ENNIS AND NANCY HAM LIBRARY
ROCHESTER COLLEGE
800 WEST AVON ROAD
ROCHESTER HILLS, MI 48307

ANIMAL MICROLOGY

ANIMAL MICROLOGY

PRACTICAL EXERCISES IN ZOÖLOGICAL MICRO-TECHNIQUE

By **MICHAEL F. GUYER**

Professor of Zoölogy in the University of Wisconsin

WITH A CHAPTER ON DRAWING BY
ELIZABETH A. (SMITH) BEAN

Former Assistant Professor in Zoölogy in the University of Wisconsin

FOURTH REVISED EDITION

THE UNIVERSITY OF CHICAGO PRESS
CHICAGO, ILLINOIS

THE UNIVERSITY OF CHICAGO PRESS, CHICAGO 37
Cambridge University Press, London, N.W. 1, England
W. J. Gage & Co., Limited, Toronto 2B, Canada

*Copyright 1906, 1917, 1930, and 1936 by The University
of Chicago. All rights reserved. Published November 1906
Second Edition February 1917. Third Edition January
1930. Fourth Edition April 1936. Eleventh Impression
1950. Composed and printed by* THE UNIVERSITY OF
CHICAGO PRESS, *Chicago, Illinois, U.S.A.*

PREFACE TO THE FIRST EDITION

For the past ten years it has been a part of the writer's duties to give instruction in microscopical technique, and it has seemed to him that there is need for a series of practical exercises which will serve to guide the beginner through the maze of present-day methods, with the greatest economy of time, by drilling him in a few which are thoroughly fundamental and standard. The book is intended primarily for the beginner and gives more attention to the details of procedure than to discriminations between reagents or the review of special processes. The student is told what to do with his material, step by step, and why he does it; at what stages he is likely to encounter difficulties and how to avoid them; if his preparation is defective, what the probable cause is and the remedy. In short, the book attempts to familiarize the student with the little "tricks" of technique which are commonly left out of books on methods but which mean everything in securing good results.

A very brief, non-technical account of the principles of the microscope is inserted (Appendix A) with the idea of giving the student just enough of the theoretical side of microscopy to enable him to get satisfactory results from his microscope. The microscope is so ably treated in the excellent works of Gage (*The Microscope*) and Carpenter (*The Microscope and Its Revelations*) that the writer feels himself absolved from any further responsibility in this matter.

The aim of the entire book is to be practical: to omit everything that is not essential; and, above all, to give definite statements about things. Appended to each chapter is a series of *memoranda* which serve to supply additional information that is more or less pertinent without obscuring the main features of the method under consideration.

In Appendix B the formulae for a number of the most widely used reagents are given with comments upon their uses and manipulation. Following this (Appendix C) is a concise table of a large number of tissues and organs with directions for properly preparing them for microscopical study.

Inasmuch as every experienced worker has his own "best" method for the preparation of almost any tissue, it is manifestly impossible to give all "best methods" in such a table. The writer believes, however, that the student will find the methods recommended all good ones which will yield satisfactory results.

In Appendix D some directions are given for collecting and preparing material for an elementary course in zoölogy.

It is hoped that the volume will prove of use: (1) as a class textbook; (2) as a guide to the independent individual worker (teacher, physician, college or medical student, or novice); (3) as a reference book for teachers, in the preparation of material for courses in elementary zoölogy, histology, or embryology.

In the matter of expressing his obligations the writer is at a loss to know just what to do. Many of the methods in microscopical technique have been handed down tradition-wise from one worker to another until their origin is unknown; they are the accumulated experiences of several generations of workers. Furthermore, many points have been absorbed, as it were, by the writer, from fellow-workers in the Universities of Chicago, Nebraska, and Cincinnati, respectively; consequently the obligation cannot be specifically expressed. Where the name of the originator of a method is known, due credit has been given. The books to which the author is most heavily indebted are the volumes of Gage and Carpenter, already mentioned, Lee's *Microtomist's Vade-Mecum*, Whitman's *Methods in Microscopical Anatomy and Embryology*, Hardesty's *Neurological Technique*, Foster and Balfour's *Elements of Embryology*, Minot's *Laboratory Text-Book of Embryology*, Huber's translation of the Böhm-Davidoff *Text-Book of Histology*, Stöhr's *Text-Book of Histology*, Mallory and Wright's *Pathological Technique*, Bausch's *Manipulation of the Microscope*, and the *Journal of Applied Microscopy*. Grateful acknowledgment is also made to the various manufacturers of microscopical instruments and appliances for the loan of most of the cuts which have been used in this volume.

M. F. G.

PREFACE TO THE SECOND EDITION

The favorable reception accorded the first edition of *Animal Micrology* has encouraged the author to believe that a second edition, incorporating some of the many new methods which have appeared during the past ten years, would be equally welcome. The general plan of the book has not been altered (see Preface to the First Edition, on a preceding page), although changes have been made on nearly every page, many sections have been entirely rewritten, and two new chapters, one on "Cytological Methods," the other on "Drawing," have been added. The chapter on drawing has been prepared by Dr. Elizabeth A. Smith.

In spite of a determined effort to limit the book to its former size, it has expanded by over fifty pages. For every method dropped there seemed to be a host of good new ones demanding recognition. These in the main, however, have been left to the encyclopedia and the various technical books and journals listed at the end of the volume. As in the first edition, the policy has been, not to attempt to give all "best" methods, but rather to select representative good ones which have proved their worth by satisfactory tests in American laboratories.

Whatever merit the new edition may prove to have over that of the earlier one is due in no small measure to the many helpful suggestions of my colleagues in other colleges and universities. I am particularly indebted in this respect to Professors C. E. McClung, R. R. Bensley, H. McE. Knower, F. L. Landacre, F. C. Waite, B. M. Allen, George R. La Rue, Edward L. Rice, F. D. Barker, R. M. Strong, and H. L. Wieman, and to Doctors Elizabeth A. Smith and C. H. Heuser.

M. F. G.

PREFACE TO THE THIRD EDITION

The continued use of this book in both American and foreign laboratories encourages the author to believe that it still satisfactorily meets the needs of beginners in microscopical technique. That they may have the benefit of the more important advances of recent years this third edition has been prepared. As in former editions the effort has been to make it, not an encyclopedia of technique nor a catalogue of all possible procedures, but a representative series of reliable standard methods

For helpful suggestions in connection with the present edition, I am particularly indebted to Harold W. Beams, Joseph B. Goldsmith, Christopher J. Hamre, Frederick L. Hisaw, Steven J. Martin, Harvey M. Smith, Opal M. Wolf, and Chao-Fa Wu.

M. F. G.

PREFACE TO THE FOURTH EDITION

If the author may judge from the continued use of this book in many laboratories for some thirty years, together with comments he has had on it from time to time, it has become a sort of stand-by for beginners in animal microscopy, for many physicians who wish to do microscopic investigation, and for teachers who are their own technicians. He feels obligated, therefore, to pass on to such users any knowledge of outstanding advances in technique that comes his way which may save their time materially or make for greater precision. Because of its time-saving advantages and also its value with tissues which become unduly hard, brittle, or friable under ordinary treatment, introduction of the dioxan method (chap. vii) would alone warrant a new edition. Minor changes, mainly additions, have been made on many pages, however; and such new special techniques as have proved to be exceptionally valuable in the author's own laboratory or in those of his immediate associates have been included.

Grateful acknowledgement is made to Harland W. Mossman, Pearl E. Claus, Nellie M. Bilstad, Ernst J. Dornfeld, Melvin Doner, James O. Foley, and Merlin L. Hayes for helpful suggestions.

M. F. G.

CONTENTS

	PAGE

INTRODUCTORY 1
  Apparatus and Supplies Required, 1; General Rules, 5.

CHAPTER I. PREPARATION OF REAGENTS 7
  Practical Exercises, 7; Memoranda, 11–14.

CHAPTER II. GENERAL STATEMENT OF METHODS 15
  Killing, Fixing, and Hardening, 15; Washing, 17; Dehydrating, 18; Preserving, 19; Staining, 19; Clearing, 22; Mounting, 23; Imbedding, 23; Affixing Sections, 24; Decolorizing, 25; Bleaching, 25; Corrosion, 25; Decalcification and Desilicidation, 25; Injection Methods, 26; Isolation of Histological Elements, 26; Normal or Indifferent Fluids for Examining Fresh Tissues, 27; General Scheme for Mounting Whole Objects (*In Toto* Preparations) or Sections, 27.

CHAPTER III. KILLING AND FIXING 28
  Cautions, 28; Fixing with Zenker's Fluid, 29; Fixing with Bouin's Fluid, 30; Formalin as a Fixing Reagent, 30; Memoranda, 31–34.

CHAPTER IV. SIMPLE SECTION METHODS 35
  Free-hand Section Cutting, 35; Memoranda, 36, 37.

CHAPTER V. THE PARAFFIN METHOD: INFILTRATION AND SECTIONING . 38
  The Method, 38; Memoranda, 44–48; Difficulties Likely To Be Encountered in Sectioning in Paraffin, and the Probable Remedy, 48.

CHAPTER VI. THE PARAFFIN METHOD: STAINING AND MOUNTING . 52
  Staining with Hematoxylin, 52; Double Staining in Hematoxylin and Eosin, 54; Double Staining in Cochineal and Lyons Blue, 54; Staining with Heidenhain's Iron-Hematoxylin, 54; Iron-Hematoxylin with Other Stains, 56; Staining in Bulk before Sectioning, 56; Paraffin Method for Delicate Objects, 57; Euparal as a Mounting- and Preservation-Medium, 58; Memoranda, 58–63.

CHAPTER VII. THE DIOXAN METHOD 64
  Why Valuable, 64; The Method, 64; With the Freezing Method, 67; Memoranda, 65–67.

CHAPTER VIII. THE CELLOIDIN METHOD 68
  The Method, 68; Staining Celloidin Sections in Hematoxylin and Eosin, 71; Memoranda, 71–76.

	PAGE

CHAPTER IX. THE FREEZING METHOD 77
  The Method, 77; Memoranda, 79–81.

CHAPTER X. METALLIC SUBSTANCES FOR COLOR DIFFERENTIATION . 82
  A Golgi Method for Nerve Cells and Their Ramifications, 82; Memoranda on Golgi Methods, 83, 84; Other Silver-Nitrate Methods, 84; Memoranda on Silver Methods, 85–87; Gold-Chloride Method for Nerve Endings, 88.

CHAPTER XI. ISOLATION OF HISTOLOGICAL ELEMENTS. MINUTE DISSECTIONS 89
  Dissociation by Means of Formaldehyde, 89; Isolation of Muscle Fibers by Maceration and Teasing, 89; Maceration by Means of Hertwig's Fluid, 90; Mall's Differential Method for Reticulum, 90; Minute Dissection and Mounting of Various Parts of Insects, 91; Memoranda, 92, 93.

CHAPTER XII. TOOTH, BONE, AND OTHER HARD OBJECTS 94
  Sectioning Decalcified Tooth, 94; Sectioning Decalcified Bone, 94; Sectioning Bone by Grinding, 94; Memoranda, 95

CHAPTER XIII. INJECTION OF BLOOD AND LYMPH VESSELS . . . 97
  Red Injection Mass, 97; Blue Mass, 97; Yellow Mass, 98; Injecting with a Syringe, 98; Micro-Injection of Embryonic Vessels, 101; Corrosion Methods, 104; Memoranda, 100–106.

CHAPTER XIV. OBJECTS OF GENERAL INTEREST: CELL-MAKING, FLUID MOUNTS, "IN TOTO" PREPARATIONS, ETC. 107
  Turning Cells, 107; Mounting in Glycerin (Water Mites, Transparent Larvae), 108; Killing and Mounting Hydra, 109; Mounting in Glycerin-Jelly (Small Crustacea, etc.), 109; Mounting in Balsam (Flat Worms, Mosquito, Gnat, Aphid), 110; Opaque Mounts (Beetles, Wings of Moths and Butterflies, Head of Fly, Foreleg of Dytiscus), 111; Dry Mounts, 112; Spalteholz Method of Clearing Total Specimens, 117; Memoranda (Including Directions for Mounting Other Forms), 113–20.

CHAPTER XV. BLOOD 121
  Examination of Fresh Blood, 121; Effects of Reagents, 121; To Demonstrate Blood Platelets, 121; Stained Preparation of Fibrin, 121; Crystals of Blood, 121; Cover-Glass Preparations (Dry), 122; Rapid Method, 123; Enumeration of Blood Corpuscles, 124; Observation of the Blood Current, 126; Inflammation, 126; Memoranda, 127–29.

Contents

CHAPTER XVI. BACTERIA 130
Bacterial Examination, 130; Cover-Glass Preparations from Fluid Media, 130; Staining and Mounting, 131; Bacteria in Tissues, 131; Methylen Blue Stain for Bacteria in Tissues, 132; Gram's Method for Bacteria in Tissues, 132; Hanging-Drop Preparations, 132; Memoranda, 133–36.

CHAPTER XVII. SOME EMBRYOLOGICAL METHODS: SECTIONS AND WHOLE MOUNTS OF FROG AND CHICK; OTHER FORMS . . . 137
The Frog, 137; Section Method, 137; Whole Mounts, 138; Memoranda on Amphibian Material, 139–41; The Chick, 141; General Memoranda on Embryological Methods and Materials, including *In Vitro* Cultures, 143–57.

CHAPTER XVIII. SOME CYTOLOGICAL METHODS 158
General Remarks, 158; Mitosis, 158; Testis of Crayfish, Sections, 159; Smears, 160; Blastodisk of Whitefish, 160; Testis of Necturus, 160; Somatic Cells of Ambystoma, 161; Living Cells, 161; Mitochondria, 162; Golgi Apparatus, 165; Staining of Living or Fresh Tissues, 167; Tests for Certain Cellular Structures, 167; Special Methods, 170; Allen's B-15 and B-20 Methods, 170; Hance's Cold Method, 171; Photographing Cellular Structures, 172; Accessory Chromosomes, 173; Aceto-Carmine Preparations, 173; Protoplasmic Currents, 174; Celloidin instead of Paraffin, 174; Urea in Fixing Fluids, 174; Drop-Method of Changing Fluids, 174; Dehydration by Dialysis, 175; Tissues of Young Adults Desirable, 175; Dissection of Living Cells, 175; Estimation of Carbon Dioxide, 176; Euparal, 176; Memoranda, 173–78.

CHAPTER XIX. RECONSTRUCTION OF OBJECTS FROM SECTIONS . . . 179
Reconstruction in Wax, 179; Geometrical Reconstruction, 181; Blotting Paper Method, 181; Rolling Drawings into the Wax, 181; Photography in Reconstruction Work, 182; Cutting Out Wax Plates, 182; Memoranda, 181, 182.

CHAPTER XX. DRAWING 183
Materials for Class Work, 183; Methods of Representation, 184; Outline, 184; Depth, 184; Ink Drawings, 184; Pencil Drawings, 185; Wash Drawings, 186; Size and Arrangement of Drawings, 187; Labeling, 187; Modes of Representation for Special Courses, 188; Embryology, 188; Histology, 189; Cytology, 189; Drawings for Publication, 190; Materials for Manuscript Drawings, 190; Camera Lucida, 191; Reduction, 191; Line Process, 191; Half-tone, 192; Wash and Combination Drawings for Reproduction, 193; Lithography, 194; Arrangement for Reductions, 194; Lettering, 194.

Contents

APPENDIX A. THE MICROSCOPE AND ITS OPTICAL PRINCIPLES . . . 197
  Optical Principles, 197; Images, 199; The Simple Microscope, 200; The Compound Microscope, 200; Defects in the Image, 202; Nomenclature or Rating of Objectives and Oculars, 206; Some Common Microscopical Terms and Appliances (Alphabetically Arranged), 206; Manipulation of the Compound Microscope, 226.

APPENDIX B. SOME STANDARD REAGENTS AND THEIR USES . . . 230
  Fixing and Hardening Agents, 230; Stains, 243; Normal or Indifferent Fluids, 263; Dissociating Fluids, 264; Decalcifying Fluids, 265.

APPENDIX C. TABLE OF TISSUES AND ORGANS WITH METHODS OF PREPARATION 267

APPENDIX D. PREPARATION OF MICROSCOPICAL MATERIAL FOR A GENERAL COURSE IN ZOÖLOGY 286

APPENDIX E. TABLE OF EQUIVALENT WEIGHTS AND MEASURES . . 310

APPENDIX F. REFERENCES 312

INDEX 313

INTRODUCTORY

APPARATUS AND SUPPLIES REQUIRED

The student should provide himself with the following supplies:

One half-gross box best grade glass slides, standard size (25×75 mm.).
One-half ounce, 18 mm. or ¾ in., round cover-glasses, medium thickness, (0.18 mm.).
Thirty 25×50 mm. cover-glasses, medium thickness.
Two or three Pillsbury slide boxes (Fig. 1).
One box of labels for slides.
Three to six camel's hair brushes (Fig. 2).
Six pipettes (Fig. 3).
One package Gillette razor blades.
One set of dissecting instruments as follows:
    One large scalpel or cartilage knife (Fig. 4).
    One small scalpel (Fig. 5).
    Two needles (Fig. 6).
    One fine straight scissors (Fig. 7).
    One fine straight dissecting forceps, file-cut points (Fig. 8).
    One blow-pipe (Fig. 9).
    One section lifter (Fig. 10).
To which may well be added:
    One heavy scissors (Fig. 11).
    One curved scissors (Fig. 12).
    One heavy forceps (Fig. 13).
    One fine forceps, curved tips (Fig. 14).
One horn spoon.
One desk memorandum calendar.
Blank cards (about 75×100 mm.) for keeping records of experiments. The kind of card used for library card catalogue will do.
One section razor (Fig. 15).
A piece of moderately heavy copper wire with one end hammered out to a width of 7 to 10 mm.
Towels.
A glass-marking pencil (wax) or writing diamond will be found useful.
Coarse carborundum "engraver's pencil points," which may be purchased for seventy-five cents a dozen, are very satisfactory for marking glass, according to Professor C. E. McClung.

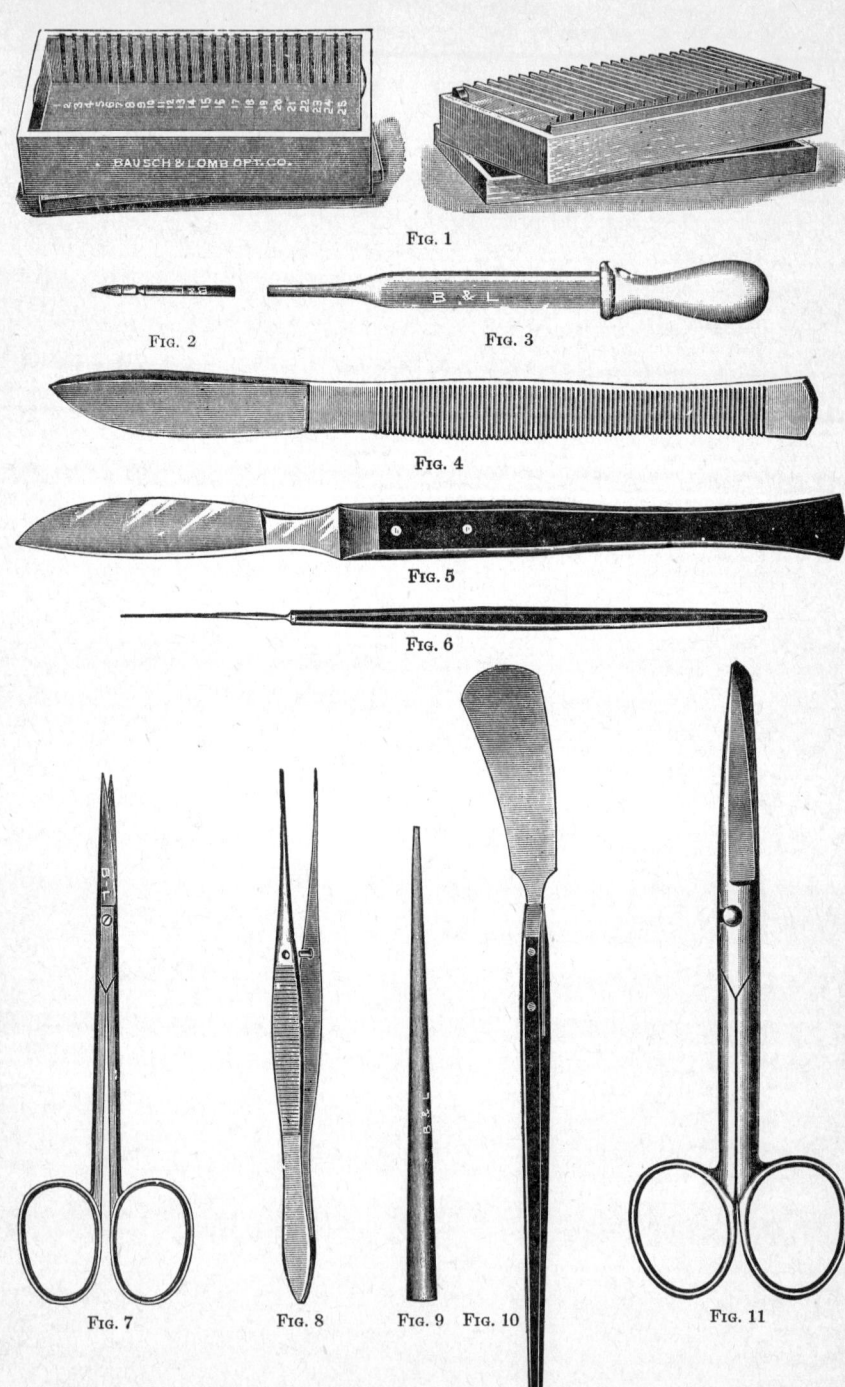

Fig. 1
Fig. 2
Fig. 3
Fig. 4
Fig. 5
Fig. 6
Fig. 7
Fig. 8
Fig. 9
Fig. 10
Fig. 11

Introductory 3

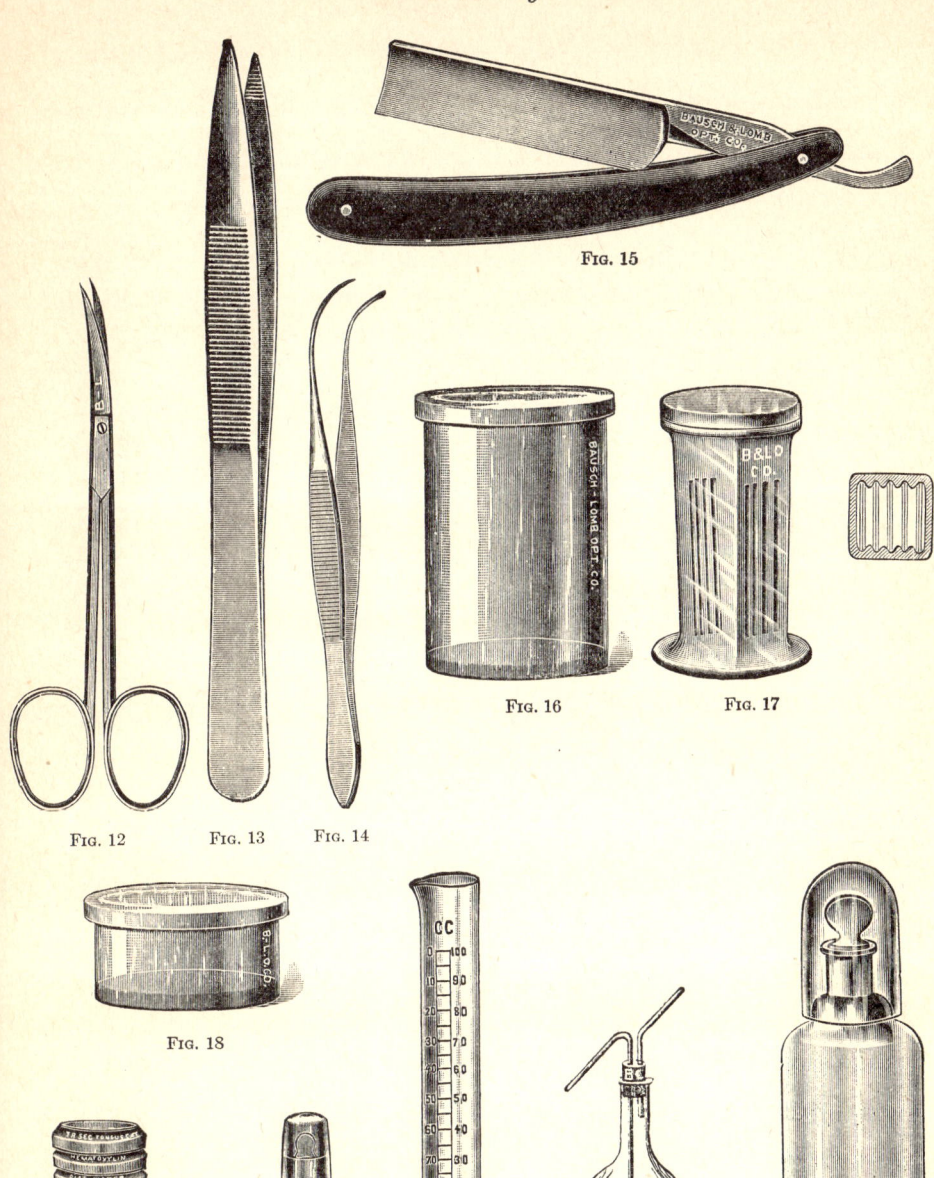

Fig. 15

Fig. 12   Fig. 13   Fig. 14   Fig. 16   Fig. 17

Fig. 18

Fig. 19   Fig. 20   Fig. 21   Fig. 22   Fig. 23

Apparatus ordinarily supplied by the laboratory:

Desk with drawers.
Locker for microscope.
Compound microscope and accessories (Appendix A).
Dissecting microscope (Figs. 62, 68).
Microtomes (Figs. 26, 27, 28, 31).
Paraffin oven (Figs. 24, 25).
Tall stenders (about 85 mm. deep). Each student should have at least eight (Fig. 16).
Coplin staining jars (Fig. 17). Tall stenders may be used instead. About 14 are needed for each student.
Flat stenders (Fig. 18); half a dozen for each student.
Syracuse watch-glasses (Fig. 19); eight to each student.
Balsam bottle (Fig. 20).
Graduated cylinders for measuring liquids (Fig. 21).
Wash-bottle (Fig. 22).
Celloidin bottle (Fig. 23).
Turntable (Fig. 36).
Injecting apparatus.
Reagent bottles and vials.
Other apparatus and supplies such as bone-forceps, bone-saws, glass tubing, glass rods, beakers, burners, filter-paper, funnels, evaporating-dishes, sand bath, dropping-bottles, balances, mortar and pestle, etc.

For apparatus or supplies not listed in this book the student is referred to the illustrated catalogues of dealers and manufacturers such as: The Bausch & Lomb Optical Co., Rochester, N.Y.; The Ernst Leitz Optical Works, American branch, 60 E. Tenth St., New York City; The Spencer Lens Co., Buffalo, N.Y.; Carl Zeiss Optical Works, American branch, 485 Fifth Ave., New York City; R. & J. Beck, 68 Cornhill, London; Eimer & Amend, New York City; Arthur H. Thomas Co., Philadelphia, Pa.; Central Scientific Co., Chicago, Ill.; Chicago Apparatus Co., Chicago, Ill.; General Biological Supply House, Chicago, Ill.

IMPORTANT GENERAL RULES

1. Keep everything clean!
2. Have a definite place in your desk for each piece of apparatus and arrange reagents in order on top of it.
3. Use cards for keeping records of materials. Each card should have a number corresponding to that of each special object or piece of tissue, and should show the name of the preparation, date, reagents used, time left in each reagent—in short, all data concerning the manipulation of the material.
4. Jot down in a blank calendar the various things to be done at future dates, such as changing of reagent on tissues, etc., and then go over this memorandum carefully each day when you first come into the laboratory.
5. Use only clean vessels in preparing reagents, and clean up all glassware while it is yet moist.
6. Reserve and mark a separate pipette for each of the chief reagents (absolute alcohol, oils, acids, etc.).
7. In making up solutions, 1 gram of a salt in 100 c.c. of liquid is reckoned ordinarily as a 1 per cent solution, 3 grams as a 3 per cent solution, etc. But if solutions are to be of 10 per cent strength or over, it is better to weigh out the dry material to the desired percentage and then add enough of the liquid to make the whole weigh 100 grams. For example, to make a 25 per cent aqueous solution of caustic potash, add 25 grams of caustic potash to 75 c.c. of water. A saturated solution contains all of a given substance that the liquid will take up. When a solution is called for without specifying the solvent, an aqueous solution is meant.
8. In weighing salts always first put paper in the scale pans to protect them.
9. In making solutions or mixtures in which only a small amount of one reagent is used, after mixing, pour back some of the mixture into the small vessel and rinse it thoroughly in order to get all of the original contents out.
10. When pouring liquids from bottles keep the label of the bottle

turned toward the palm of the hand. Do not lay down **stoppers but** hold them by their tops between the knuckles.

11. Before leaving the laboratory put away your instruments and clean and put in its place whatever laboratory apparatus you may have been using.

12. All solid waste materials, acids, stains, etc., should be **thrown** into stone jars, not into the sink.

CHAPTER I
PREPARATION OF REAGENTS

The following reagents should be prepared by each student.

1. Grades of Alcohol.—To obtain a given percentage of alcohol through dilution of a higher percentage with distilled water, subtract the percentage required from the percentage of the alcohol to be diluted; the difference is the proportion of water that must be added. Thus, if 35 is the percentage required, and 95 the percentage to be diluted, then $95-35=60$; hence, 60 parts of water and 35 parts of 95 per cent alcohol are the proportions for mixing.

This means that in practice one needs only to fill the graduated measuring cylinder to the same number as the percentage required (e.g., 35) with the alcohol to be diluted (e.g., 95) and then fill up to the percentage of the latter with distilled water. In this way one would obtain 95 c.c. of alcohol of the percentage required, if the measuring cylinder is graduated in cubic centimeters.

Prepare about 250 c.c. of 35, 50, 70, and 83 per cent alcohols, respectively, from 95 per cent alcohol and distilled water. The commercial alcohol used, though really about 96 per cent, may be figured on the basis of 95 per cent.

Owing to the differences in the specific gravities of the different percentages of alcohol, the foregoing method gives only approximate results; they are sufficiently accurate, however, for most biological work.

2. Absolute Alcohol.—It is customary in most laboratories to purchase so-called absolute alcohol specially prepared for laboratory purposes. Squibb's absolute alcohol (99.8 per cent) is commonly used. Inasmuch as such alcohol is an expensive reagent, economy sometimes necessitates that the student undertake the more tedious process of making his own absolute alcohol. Crystals of copper sulphate are heated until the water of crystallization is driven off and the sulphate is left as a white powder. Such anhydrous sulphate is added to a bottle of commercial (96 per cent) alcohol. The water in the alcohol immediately unites with it, turning it blue. Anhydrous sulphate should be added until it no longer turns blue. The alcohol is then filtered into a clean, dry bottle which must have a tight-fitting cork or ground-glass stopper. It is well to smear the glass stopper with vaseline, so that when it is placed in the bottle all moisture from the air may be completely excluded. Any laboratory using considerable quantities of absolute alcohol should have its own still.

3. Acid Alcohol.—

Alcohol (35 per cent)	50 c.c.
Hydrochloric acid (pure)	3 drops

For sections use the mixture only a few seconds or minutes. For material stained in bulk, add twice as much 35 per cent alcohol and leave the object in it until sufficiently decolorized (2 to 24 hours).

4. Ether and Alcohol.—Absolute alcohol and sulphuric ether equal parts. Quantity, 400 c.c. Keep the ether distant from all flames.

5. Normal Saline.—Prepare a 0.75 per cent solution of sodium chloride in distilled water. This is termed a normal salt solution because it is a solution of about the same density as natural lymph and is much less harmful to living tissues than is distilled water. Quantity, 500 c.c.

6. Formalin (also termed formal, formol, formolose).—Commercial formalin is a 40 per cent solution of formaldehyde in water. A 4 per cent solution of formalin would be made by taking 4 volumes of commercial formalin and 96 volumes of water. This is, however, only a 1.6 per cent solution of formaldehyde. Make a 10 per cent solution of formalin. Quantity, 250 c.c.

7. Zenker's Fixing Fluid.—

Bichromate of potassium	2.5 grams
Bichloride of mercury (corrosive sublimate)	5.0 grams
Sodium sulphate	1.0 gram
Water	100 c.c.
Glacial acetic acid	5 c.c.

Dissolve the bichromate and the sublimate in the water with the aid of heat. Keep the acetic acid in a separate bottle until the fixing fluid is to be used, as it will produce changes in the chrome salt if added at once and allowed to stand.

CAUTION.—In handling corrosive sublimate do not use metal instruments because it corrodes metal. Use a glass or horn spatula.

8. **Bouin's Picro-Formol.—**
   Picric acid, saturated aqueous solution........ 75 parts
   Formalin................................... 25 parts
   Acetic acid (glacial)....................... 5 parts

   One gram of picric-acid crystals will saturate about 75 c.c. of water.

9. **MacCallum's Macerating Fluid.—**
   Nitric acid................................. 1 part
   Glycerin.................................... 2 parts
   Water....................................... 2 parts

10. **Decalcifying Solution.—**
    Nitric acid (strong)........................ 10 c.c.
    Alcohol (70 per cent)....................... 90 c.c.

11. **Alum-Cochineal.—**
    Potassic alum............................... 6 grams
    Powdered cochineal.......................... 6 grams
    Distilled water............................. 90 c.c.

Boil for half an hour; after the fluid has settled, decant the supernatant liquid, add more water to it, and boil it down until only 90 c.c. of the decoction remains. Filter when cool, and add a bit of thymol or a little salicylic acid to prevent the growth of mold. For beginners, some instructors prefer borax-carmine (reagent 39, p. 246) to alum-cochineal.

12. **Delafield's Hematoxylin.—**Prepare 100 c.c. of a saturated aqueous solution of ammonia alum (about 20 grams of aluminum ammonium sulphate to 100 c.c. of water). Dissolve 1 gram of hematoxylin crystals in 10 c.c. of absolute alcohol, and add it, drop by drop, to the first solution. Expose this mixture to air and light for several weeks (two months is not too long) to "ripen." Ripening consists in an oxidation of the hematoxylin to form hematein. This may be accomplished at once with some degree of success through the addition of a few cubic centimeters of a neutralized solution of peroxide of hydrogen, or, better, by exposure in a shallow dish to a powerful quartz mercury vapor light (ultra-violet light) for from 20 to 30 minutes, at a distance of about 2 feet. When ripe, filter the solution and add 25 c.c. of glycerin and 25 c.c. of methyl alcohol (see memorandum 1). It is well to have a stock solution of this stain already prepared to be used in case the student's preparation is not ready in time. For many tissues, Harris' modification may well be substituted for this one (see reagent 59, p. 252).

Most laboratories keep on hand a stock solution of hematoxylin made by dissolving 1 part of hematoxylin crystals in 10 parts of absolute alcohol. In the course of several months or a year this solution ripens to a dark wine-red color. It may then be used in making up the various hematoxylin solutions and, being ripe, will stain at once.

>   NOTE.—*At this point the student should begin chap. iii in order that no time may be lost. The present chapter may then be completed while the tissues are becoming fixed and hardened.*

13. Orange G.—

Orange G (Grübler's)	1	gram
Distilled water	100	c.c.

14. Congo Red.—

Congo red	0.5	gram
Distilled water	100	c.c.

15. Lyons Blue.—

Absolute alcohol	100	c.c.
Bleu de Lyon	0.3	gram

16. Eosin.—

Eosin	0.5	gram
Alcohol, 95 per cent	100	c.c.

17. Iron-Hematoxylin (Heidenhain's).—Two solutions are used. They are not to be mixed.

Solution I:

Ferric alum (clear violet crystals)	2.5	gram
Distilled water	100	c.c.

Solution II:

Hematoxylin	0.5	gram
Distilled water	100	c.c.
Alcohol, 95 per cent	10	c.c.

First dissolve the hematoxylin in the alcohol, then add the water.

The hematoxylin should ripen (see reagent 12, p. 9) for some three or four weeks. The ferric alum of the histologist is always ammonio-ferric sulphate. For variants, see note on page 55.

18. Canada Balsam.—Dry 2 grams of Canada balsam on a sand bath or in a warm chamber until it becomes hard (1 to 2 hours at 65° C.) Do not overheat. When cool add enough xylol to make a very thin syrupy fluid. Roll a sheet of paper into a cone to serve as a funnel, and filter the fluid through absorbent cotton. Thicken the solution slightly by leaving the cap off the bottle in a place free from dust, and allowing some of the xylol to evaporate. Or, fill your balsam bottle one-third full of the liquid xylol-balsam now on the market and dilute to the proper consistency.

19. **Mayer's Albumen Fixative.**—Beat the white of an egg with an eggbeater and pour it into a tall cylinder. Let stand until the air brings all suspended matter to the top. Skim off the latter and filter through a suction filter. To the remainder add an equal volume of glycerin, and a bit of salicylate of soda (1 gram to 50 c.c.) or thymol to prevent putrefaction.

20. **Celloidin.**—Soak 15 grams of dry celloidin (see n. 11, p. 73) overnight in just sufficient absolute alcohol to cover it. Then dissolve it in 200 c.c. of ether-alcohol (see reagent 4). In a second bottle prepare a thinner solution by taking about one-third of the original and diluting it with its own volume of ether-alcohol. The solutions are best kept in bottles with glass stoppers and ground-glass caps. Label the bottles thick and thin celloidin, respectively.

Fig. 24.—Warming Table
There should be two rectangular boxes (about 3×3×16 cm.) to contain paraffin if the table is to be used for paraffin imbedding. When in use the boxes are so placed on the imbedding-table that the paraffin in one end remains melted; in the other, solid. Regulate the temperature by placing the flame at the proper distance under the acute angle of the table. It is best, when gas is used, always to turn on the gas completely and then regulate the height of the flame by means of a clamp on the rubber tubing which conducts gas to the burner.

21. **Paraffin.**—In one of the cups of a warm paraffin oven (Fig. 24 or 25) put 75 grams of paraffin, melting at 52° C. (see memorandum 5, p. 12). A supply of softer and of harder paraffin (e.g., melting at 43° and 58° C.) should also be at hand.

Other Reagents.—Provide yourself with 500 c.c. of distilled water, 200 c.c. of xylol, 25 c.c. of clove oil, 25 c.c. of glacial acetic acid, 50 c.c. of a cedar-wood oil, 25 c.c. of a saturated solution of iodine in 70 per cent alcohol, 75 c.c. of chloroform, 30 c.c. of glycerin, and 250 c.c. of absolute alcohol if it has not been already prepared. Keep the absolute alcohol and the xylol carefully corked to exclude moisture. Before measuring out any of these reagents, see that both the graduate and the bottle are perfectly clean and *dry*.

MEMORANDA

1. **Ethyl Alcohol** (grain alcohol) is the kind commonly used in histological laboratories. Upon presentation of the proper credentials to the internal revenue officers, it may be purchased by the barrel from distillers, tax

free, by educational institutions. Such commercial alcohol is of about 96 per cent strength. When the strength is unknown, it should be tested by means of an alcoholometer (see 2, below).

Methyl Alcohol (called also wood alcohol or wood spirits) is cheaper than ethyl alcohol in case the latter cannot be had tax free, and is fairly satisfactory in most cases. It is poisonous and must be carefully handled. It is of about 90 per cent strength. The "methylated spirits" of English microscopists is grain alcohol containing 10 per cent of methyl alcohol.

Rectified Spirit is a 91 per cent alcohol (84 per cent in England).

2. **The Alcoholometer** is a convenient instrument for determining the strength of alcohol, or the percentage of absolute alcohol in a spirituous mixture. It is a kind of hydrometer with a scale marked to indicate the percentages of alcohol. Different strengths of alcohol have different specific gravities; consequently the instrument will float higher or lower in the liquid according to the percentage of alcohol present. The number on the scale just at the surface of the liquid indicates its strength.

3. **Rule for Dilution** of a given strength of a solution with a lower percentage of the same solution (for where the diluent is water, i.e., zero per cent, see rule under reagent 1): Subtract the percentage required from the percentage of the solution to be diluted; also subtract the percentage of the diluent from that of the strength required. The differences are the relative proportions of the diluent and the solution to be diluted that must be used. Thus, to prepare a 35 per cent solution from 95 and 20 per cent solutions: $95-35=60$; $35-20=15$; hence, 60 to 15 or 4 to 1 are the proportions desired. That is, 4 parts of the 20 per cent and 1 part of the 95 per cent solution must be used to obtain a 35 per cent solution.

4. "**To Remove Fixed Stoppers,** take the bottle in the left hand with the thumb applied to one side of the stopper, then tap the other side of the stopper with some heavy instrument, such as the handle of a pocket-knife, pressing the thumb against the direction of the tap. Turn the bottle round, gradually tapping until the stopper loosens. Should this device prove of no avail (which is very rarely), hold the neck of the bottle in a spirit flame, and quickly withdraw the stopper as the glass of the neck expands. This is a somewhat risky procedure, but is very effectual if done smartly" (*Journal of Applied Microscopy*, VI, 2116). The glass of the neck may be more safely heated by looping a heavy cord about it and sawing the cord back and forth until the friction warms the glass.

5. **A Simple Paraffin Bath,** recommended by several workers, may be made by suspending an electric-light bulb over a tumbler of paraffin. The height of the bulb should be so adjusted that some unmelted paraffin remains at the bottom. Tissues will thus come to lie where the paraffin is just at the melting-point.

McClung suspends a 150-watt nitrogen-filled bulb, provided with a lamp shade, over a beaker or tumbler three-fourths full of paraffin. The bulb should be so adjusted as to keep the paraffin melted to the depth of about 1 inch. Rapid evaporation of the dealcoholizing agent is facilitated by such an arrangement and overheating is avoided. The same lamp may be used for spreading the paraffin ribbon on slides, and for drying.

In our own laboratory individual baths made as shown in Figure 25 are in general use. Ideas from various workers are embodied. Such baths are inexpensive and efficient. All that is required is: (1) a chemist's ordinary iron stand with adjustable clamp-holder and clamp for electric light; (2) a

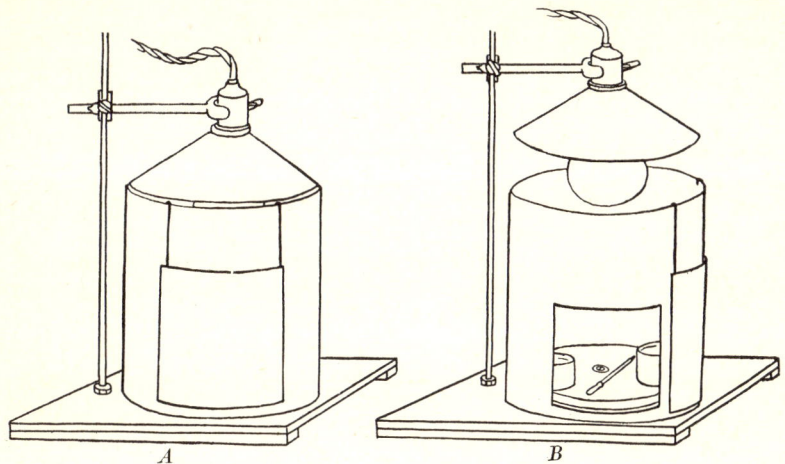

FIG. 25.—Wisconsin Paraffin-melting Oven

A, with slide closed and lamp lowered; *B*, with slide open, showing pipette, glass dishes for paraffin, and the circular floor which can be rotated. The lamp should be so adjusted that some unmelted paraffin always remains at the bottom of each dish; tissues will thus lie where the paraffin is just at the melting-point.

60-watt mazda bulb and 8-inch tin shade; (3) a cylinder of galvanized iron open at each end, $8\frac{1}{2}$ inches in diameter and 8 inches high; and (4) a wooden base on top of which is a wooden rotary disk over which the cylinder fits. In one side of the cylinder a door 4 inches square is cut which can be closed by a strip of galvanized iron suspended by two wire hooks from the upper edge of the cylinder. Small vessels containing paraffin are placed on the disk, which can be rotated, and the lamp is raised or lowered to a height which keeps the paraffin properly melted. Paraffins of different melting-points may be kept on the disk by covering those of higher melting-point with glass covers. To keep out dust a strip of cheesecloth may be tied over the crack between the shade and the top of the cylinder. Because of the necessity of raising or lowering the lamp at times, it has been found most practical to have the shade of smaller diameter than the cylinder.

6. A Convenient Warming Table can easily be made by lining a light, flat box, such as a starch box, with sheet asbestos, installing two electric light bulbs to provide heat, and replacing the top with a sheet of galvanized iron, or, better, a stiff sheet of asbestos. The bulbs should be of just sufficient wattage to keep the top of the box warm enough to cause the sections to flatten without melting the paraffin. Slides may be left to dry on such a table. See Figure 24 for table to be used with gas flame.

CHAPTER II

GENERAL STATEMENT OF METHODS

Each of the reagents which has been prepared is used for one or more of the purposes to be discussed in this chapter.

All methods of preparation in microscopy are to enable us to learn more of the structure and functions of objects than would otherwise be apparent. We endeavor to study them in as near their natural condition as possible. While the study of living or of fresh material is desirable it can be carried on only to a very limited extent. Most structures of the animal body, though opaque, must be examined largely by transmitted light, hence special preparation is necessary to put them into suitable condition. This is accomplished—

1. By cutting them into thin slices (*section method*).
2. By separating them into their elements (*isolation*)—
   a) Mechanically (*teasing*), or
   b) With the aid of fluids which remove the cement substance (*dissociation* or *maceration*).

In most instances the minute structure of a tissue or of an organism can be studied to the best advantage only after the application of certain agents which serve to emphasize the various structural elements. A tissue so prepared is an artificial product in that it is not exactly the same as it was in the living organism, but recent studies of protoplasm in the living condition by competent investigators strengthen the belief that many reagents preserve very faithfully the actual structure of the cell contents. The liquid albuminoids are apparently the materials which suffer the greatest modifications. Since alterations do occur, however, it is clear that in our interpretations of prepared material we must reckon carefully both with the original nature of the object and with the factors introduced by ourselves.

KILLING, FIXING, AND HARDENING

The first step in the preparation of tissues ordinarily is the employment of some reagent which will kill the tissues and fix their various components in the characteristic stages of their activities. Such material may then be preserved indefinitely for future use.

It is customary to discriminate between killing, fixing, and hardening, although the same reagent may fulfil all three requirements. Killing refers particularly to the destruction of the life of the tissue, a process which may be either slow or instantaneous. In slow killing it is usual to employ narcotics such as ether, chloroform, chloral hydrate, chloretone, carbon dioxide, nicotin, cocain, or weak alcohol. Ice is also used sometimes. Such methods are of particular value with highly contractile animals which are desired in the extended condition. Such forms are narcotized completely or until they are unable to contract, and then frequently fixed and hardened in other or stronger fluids. Where practicable, instantaneous killing and fixing is preferable because tissues have then no time to undergo postmortem changes. The same fluid ordinarily is employed for killing and fixing.

The purpose of fixation is—

a) To preserve the actual form of tissue elements.

b) To produce optical differences in structure, or so to affect the tissues that such differences will be brought out through subsequent treatment with stains or other reagents.

To accomplish this the fixing agent must possess the following properties:

1. It should kill the tissue so quickly that few structural changes can occur.

2. It should neither shrink nor distend the tissue.

3. It should be a good preservative; that is, it must render the tissue element insoluble and prevent postmortem changes.

4. It should penetrate all parts equally well.

5. It should put the tissue in condition to take stains unless it of itself produces sufficient optical differences in the various parts of the tissue.

No ideal single reagent has been discovered which meets all of these requirements, hence it is customary to combine two or more reagents which individually possess certain of these desirable qualities. All of the best fixing reagents are mixtures. For example, acetic acid is very generally used in fixing mixtures because it penetrates well, produces good optical differentiation, and counteracts the tendency of some reagents (e.g., corrosive sublimate) to shrink tissues. Unfortunately it is destructive of various cytoplasmic structures such as mitochondria and Golgi apparatus, and even to parts of the resting nucleus. Where cell structures other than chromosomes are to be studied, cytologists recommend omission of the acetic acid from the standard cytological reagents (see pp. 158 ff.). A few workers report successful substitution of formic for acetic acid. It is said to penetrate more rapidly, to dissolve fatty substances less, and to be more toxic, all three of which characteristics are usually desirable. Other workers, however, report indifferent success in such substitution. Again, osmic acid, which is an excellent fixing agent for very small pieces of tissue, penetrates very poorly; consequently for most objects it must be mixed with reagents which penetrate rapidly and thoroughly.

Some fixing agents (corrosive sublimate, chromic acid, osmic acid, etc.) enter into chemical combination with certain of the tissue elements; others (alcohol, picric acid, nitric acid, hot water, etc.) act by coagulating or precipitating various constituents of tissues.

Certain organic substances, such as urea and sugar (maltose), have been found to improve various fixing fluids for many tissues, particularly where cytological detail is desired. They seem to assist in preventing clumping of chromosomes. About 2 per cent of urea is commonly used. The proper amount for a given tissue is best learned by trial.

The chief object of *hardening* is to bring tissues to the proper consistency for cutting sections. The process, although begun ordinarily by the fixing agent, is usually completed in alcohol. Some objects are not sufficiently hardened until they have remained in alcohol for many hours, or even days. As a rule, tissues should remain in alcohol of at least 70 per cent strength for a minimum of 24 hours after the preliminary operations of fixing, washing, etc., before they are subjected to further treatment. There is always danger, however, of overhardening. The tissue should be of about the same hardness as the medium (paraffin, celloidin) in which it is to be sectioned.

WASHING

Fixing agents ordinarily, with the exceptions of alcohol and formalin, must be washed out thoroughly or they are likely to interfere with subsequent processes. Aqueous solutions are washed out usually in water or a low percentage of alcohol; alcoholic solutions, with alcohol of about the same strength as that of the fixing agent. Washing usually requires from 10 to 24 hours, with several changes of the liquid. If water is the washing agent, it is best where practicable to use running water.

Chromic acid and its compounds should be washed out in running water. This should be done in the dark in order that precipitation may be avoided.

Picric acid, or solutions containing it (except picro-formalin mixtures), must be washed in strong alcohol (70 per cent), never in water, because the latter seems to undo the work of fixation.

Corrosive sublimate and mixtures containing it are washed out in water or alcohol. A little tincture of iodine should be added to the wash from time to time to insure the removal of all corrosive-sublimate crystals. Sufficient iodine has been added when it no longer loses its reddish color after being in contact with the preparation for a short time.

Osmic acid and mixtures containing it should be washed in running water.

DEHYDRATING

While under certain circumstances objects may be mounted in aqueous media for examination, in the majority of cases, especially where the preparation is to be a permanent one, it has been found best to remove all water from the tissues, that is, to *dehydrate* them. This renders preservation more certain, and it is a necessity, moreover, if the object is to be imbedded later in paraffin or celloidin, for neither of these substances is miscible with water. Because of its strong affinity for water and the ease with which it may be manipulated, alcohol has come to be used universally for this purpose. It completes the process of hardening at the same time. The dehydration must be gradual. In tissues transferred from water or aqueous solutions directly to strong alcohol (or vice versa) violent diffusion currents are set up which produce serious distortion of the tissue elements. For this reason a series of alcohols of gradually increasing strength (e.g., 35-50-70-83-95 per cent) is used. The more delicate the object, the closer should be the grades of alcohol.

Professor C. E. McClung recommends a "drop" method for all purposes. If, for example, an object in water is to be carried up into the higher alcohols, he places a vessel containing 95 per cent alcohol and the vessel containing the object in water under a bell-jar. The vessel containing alcohol is raised above the level of the other vessel and a string or a capillary siphon is set to carry over the alcohol drop by drop into the water. The amount of 95 per cent alcohol must be apportioned to the amount of water so that the final desired strength will be reached by the time the alcohol has all passed over into the water. The vessel containing the specimen should be agitated frequently to secure thorough mixture. Obviously other fluids may be changed by the same method. For a more elaborate form of drop apparatus see memorandum 6, page 174.

D. A. Johansen points out (*Science*, LXXXII [1935], 253) that biological technicians have not discriminated as they should between *dehydration*, which should be merely the removal of the free water of tissues, and *desiccation*, which also causes the removal of the combined water. A dehydrating fluid such as absolute alcohol, which is also a powerful desiccator, commonly makes tissues excessively hard and causes various shrinkage phenomena. Removal of all free water but retention of water-absorption capacity is the end to be attained in the preparation of tissues for sectioning. Dioxan (see chap. vii) and tertiary butyl alcohol give much promise of aid in this respect.

PRESERVING

After fixing and washing, the process of dehydration is begun ordinarily and tissues are carried as far as 70 per cent alcohol. It is customary to leave them in alcohol of from 70 to 83 per cent strength until they are needed. They may remain here indefinitely. If they are to be preserved for a long time (for months), however, it is better to keep them in a mixture of equal parts of glycerin, distilled water, and 95 per cent alcohol.

STAINING

A few fixing agents produce sufficient optical differentiation in tissues, but as a rule this must be accomplished through the addition of certain stains. Most of the stains used have more or less of a selective action; that is, they pick out certain elements of the tissue, and thus enable one to see details of structure that would otherwise be invisible. Their action, however, depends largely upon the nature of the fixing agent which has previously been used. The secret of good staining, indeed, lies largely in proper fixation.

There are large numbers of stains of very different chemical constitution (acid, neutral, and alkali) and they may act in very different ways upon the material to be stained. For example, some show affinity only for certain elements of the nucleus, others for the cytoplasm of cells, and some are present in tissues only physically as deposits, while others enter into chemical combination with certain of the cell constituents. A few, such as borax-carmine, are general stains, and affect to a greater or less degree practically all the tissue elements.

It is not the purpose of the present book to enter into a prolonged discussion of the theory of staining or to undertake a description and classification of stains. For this the reader is referred to the excellent compendium of Lee (*The Microtomist's Vade-Mecum*), or the *Handbook of Microscopical Technique*, edited by C. E. McClung.

The stains of widest application are (1) the **Carmines,** (2) the **Hematoxylins,** (3) the **Anilins,** and (4) **Metallic substances.**

Carmine is a brilliant scarlet or purplish coloring matter made from the bodies of the cochineal and kermes scale insects. The carmine stains, including cochineal, have been largely used in the past for all kinds of work, but at present they are used more particularly for staining objects in bulk before sectioning, or objects which are not to be sectioned. They are easy to use, and will follow almost any fixing agent. In case of overstaining, weak hydrochloric acid (0.1 to 1 per cent) is used to decolorize the tissues. For formulae see Appendix B, page 230.

Hematoxylin is a compound containing the coloring matter of logwood. The hematoxylins follow well almost any of the fixing agents; they are especially recommended after fluids containing chromic acid or its salts. According to Mayer, the active agent in these stains is a compound of hematein with alumina. This blue-colored solution is precipitated in tissues, particularly in nuclei, by certain organic and inorganic salts, such as phosphates. The hematein is produced by the oxidation of hematoxylin. The so-called "ripening" is simply this change, which is brought about by exposing the hematoxylin solution to air. If the pure hematein is used in making the stain, therefore, the latter will be ready for use immediately, because it need not undergo the ripening process (see remarks under 12, p. 9). For formulae see Appendix B, page 230.

Anilin is a colorless coal-tar derivative, and is the base from which many of the numerous coal-tar dyes are made. The anilins are brilliant stains of all colors. They are used almost exclusively for staining sections or thin membranes, and are of great service to the microscopist, although, as a rule, they fade in time.

The basic anilin stains, such as methyl green, methyl violet, gentian violet, methylen blue, safranin, Bismarck brown, toluidin blue, and thionin, are usually nuclear stains. On the other hand, the acid anilin stains, such as acid fuchsin, eosin, light green, orange G, bleu de Lyon, nigrosin, benzopurpurin, and aurantia, are ranked as cytoplasmic stains. Some of these stains must be made up fresh every two or three months, as they frequently spoil if kept much longer.

The metallic substances used for color differentiation operate principally as *impregnations* rather than as stains. The coloring matter is held physically as a precipitate or reduction product in certain of the tissue elements. The commonest reagents of this class in use are silver nitrate and gold chloride.

The different tissue elements frequently show affinity for different stains; consequently it is a common practice to use more than one stain. Very decided contrasts may thus be produced, such as red and blue, red and green, green and orange, etc. It is not uncommon, in fact, to have triple and even multiple staining. In such staining the stains are sometimes applied consecutively; in other cases, at different points in the process of general manipulation. Sometimes all the stains may be mixed together, so that immersion of the sections in one liquid is all that is required for double or multiple staining.

A general rule in staining, especially for entire or bulky objects, is that the specimen should be transferred to the stain from a reagent in which the percentage of water is approximately the same as that of the stain. The same is true when the object is removed from the stain. For example, if the stain to be used is an aqueous solution, the object should enter it from an aqueous solution; if the stain is made up in 95 per cent alcohol, the object should enter from 95 per cent alcohol, etc. For reasons see "Dehydrating."

In the past most stains used by biologists have been known merely by trade names, and inasmuch as different manufacturers have used different methods in preparing their dyes, often those bearing the same name vary widely in actual composition and effects, due mainly to the impurities contained. In recent years, in America, a Commission on Standardization of Biological Stains has been testing and standardizing various dyes used in biological technique. Three series of tests are applied to any stain submitted by manufacturers: (1) its value and reliability for bacteriological staining; (2) for histological staining; and (3) its chemical composition in so far as present knowledge makes this possible. The results of this investigation up to 1929 are set forth in Conn's *Biological Stains, A Handbook on the Nature and Uses of the Dyes Employed in the Biological Laboratory* (2d ed.; published by the Commission, Geneva, N. Y.). *Stain Technology*, a quarterly journal issued by the Commission (ed. H. J. Conn, New

York Agricultural Experiment Station, Geneva, N. Y.), reports the latest findings about stains and staining procedure.

The following stains have been put on a certification basis by the Commission:

Anilin blue, water soluble	Fuchsin, acid	Neutral red
Azure A	Fuchsin, basic	Nile blue A
Azure C	Gentian violet	Nigrosin
Azure Erie garnet	Giemsa stain	Orange G
Bismarck brown Y	Hematoxylin	Orange II
Brilliant cresyl blue	Hematein	Phloxine
Brilliant green	Indigo carmin	Pyronin
Carmin	Janus green B	Rose Bengal
Congo red	Jenner's stain	Safranin O
Cresyl violet	Light green, S. F., yellowish	Sudan III
Crystal violet		Sudan IV
Eosin, alcohol soluble	Malachite green	Tetrachrome stain (MacNeal)
Eosin, bluish	Martius yellow	
Eosin, yellowish	Methyl green	Thionin
Erie garnet B	Methyl orange	Thionin eosinate mixture
Erythrosin	Methyl violet	Toluidine blue
Ethyl eosin	Methylene blue	Wright's stain
Fast green FCF	Methylene violet	

These stains may be obtained through dealers in general laboratory supplies. In ordering, specify "Commission certified stains."

CLEARING

In the vast majority of cases tissues are too opaque for satisfactory examination until they have been treated with certain clarifying reagents or *clearers* which render them more transparent.

Such reagents as glycerin, glycerin-jelly, etc., are used when the object is to be cleared, without alcoholic dehydration, directly from water. Usually, for permanent preparations, the alcoholic dehydration method is employed and it then becomes necessary to use a clarifying reagent which will replace the alcohol and facilitate the penetration of the final mounting-medium (balsam or damar).

Perhaps the most useful and rapid clearer is xylol. Xylol, however, is very sensitive to moisture, and if the preparation has not been thoroughly dehydrated the final mount will appear milky. See page 63, memorandum 25, for *carbol-xylol*. Cedar-wood oil, though somewhat slower than xylol, is one of the best clearers. It is also one of the safest, because tissues may be left in it indefinitely. Other good clearers after alcohol are oil of origanum, sandal-wood oil, oil of cloves, cassia oil (cinnamic aldehyde) toluol, oil of bergamot, anilin oil (for watery specimens), carbolic acid (for watery specimens), and beech-wood creosote. Anilin oil will clear from 70 or 80 per cent alcohol. It should be followed by oil of bergamot, cassia, or wintergreen, according to McClung. Clove oil should not be used for celloidin sections because it dissolves celloidin. It is also inapplicable ordinarily after most anilin dyes because of its tendency to extract them. Among the best reagents for celloidin sections are cedar-wood oil, oil of origanum, creosote, and Eycleshymer's clearer (memorandum 4, p. 72).

While "clearing" refers especially to the rendering transparent of tissue elements, and *dealcoholization* to the removal of alcohol previous to imbedding in paraffin, very frequently the same reagent is used for either purpose and the term "clearing" has come to be used in either sense.

MOUNTING

After tissues have been cleared the final step is to mount them in some suitable medium for preservation and inspection.

If tissues are to be mounted directly from water or aqueous media, glycerin, glycerin-jelly, or Farrant's solution (see p. 120, memorandum 18) is used ordinarily. If the alcoholic dehydration method is employed, balsam, gum damar, hyrax or euparal is the final mounting-medium. The balsam or damar is dissolved commonly in xylol, although turpentine, chloroform, or benzol may be used as the solvent. In my experience xylol-balsam is the most satisfactory for ordinary purposes. However, some of our best American technicians prefer gum damar dissolved in xylol.

IMBEDDING

In order to section tissues or objects satisfactorily it is frequently necessary to imbed them in a suitable matrix. *Simple imbedding* consists in merely surrounding the object by an appropriate medium to hold it in place while it is being cut. In *interstitial imbedding* the object is saturated (*infiltrated*) with the imbedding-substance which, when all cavities and interstices are filled, is caused to set; thus it supports all parts of the tissue and holds the components in place when sections are made. Infiltration imbedding is of great importance to microscopists and much of the space of the present book is given up to drilling the student in the details of the two chief infiltration methods, viz., the *paraffin method* and the *celloidin method*. Infiltration with gum is also not infrequently resorted to, especially for tissues which would be injured by alcohol, or for sectioning by the freezing method.

Paraffin is a translucent, waxy material derived from various sources, one of the commonest of which is crude petroleum. Paraffins of low and of high melting-points, termed respectively soft and hard paraffin, should be kept on hand so that mixtures of different degrees of hardness may be made up as necessity demands.

Celloidin is a form of pyroxylin (guncotton or collodion cotton) specially prepared for interstitial imbedding. It is dissolved in a mixture of ether and alcohol (p. 8, reagent 4) and solutions of two or three strengths are used for infiltration. For details see the method, page 68. Parlodion instead of celloidin is now commonly used by American workers (see memorandum 11, p. 73).

AFFIXING SECTIONS

When mounting sections upon a slide, especially if they are yet to be stained, it is usually necessary to affix them firmly to the slide to prevent later displacement. For paraffin sections Mayer's albumen fixative (reagent 19, p. 11), or a combination of this method with the water method, is most widely used. The water method alone often proves adequate, particularly with thin sections. The slide is flooded with water, or, better, albuminized water made by adding 3 drops of albumen fixative to 30 c.c. of distilled water, and the sections are floated upon its surface. The paraffin ribbon should be gently heated until it becomes translucent but not melted, in order to make it spread and flatten properly. As the layer of water evaporates, the sections are slowly drawn down into close contact with the slide. When perfectly dry they are usually so firmly affixed that they will not become detached even after the removal of paraffin from them. It is common, however, and safer to use a thin film of albumen fixative as a cementing substance between the water and the surface of the slide.

In the case of celloidin sections, if only one or a few sections are to be mounted on one slide, it is a common practice to stain the sections and transfer them through the various reagents, even to clearing, before mounting them on the slide. In such cases the sections need not be fixed to the slide. With serial sections, however, the sections must be

General Statement of Methods

held in place in some way during their transition through the reagents (see memoranda 12 and 13, p. 74). Unlike paraffin, the celloidin is not ordinarily removed from the tissues.

DECOLORIZING

Not infrequently in staining, the tissue becomes overstained and requires that some of the color be extracted from certain of the elements to bring about a proper differentiation. The fact that certain tissue elements retain stain more tenaciously than others is sometimes taken advantage of and overstaining followed by decolorization is practiced intentionally. Alcohol slightly acidulated with hydrochloric acid (0.1 to 1 per cent) is commonly used for the extraction of surplus color. In special cases other decolorizers are used: for example, iron-alum in the iron-hematoxylin method (reagent 17, p. 10).

Overstaining tissue and then partially decolorizing it is sometimes designated as *regressive staining* in contradistinction to *progressive staining* in which the dye, once taken up by the tissue, is not removed. In progressive staining differentiation is accomplished through the selective affinity of dyes for different elements.

BLEACHING

In some cases tissues are obscured because of the presence of natural pigments or on account of blackening caused by the fixing reagent. Such tissues must be bleached. Chlorine, peroxide of hydrogen, or sulphurous acid are commonly employed. A method is given in memorandum 13, page 47. Also see II, page 178.

CORROSION

To obtain skeletal structures, as, for example, the spicules of sponges or the hard parts of insects, various methods of corrosion are employed. Nitric acid, caustic potash, caustic soda, eau de Javelle are reagents often used for this purpose. Corrosion preparations of injected vessels and cavities may also be made.

DECALCIFICATION AND DESILICIDATION

Tissues impregnated with lime salts or with silica must have such hard parts removed usually before they can be sectioned. For decalcification, one of several acids may be used. The details are given in the chapter on bone, tooth, etc. (chap. xii). For decalcifying reagents, see Appendix B, v.

Where desilicidation is necessary hydrofluoric acid may be employed, although, because of its property of attacking mucous membranes, its use is attended with more or less danger for the operator. It is added drop by drop to the tissue which has previously been placed in a paraffin-coated vessel (the acid attacks glass). If the tissue is not too heavily impregnated with silica, it is safer to use an old section razor and try to cut sections without previously treating them with hydrofluoric acid.

INJECTION METHODS

The injection of colored masses into the blood vessels and other vessels of the body is frequently practiced to aid in determining their distribution and their relation to the surrounding tissues. The dye is termed the *coloring mass* and the substance to which it is added the *vehicle*.

ISOLATION OF HISTOLOGICAL ELEMENTS

Isolation is one of the most valuable means of forming a correct conception of cells and fibers. It has the advantage over sections that the elements may be inspected in their entirety and from all sides. The separation is accomplished, as already noted, by (1) reagents which dissolve or soften cell cement and interstitial material without seriously affecting the cells (*maceration* or *dissociation*), or (2) mechanically by means of dissecting needles (*teasing*) or both. Hardening and fixing reagents in general if diluted to about one-tenth are efficient for dissociation. Gage recommends normal saline as preferable to water for dilution. The dissecting microscope or some kind of lens-holder and lens are valuable aids in isolating tissue elements. For practical methods consult chapter xi; for reagents, Appendix B, iv.

NORMAL OR INDIFFERENT FLUIDS FOR EXAMINING FRESH TISSUES

It is desirable frequently to examine fresh material in as near a natural condition as possible, hence recourse is had to the so-called indifferent fluids. While not wholly indifferent, they ordinarily produce but slight changes in tissues and their elements from the viewpoint of the microscopist. The liquids most commonly used for this purpose are discussed in Appendix B, iii.

GENERAL SCHEME FOR MOUNTING WHOLE OBJECTS (*IN TOTO* PREPARATIONS) OR SECTIONS

Whole Objects (for balsam mounts)	*Section Methods* (paraffin and celloidin)
Killing and fixing	Killing and fixing
Washing	Washing
Staining	(Staining, if to be stained in bulk)
(Decolorizing if necessary)	Hardening and dehydrating
Dehydrating	Absolute alcohol
Clearing	
Mounting	

	Paraffin Method		*Celloidin Method*
	Dealcoholization (xylol)		Ether-alcohol
	Melted paraffin		Thin celloidin
	Imbedding		Thick celloidin
If not stained in bulk	Sectioning	*If not stained in bulk*	Imbedding
Through alcohols to stain	Affixing sections	Staining	Sectioning*
Staining	Removal of paraffin	Washing (and decolorizing if necessary)	Dehydrating to 95 per cent alcohol
Washing	Absolute alcohol		Clearing
Dehydrating (and decolorizing if necessary)	Clearing		Mounting
	Mounting		

* If sections are to be arranged serially they are best mounted in sheets or affixed to the slide as soon as cut.

CHAPTER III

KILLING AND FIXING

CAUTIONS.—1. *Use only fresh tissues and work rapidly so that the tissue elements will not have time to undergo postmortem changes.*

2. *Remove organs carefully and avoid crushing or pressing the parts to be prepared.*

3. *Tissues should never be allowed to dry from the time they leave the animal until they are finally mounted for microscopical examination except at one point in the paraffin method.*

4. *Use only small pieces (2 to 6 mm. cube) of tissue whenever possible, or penetration of the reagent will be insufficient. Embryos and small objects up to 4 cm. in size may be placed entire in certain of the fixing fluids.*

5. *For fixing and hardening, the bulk of the fluid should be from 10 to 90 times that of the object. Too many pieces should not be placed in the same vial.*

6. *Use only clean reagents. It is well to let the object rest on a bit of cotton in the bottom of the vial or have it suspended from the vial mouth so that the reagent may penetrate equally from all sides. Penetration is aided by heat.*

7. *When necessary to wash fresh tissue, it is usually best to use normal saline, and not water. Let it flow gently over the surface of the object or slowly twirl the latter in the fluid. Do not scrape off foreign matter.*

8. *In many cases the killing and fixing reagent does not harden the tissue sufficiently and the hardening process must be completed in alcohol.*

9. *Keep the reagents and preparations from direct sunlight.*

10. *Carefully lable each vessel containing tissue. State the contents, the fluid used, and the date. Label on the side.*

11. *Keep on cards a careful record of the reagents used, and the time when changed, for each separate piece of tissue.*

PRACTICAL EXERCISE

In our own laboratory after many years of experience we have settled upon the gill of the fresh-water mussel and the tissues of *Amphiuma* (the "Congo eel") as the best material we can find with which to start beginners in technique. The cells of *Amphiuma* are so remarkably large and the tissues so diagrammatically clear in microscopic preparations that any extra trouble the instructor may be put to in securing this amphibian is well worth while. One animal (which may be secured alive from the Southern Biological Supply Company of New Orleans, La., for from one to three dollars, depending on size) will supply a very large class. In any event, tissues from tailed amphibia are preferable to those from frogs or toads for manipulation by beginners.

Kill the animal by beheading with stout sharp scissors. Open the body cavity as soon as possible and secure the tissues specified. Inasmuch as the intestinal lining begins to digest itself promptly, tissues from the alimentary tract should be the first to be removed and placed in the fixing fluid. The valves of the clam or mussel should be pried apart and bits of the gills quickly clipped off and placed in the fixer.

1. **Fixing with Zenker's Fluid.**—Place small bits of mussel or clam gill in a vial or small bottle of Zenker's fluid.

Kill a urodele amphibian (preferably *Amphiuma*) as directed above. Remove a piece of the intestine about 12 mm. long, and after washing it thoroughly in normal salt solution place it in a vial of the fixing mixture. Also, in separate vials place small pieces of liver, kidney, spleen, cardiac, and pyloric ends of the stomach, bladder, spinal cord, and brain. In every case use by bulk at least ninety times as much fluid as tissue. After fixation, which requires from 6 to 24 hours, wash the objects in running water for from 12 to 24 hours. Then transfer them through 35 and 50 per cent alcohol (20 minutes each) into 70 per cent alcohol. Add sufficient iodine solution to give the alcohol a port-wine color. The iodine will remove any mercuric crystals which may have formed in the tissues. As often as the color disappears the iodine must be renewed. After from 12 to 36 hours of this treatment, the color persists and the objects should then be transferred to fresh 70 or 80 per cent alcohol, which must be renewed until it no longer extracts iodine from the specimens. Too prolonged washing with iodine solution tends to undo the work of fixation and to hinder staining. Many workers prefer to omit the treatment with iodine until the tissue has been cut into sections. In such cases slides bearing sections are treated with dilute iodized alcohol for 30 minutes and then washed thoroughly in 70 per cent alcohol.

Zenker's is one of the best general fixing agents known, particularly for ordinary vertebrate histological work. Any of a great variety stains may be used after it, and it fixes satisfactorily almost any kind of tissue. The time during which objects should be left in the fluid varies from 30 minutes for delicate ones to 24 to 36 hours for larger or denser tissues, although many objects may be left a longer time without injury.

2. **Fixing with Bouin's Fluid.**—Place small pieces of trachea, tongue, cornea, intestine, and testis or ovary in Bouin's fluid for from 4 to 16 hours. After fixation, wash the tissues in several changes of 50 per cent alcohol, then in several changes of 70 per cent alcohol. Preserve in 70 or 80 per cent alcohol. Bouin's fluid is an excellent reagent which gives a very delicate fixation. It is one of the safest for the beginner because it is almost impossible to go wrong in its use. Objects may be left a considerable time in it without injury. For staining after Bouin see comment, p. 235.

3. **Formalin as a Fixing Reagent.**—Place a piece of spinal cord, sciatic nerve, liver, and fragments of muscle in which nerves terminate, in 10 per cent formalin and leave until needed for work later. Formalin in varying percentages is widely used for the preservation and fixation of specimens for dissection. It has been employed especially for the central nervous system. However, Miss Helen D. King (*Journal of Comparative Neurology*, XXIII, No. 3 [August, 1913]), who has made a careful study of the effects of this fluid on the brain of the white rat pronounces it unfit for cytological work because of its tendency to swell brain tissue. On the other hand, she finds that nerve tracts are apparently not affected adversely by it. If a formalin-fixed brain is to be used for histological purposes, she advises its transfer to alcohol as soon as it is fixed and hardened. Formalin ordinarily has a slightly acid reaction due to the presence of formic acid. Miss King finds such formalin less harmful than that which has been neutralized. For faithful preservation of cells, however, she prefers some other fixing fluid, such as Bouin's. For simple preservation, solutions ranging from 2 to 5 per cent are adequate, but for fixation the solution should be stronger (10 per cent). Entire human brains may be fixed and hardened in a 10 per cent solution.

Formalin is much used for fixing and preserving when frozen sections are to be made, and it is particularly serviceable where a study of fat is desired. It also interferes less with microchemical tests than most other reagents.

MEMORANDA

1. **Tissues Are Preserved in Alcohol** of from 70 to 80 per cent strength, but if they are to remain several months it is better to preserve them in a mixture of equal parts of glycerin, distilled water, and 95 per cent alcohol. Tissues left long in alcohols of grades higher than 80 per cent tend to become hard and brittle.

2. **Hardening.**—Read carefully the remarks on hardening in chap. ii.

3. **Tissues Should Not Be Left in the Fixing Agent** longer, ordinarily, than is necessary to get results. Some, however, require a long time to bring out the optical differences of their elements. Experience alone can teach the time required in a given case. Such a reagent as formalin kills, fixes, hardens, and preserves, all at the same time. However, see remarks under 3, p. 30.

4. **Fixation by Injection** is highly advantageous with many tissues because the fixing fluid is brought quickly into contact with all parts of the body. The vascular system is first washed out with normal salt solution and then filled with the fixer. For fluids containing corrosive sublimate a glass syringe and cannula, instead of a metal one, should be used.

5. **Hollow Organs** should be filled with the fixing fluid and then suspended in a vessel of the same.

6. **For Transferring Small Objects** through reagents the method of Chao-Fa Wu has proved most practicable in our laboratory. A cone is made by folding twice a square of ordinary lens-paper. Small objects may be carried through the alcohols, clearing fluid, and paraffin baths in this lens-paper cone, thus avoiding injurious handling with forceps or other instruments. In imbedding, unfold the cone, lower the surface of the lens-paper to which the objects are attached into the melted paraffin. The hot paraffin loosens the objects from the paper. They may then be arranged in the proper positions with a warm needle.

Walton uses shell vials which measure about 10 cm. in height by 3 cm. in diameter. Through the center of a flat cork which fits the vials a hole is made and a glass tube (about 9 cm. $\times$ 1.5 cm.) is inserted so that its lower end dips well into the reagents in the vials. The lower end of the tube is closed with fine-meshed cloth and the objects are placed within the tube. To transfer the objects one simply removes the cork bearing the tube and inserts it in the vial containing the desired reagent. The upper end of the tube may be closed with a cork of the proper size. To avoid disturbance from changes in air pressure a small hole should be bored in the side of the tube just below the lower level of the larger cork. The vials are supported as indicated in memorandum 7.

Very small objects (e.g., small eggs) may also be made up into little packets in bits of the cast-off epidermis of the frog or salamander, according to the method of Professor Boveri. The epidermal film is spread over the concavity of a hollow ground slide and saturated with alcohol of the same strength as that which surrounds the objects. After the latter have been transferred by means of a pipette to this sheet of epidermal tissue, two opposite edges of the sheet are folded over the objects, then the other two are brought together, twisted about each other, and pinned with a fine pin. The pin, bearing a label, is used as a handle to transfer the little bag of objects from one reagent to another. Since the epidermal tissue cuts readily, it need not be removed if the objects are to be imbedded for sectioning.

Professor C. E. McClung places fresh bits of tissue on small strips of paper with the proper index number on the opposite side. The paper is then inverted on the surface of the fixing fluid. The tissue will adhere to the paper through all subsequent processes. For washing, dehydrating, etc., simply float the paper on the proper fluid. A number of objects may thus be handled together. The plane of section may also be marked by the paper.

Gelatin capsules, such as are used for medicines, have been recommended for small objects by various workers. A hole, of sufficient size to let through the reagent but hold back the objects, is pricked in each end of the capsule.

7. **Shell Vials, Small Bottles,** etc., when in use are best supported in shallow auger holes of proper size in thick blocks of wood.

8. **Material Which Is To Be Kept Indefinitely** should be put in tightly stoppered vials in a place away from strong light. Glass stoppers should be used, since cork, besides shrinking and disintegrating with age, may give off extractives which injure delicate tissues. It is best to pack the vials in a museum jar on cotton and then seal the jar securely to prevent evaporation. Material is even more secure if the museum jar is partly filled with alcohol; in such a case each small vial should have a label of the contents placed within it.

Another way to prevent evaporation from vials or bottles is to "cap" them with a suitable varnish (see memorandum 9).

9. **To Seal Bottles and Preparation Jars** ("bottle-capping"), dip the stopper and part of the neck in collodion varnish made as follows:

Pyroxylin (e.g., collodion or celloidin)............	1 oz.
Ether..	6 oz.
Alcohol......................................	8 oz.

When the pyroxylin has completely dissolved add 2.5 drams of camphor (from *Pharmaceutical Era*, XXX, 528).

10. **For the Preservation of Anatomical Material** for other than cytological or histological dissection, Professor George Wagner of our own laboratory finds nothing as serviceable as Keiller's fluid. The formula is as follows:

Formalin.............................	1.5	parts
Carbolic acid.........................	2.5	parts
Glycerin.............................	10	parts
Water...............................	86	parts

If a good grade of carbolic acid is used, disagreeable odors will largely be avoided. For embalming, a suitable reservoir is filled with the fluid and suspended some 3 or 4 feet above the body to be preserved. About 6 feet of rubber tubing, provided with clamps and a glass cannula of proper size to fit the femoral artery, is attached to the reservoir. After killing the animal with illuminating gas or chloroform, the fluid is injected through the femoral artery. An important precaution is to have the column of fluid in the tube and cannula free from air bubbles or foreign materials. The pressure should be such at all times as to prevent blood from running back through the cannula. The animal should be subjected to this embalming process for at least two hours. Obviously, by increasing the size of the reservoir and the number of tubes, several animals may be treated at the same time.

At the conclusion of the embalming process, if the arteries are to be injected with a colored mass, the rubber tube should be disconnected from the cannula and the latter be left in place in the artery. After the animal has remained for 24 hours in an upright position the injection may be undertaken. For study of the blood vascular system, however, some workers prefer to inject fresh animals and preserve them in formalin. For a satisfactory starch injection mass see memorandum 16, p. **97.**

In our own laboratory it is the practice to skin the embalmed animal and wrap the body in cheesecloth saturated with the embalming fluid. It is kept in a zinc box of suitable size which has a close-fitting lid. We find a box 25 inches long, 8 inches wide, and 9 inches deep a very good size for cats. For a more detailed account of embalming with Keiller's fluid see Bensley, *Practical Anatomy of the Rabbit*, pp. 194–98.

The following mixture, recommended to the author by Professor Kincaid of the Washington State University, has given most excellent results. To a mixture of equal parts of glycerin and 95 per cent alcohol sufficient formalin is added to make the whole about a 2 per cent formalin. Specimens remain perfectly flexible in this mixture, and, indeed, after they have become thoroughly saturated, many forms (crustacea, insects, etc.) may be removed and kept as dry specimens which still retain their flexibility.

11. **Material Which Has Been in Formalin** and is to be dissected may be rendered more pliable, and **harmless to the skin**, by soaking in a 3 per cent solution of carbolic acid.

12. **For Washing Out Specimens** it is advisable to have the laboratory water-pipe provided with numerous small cocks about 10 cm. apart, so that each student may have the use of one or more. A piece of rubber tube, long enough to reach to the bottom of the vessel containing the specimen and fitted with a bit of glass tube in the free end, may be attached to each outlet. If the objects are very small they may be placed in perforated porcelain thimbles which may be purchased from instrument dealers. For minute objects the

thimbles may need to be lined with fine gauze. Bits of glass tubing with fine gauze over the two ends are also useful for small objects. Chi-Hsiun Chu (*Science*, LXXVIII [1933], 510) puts bits of tissue in small glass tubes separated by cheesecloth and connected to one another and to the faucet by rubber tubes.

13. **Alcohol Fixation** though in most instances unsatisfactory is sometimes employed because of its ease of manipulation or for the want of other fixers. Small pieces of tissue should be hardened in absolute alcohol (at least not less than 95 per cent) for at least two days. Larger pieces of tissue require longer time. They should be thin. The alcohol should be changed every day for the first three days. Hot alcohol is often used for insects. Acetic acid (p. 230) is used with alcohol sometimes to increase penetration and to counteract its tendency to shrink tissues. This mixture is usually preferable to alcohol alone. Since alcohol is a reducing agent, it should not be used in a mixture with chromic acid, osmium tetroxide (osmic acid), or potassium bichromate.

14. **For Tissues Which Harden Unduly in the Usual Fixing Agents,** and for tissues in general, Petrunkevitch (*Science*, Vol. LXXVII, No. 1987 [January 27, 1933]) recommends a cupric-paranitrophenol fixing fluid. The formula is as follows:

60 per cent alcohol	100 c.c.
Nitric acid (c.p., sp. gr. 1.41 to 1.42)	3 c.c.
Ether	5 c.c.
Cupric nitrate (c.p. crystals $Cu(NO_3)_2$)	2 grams
Paranitrophenol (c.p. crystals)	5 grams

The mixture remains stable for months in glass-stoppered bottles. The rate of penetration is about 0.5 mm. an hour. After fixation wash in several changes of 70 per cent alcohol. Fixation in this fluid is also recommended as preliminary to macroscopic dissections of invertebrates and of mammalian embryos.

CHAPTER IV

SIMPLE SECTION METHODS

FREE-HAND SECTION CUTTING

This method is important because it requires no costly appliances; although the sections are not as accurately cut as when mechanical aids are used, the method is simple, rapid, and adequate for the more general histological and pathological work.

1. The section razor is flat on one side (the lower) and hollow ground on the other (Fig. 15). *It must be sharp.*

2. A shallow glass dish or watch-glass partly filled with water is also necessary. Before making a section, dip the razor flatwise into the liquid, or use a camel's hair brush; see that the upper surface is well flooded.

3. Sit in such a way that the forearm may be steadied against the edge of the table.

4. Use a piece of liver which was fixed in formalin, first rinsing it in water. Take the tissue between the thumb and forefinger of the left hand, and hold it in such a way that a thin slice may be cut by drawing the knife along the surface of the forefinger.

5. Rest the flat surface of the knife upon the forefinger, and, beginning at the heel of the knife, carefully draw the blade toward you diagonally through the tissue, slicing off a thin section of as uniform thickness as possible.

6. As each section is cut, float it off into the water; if it adheres to the blade, remove it by means of a wet camel's hair brush.

7. Practice until very thin sections are obtained, then place the dish upon a black surface, and with a needle or section-lifter transfer the thinnest and best sections, if only fragments, to a watch-glass containing water.

NOTE.—In case the tissue has been preserved in alcohol, cut the sections under 70 per cent alcohol instead of water, then transfer them to 50 and 35 per cent alcohol successively and finally to water, leaving them in each liquid from 3 to 5 minutes.

8. Next, place the sections in about 3 c.c. of Delafield's hematoxylin diluted with an equal volume of water, and leave them for various

lengths of time (3, 7, 12, or more minutes) to determine the time for successful staining.

9. Transfer the sections from the stain to tap water, and gently move them about for from 5 to 10 minutes to wash out the excess of the stain. If the sections are still overstained, place them in 5 c.c. of distilled water to which 3 drops of acetic acid have been added. Leave for 5 minutes, or until they become lighter in color, then wash in several changes of tap water until they have again become blue.

10. Remove the sections from the water and transfer them through 35, 50, 70, 83, and 95 per cent alcohol successively, leaving them from 3 to 5 minutes in each, and then transfer them to absolute alcohol for 10 minutes, and finally to xylol for 10 minutes, or until clear.

11. Select one or two of the best sections and transfer them to the center of a clean glass slide. After straightening them out properly, drain off the excess of the clearer and before the sections can become dry, add a drop of Canada balsam. Carefully lower a *clean* cover-glass (for cleaning see memorandum 14, p. 60) on to the balsam. There should be just sufficient balsam to spread evenly under the cover without exuding around the edges.

12. Label, stating card number, name of the preparation, and other data that it is desired to add (see p. 53, step 10).

13. Carry one of the pieces of stomach prepared in Zenker through the same treatment. The sections should be transverse sections of the stomach wall.

14. Clean up all dirty glassware *immediately*.

MEMORANDA

1. **The Thinnest Sections** are not always the best. For a general view of an organ, large, comparatively thick sections are usually better; for details of structure, thin sections.

2. **Small Pieces of Tissue** may be cemented to a cork if too small to hold conveniently between thumb and forefinger. A piece of stout copper wire is heated for a moment in the flame and touched to a bit of paraffin. As the paraffin melts, transfer drops of it to the edge of the tissue, which has been previously placed on the cork. The paraffin cools and holds the tissue fast.

Another and better method of handling a small object is to imbed it in a piece of hardened liver. In sectioning, the liver as well as the object is sliced, but they readily separate when placed in alcohol. Beef liver or dog liver is prepared for such purposes by hardening pieces about $5 \times 2 \times 2$ cm. in size in 95 per cent alcohol for 24 hours, and then transferring to fresh 95 per cent alcohol until needed. When much hand sectioning is to be done, a supply of hardened liver should be kept on hand. Many small objects may be held between pieces of pith and successfully sectioned.

3. **Well Microtomes** (Fig. 26) are inexpensive instruments which are used for simple sectioning. Such a microtome consists of a tube in which the object is placed, and at one end of which is a plate to guide the razor. The other end is provided with a screw, which, when turned, pushes the contents of the tube above the plate, thus making it possible to cut sections of a uniform thickness. The object to be cut must be firmly fixed in the well. Such tissues as kidney, liver, spleen, hard tumors, cartilage, etc., may be held sufficiently rigid by wedging small slabs of carrot, turnip, pith, or hardened liver in about them. These supporting substances must, of course, rest squarely against the bottom of the well. Soft tissues, such as soft tumors or brain, must be imbedded. Three parts of paraffin and one part of vaseline melted together and thoroughly mixed makes a very good imbedding-mass for a well microtome. To imbed, warm the microtome slightly and fill the well with the imbedding-mixture. Remove all liquids from the surface of the tissue, and pass it below the surface of the mixture just as it begins to harden around the edges.

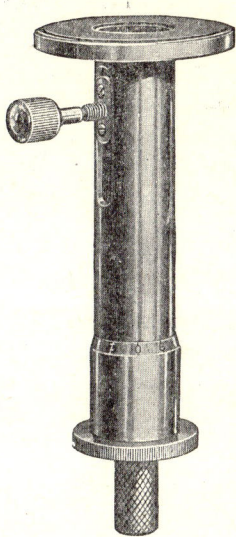

Fig. 26.—Well Microtome

When the imbedding-mass has become cold the sections are cut in the ordinary way.

4. **Temporary Mounts** may be made directly from water after staining by using glycerin as a mounting-medium. Transfer the section to the slide, add a drop or two of glycerin, and a clean cover-glass.

CHAPTER V

THE PARAFFIN METHOD: INFILTRATION AND SECTIONING

1. From 70 per cent alcohol take a small piece of mussel gill or of intestine (6 mm. long) fixed in Zenker's fluid, and also pieces of other tissues fixed in this fluid, and proceed according to the following schedule. Keep accurate records on your cards.

2. Immerse in 95 per cent alcohol for 20 to 30 minutes. A longer time will do no harm.

3. Transfer to absolute alcohol, ½ hour. Before transferring to absolute, remove the excess of 95 per cent alcohol from the object by touching it with a piece of blotting paper or a clean cloth, but do not let it become dry.

4. Cedarwood oil and absolute alcohol equal parts, 15 minutes.

5. Cedarwood oil, 1½ hours, or until the object looks clear. It can be left indefinitely as cedar oil does not harden. Rinse in xylol for from 2 to 5 minutes. Rapidly remove all excess of xylol before proceeding with step 6, but do not allow the tissue to become dry or dull looking. (For delicate objects see p. 57.)

6. Melted paraffin (melting-point about 52° C.), 2 hours. For large objects leave 8 hours or overnight. It is best to avoid as much as possible subjecting tissues to an elevated temperature.

It is well to have three vessels of the paraffin and to shift the object from one to the other at equal intervals during the process of infiltration. This facilitates penetration and gets rid of all clearer that may have been carried over to the first dish (see memorandum 8, n. 2, p. 50).

The duration of the paraffin bath varies according to the size and density of the object. Many objects of from 3 to 5 mm. in thickness are thoroughly saturated within an hour or less; others which are more impervious or which have impenetrable coverings may require several hours or even days. Lee, assuming that melted paraffin will penetrate as quickly as cold oil, takes as a guide the length of time required to clear the object in cedar oil.

CAUTIONS.—*a*) *Do not have the bath too hot.* Cooked tissues are worse than useless.

b) To keep material clean, it is well to have a false bottom of paper in the vessel containing paraffin. Make this by swinging a strip of white paper into the cup so that the loop of the paper is submerged in paraffin and the ends attached on either side to the mouth of the cup.

7. Prepare paper boxes according to the following instructions:

A small rectangular block of wood or a stick with a flat end measuring approximately 15×20 mm. is used. Cut a strip of stiff glazed paper so that it measures about 4×7 cm. Place the flat end of the block in the center of the paper with its long diameter coinciding with the long diameter of the paper. Fold the narrow side margins of the paper up along the sides of the block first, then do likewise with the ends of the paper. Turn the ears which have been formed at each corner back over what is to be the end of the box, and then fold the long end of the paper back to hold the ears in place, and also to make the end of the box of the same height as the sides. Manifestly, any size of box may be made by varying the size of the block. With a little practice the same kind of box may be folded without the use of a wooden block. See, however, memorandum 15, p. 47.

8. With a *warm*, wide-mouthed pipette transfer sufficient melted paraffin to a paper box (or Syracuse watch crystal, memorandum 15, p. 47) to cover the bottom, then, with warm forceps, remove the tissue to the box. Do not leave the tissue in the air long enough to form a film of solid paraffin on its surface! Next, fill the box with melted paraffin. This must be added before the surface of the underlying paraffin has congealed, otherwise an awkward seam will occur. Orient the object with heated needles if necessary. As soon as the paraffin has congealed sufficiently to form a thin surface film, cool it *rapidly* by plunging it into cold, preferably running water; otherwise the paraffin will crystallize and become unsuited for sectioning. Some workers prefer cold alcohol to water for hardening the paraffin block. Waste alcohol may be saved for this purpose.

CAUTIONS.—*a)* Tissues must be oriented (i.e., placed in proper position for cutting) while the paraffin is still in liquid condition. Arrange the tissue so that it will be cut at right angles (transverse) or parallel to the surface of the organ. Avoid oblique sections as they are very puzzling. For present purposes of practice cut transverse sections.

b) If whitish-looking patches are present in the block after imbedding, they are probably due to xylol or cedar oil which has been carried over into the paraffin, or to crystallization. If they occur in the immediate vicinity of the object, the block should be placed in the bath again until melted, and the object should be reimbedded.

c) Be sure that every piece of tissue is marked after it is imbedded. Tissues are sometimes kept in paraffin for months or even years before they are finally sectioned. To mark, scratch the number of the record card in the paraffin or, better, write it on the paper box or tab and leave the latter in place.

CUTTING SECTIONS

9. Study the paraffin microtome (e.g., Fig. 27 or Fig. 28); identify the parts and learn how the thickness of sections is controlled.

10. Proceed with the block of paraffin containing the mussel gill or the intestine. Make it fast to the carrying disk of the microtome in the following manner: Remove the disk from the machine and, by means of a heated steel spatula or copper wire flattened at one end, melt a small chip of paraffin on to it. Likewise warm the end of the paraffin block and quickly press it into the melted paraffin on the disk. Cement it firmly in place by means of the heated wire or spatula and cool in water.

11. With a sharp scalpel trim the free end of the block so that it presents a perfectly rectangular outline (however, see caution *c*). The length should exceed the breadth by at least one-fourth.

CAUTIONS.—*a)* In trimming do not cut farther back than the base of the object. This leaves a wide shoulder for support.

b) Leave a margin of about 2 mm. around the object.

c) To avoid reversing sections in mounting, it is frequently advantageous to have the imbedding-mass trimmed unsymmetrically. The edge which first comes in contact with the knife is left longer than the opposite edge. One may thus readily discover when a section or part of a series has been turned over.

12. Mount the object firmly in the microtome. It should just clear the knife. The flat end-surface of the paraffin block should be parallel to the edge of the knife, and the block so oriented that, in cutting, the long edge will meet the edge of the knife squarely.

13. Place the knife in position with the handle to the side away from the wheel (if a rotary microtome is used). By means of the adjusting screws tilt the cutting edge slightly toward the object so that the side of the knife will not remain in contact with the paraffin block after a section has been cut. If the knife has a flat under surface it requires more tilt than if the surface is hollow ground. For a flat under

The Paraffin Method

FIG. 27.—Minot Automatic Rotary Microscope

The object-carrier is adjustable in three planes and is perfectly rigid. The knife-carrier is also adjustable and extra heavy and solid. The feed is controlled by an adjustable cam, giving cuts of any number of microns in thickness from 1 to 25. By means of an automatically closing split-nut the carriage is returned to the beginning position after the screw is fed out the entire length. For details of construction see catalogue of the Bausch & Lomb Optical Co.

surface the tilt should be about 9 degrees from the perpendicular. See that the knife is held firmly in place.

NOTE.—It is impractical to attempt to supply and keep in repair a sufficient number of microtome knives for a large class in microscopical technique. In our own laboratory each student is required to provide himself with a package of Gillette safety razor blades. Special holders for safety razor blades may be obtained from makers of rotary microtomes. These are substituted for the regular microtome knife. This combination is very satisfactory for routine student work.

FIG. 28.—Spencer Rotary Microtome

By means of a wheel at the back the feed may be set for any thickness from 1 to 60 microns. For details of construction see catalogue of the Spencer Lens Company.

The razor blade must be held firmly in place in the holder, with not too much of the edge exposed, otherwise the edge will spring and give unequal sections. Some students prefer to provide themselves with a "section razor" of folding type, made especially for sectioning (see catalogues of microscope dealers).

CAUTION.—The knife should be kept in its case when not in the machine. The edge is very easily injured.

14. Set the regulator so that the microtome will cut sections about 10 microns thick. A micron is one-thousandth of a millimeter.

15. Unloose the catch which locks the wheel and revolve the wheel with the right hand. A few revolutions should bring the block of paraf-

The Paraffin Method

fin into contact with the knife. As each new section is cut, it displaces the last one and if the paraffin is of the proper consistency unites by one edge with the displaced section. Thus a ribbon or chain is formed. When the ribbon becomes of sufficient length, support the free end by means of a hair brush held in the left hand. To prevent breaking the ribbon avoid pulling it taut. Various ribbon-carriers have been devised for attachment to the microtome. The best of these is the cylindrical carrier made by the Spencer Lens Co. according to the suggestions of Dr. C. E. McClung. For an inexpensive, home-made form of this see Hance, *Anatomical Record*, X, No. 8 (June, 1916).

CAUTION.—Never bring a needle or other hard object near the edge of the knife. If the paraffin does not ribbon properly, consult the table at the end of this chapter.

16. When a sufficient number of sections have been cut, carefully place the ribbon on a piece of paper. Protect it from draughts of air which will carry away or disarrange the sections.

17. Cut the ribbon into strips of such length that they may be placed in successive rows one above the other under the cover-glass that is to be used. Allow one-fourth for the expansion of the ribbon when heated. Mark out on a sheet of paper the exact size of the cover-glass so that there can be no mistake in cutting strips of the proper length. A margin of 2 or 3 mm. should be allowed for the cover.

18. Place a small drop of albumen fixative on a clean glass slide (for cleaning see memorandum 14, p. 60), and spread it evenly over the surface, except the end which is to bear the label (see step 10, p. 53). With a clean finger rub off all of the fixative that can be easily removed so that only a very thin film remains.

19. Flood the slide with a few drops of distilled or albuminized (p. 24) water until the entire surface bearing the fixative is covered by a thin layer of water, but do not put on sufficient to overflow the edge. Some workers use albuminized water alone for affixing sections.

20. Take up the first strip of paraffin ribbon with a brush or needle and float it on to the surface of the water. The first section of the series should be in the upper left-hand corner, but back at least 10 mm. from the end of the slide. In case the label is to be placed on the left end of the slide, allowance must be made for it, of course. Add the successive strips of the ribbon in the order of the lines of a printed page until as many rows are in place as will conveniently lie under the cover (see step 17, p. 43), allowing for the proper margins. See that each

section presents the same aspect to the observer as its predecessor (see step 11, *c*, p. 40).

21. Warm the slide gently until the paraffin flattens out and becomes free from wrinkles. Be careful not to melt the paraffin, for heat sufficient to do so will render the albumen useless. Remove air bubbles. It is safer to heat the slide by placing it upon a warming table for a few minutes, instead of holding it above a flame. See page 14, memorandum 6, or Figure 24.

22. Drain off the excess of water and set the slide away to dry after properly numbering it with your glass-marking pencil. As the water evaporates, the sections are drawn down tightly into the film of fixative. If, after drying, air is present under the sections it may be seen from the glass side of the slide. Such sections will float off in subsequent treatment, hence the importance of removing air bubbles (step 21). The slide is seldom sufficiently dried under 6 hours. It is well to leave it 12 hours; it may be left indefinitely. The time may be shortened by placing a few thicknesses of blotting paper under the slide and drying it on a warming table or in an incubator. It is often convenient simply to place the slides in an ordinary slide box without cover and leave in an incubator until dry. Unless the slide is perfectly dry and the ribbon fully spread, the sections will float off during subsequent treatment. Take precautions to prevent particles of dirt from settling upon the surface of the sections. This is usually accomplished by placing the slides upon some kind of a rack and covering them with a bell-jar. Prepare several other slides in the same manner as the above if sufficient of the ribbon remains.

NOTE.—As time permits, cut the other sections which are imbedded in paraffin. When, as in the present case, it is not necessary to have a complete series of sections, you may place fewer sections on a slide and use smaller covers.

When a small cover is to be used, place the sections at the center of the slide. The center may readily be determined by drawing the outline of a slide on a card and connecting the opposite corners of the figure by means of diagonal lines. When mounting, place a slide over the diagram; the intersection of the diagonals shows the center.

At this point the student should make a careful study of Appendix A if he is not already thoroughly acquainted with the optical principles involved in microscopy.

MEMORANDA

1. **If Paraffin Becomes Dirty** it should be melted and filtered through a heated metal funnel.

2. **Oil of Cedar,** if used for dealcoholization before imbedding, should be

followed by at least two changes of paraffin or the paraffin does not thoroughly replace the oil and the object is likely to drop out of the sections as they are cut. In my experience this is the commonest difficulty which beginners encounter if they use cedar oil for dealcoholization. For this reason xylol or chloroform although less favorable for the tissue, is often used.

3. **Objects Imbedded in Paraffin** may be preserved in that form indefinitely. It is one of the most convenient ways, in fact, of preserving material which is to be sectioned in paraffin.

4. **Small White Objects,** if not stained before imbedding, should be tinged with a dilute solution of Congo red to facilitate orientation. For orientation in general see p. 143, memorandum 1.

5. **With Delicate Tissues** it is necessary that the transition from alcohol to clearer be gradual, hence it is best to add the clearer, a little at a time, to the last alcohol, transferring it with a pipette *to the bottom* of the alcohol. See also "drop" method, pp. 18 and 174.

6. **The Temperature of the Laboratory** must be taken into account when sectioning in paraffin. In summer use a harder, in winter a softer, paraffin.

7. **For Thin Sections** use a hard paraffin, for thick sections, a softer paraffin.

8. **Bayberry Wax** (Pohlman's suggestion) added to paraffin with a melting-point of 52°, in the proportion of about 1 to 10, makes a good imbedding medium for routine work, various workers in our laboratories find. It decreases the melting-point while increasing hardness, and prevents crystallization. Too much makes the mass crumbly.

9. **For Valuable Tissues Which Crumble in Paraffin Alone** the following somewhat tedious process (Mark, *American Naturalist* [1885] p. 628) may be resorted to. Prepare a very fluid collodion in ether-alcohol and coat the exposed surface of the object immediately before cutting each section. If the collodion leaves a shiny surface or produces a membrane when applied to the paraffin, it is not thin enough and must be further diluted with ether-alcohol. Apply the collodion with a brush with all excess of the fluid wiped away so that the brush is just moist. The fluid should touch only the face of the block in which the object is exposed. After applying, wait a few seconds for the solution to dry before cutting. See also memorandum 10.

10. **Johnson's Paraffin-Asphalt-Rubber** Method for brittle objects is a very useful one. One part of crude India rubber cut into very small pieces is mixed with 99 parts of hard paraffin which has previously been melted and tinged to a light amber color with a small amount of asphalt ("mineral rubber"). The mixture is then subjected to a temperature of 100° C. (not higher) for 24 to 48 hours, or left in a paraffin oven at 60° C. for several days. Use only the supernatant fluid. It is allowed to cool and remain cold until needed, because the rubber separates out after a time if the mixture continues melted. Johnson (*Journal of Applied Microscopy*, VI, 2662) recommends it

as even better than paraffin for all kinds of work for which paraffin is commonly employed. Proceed as in the ordinary method, using xylol (*not* cedar oil) for dealcoholization and also for clearing sections.

Hance (*Science*, LXXVII [1933], 353) chops crude rubber with scissors and drops it into melted paraffin which is smoking hot, stirring occasionally. About 20 grams of rubber can be dissolved in 100 grams of paraffin (after 3 to 4 hours). He finds that 4 to 5 grams of this mixture and 1 gram of beeswax added to 100 grams of melted paraffin makes, after filtering, an excellent imbedding mixture, pale yellow in color.

11. **Keep All Parts of the Microtome** clean and well oiled with watch oil or pure paraffin oil of 25° C. The instrument should be covered when **not** in use.

12. **Keep the Microtome Knife Sharp.** It should receive frequent stroppings. For sharpening the knife two hones are commonly used.

Honing.—If the knife is very dull it is first honed on a Belgian yellow hone, an open-grained stone which cuts the metal of the knife rapidly. The surface of the stone is kept moist with filtered kerosene oil or lathered with palm-oil soap. After the nicks and other inequalities of the edge of the knife have been removed, the honing is best finished on a good fine-grained blue-water stone. If the knife is well cared for the yellow stone is seldom necessary.

In honing, the stone is laid flat on the table with its end toward the operator and its surface properly lubricated. A very dull knife is ground at first on the concave side only until it develops a fine "wire edge" along the full length of the blade. It is then ground on each side alternately until the wire edge has disappeared completely. In grinding, the knife must remain flat on the hone and pass lightly over the full length of the surface, edge foremost in a diagonal direction from point to heel, although itself remaining at right angles to the long axis of the hone. The honing has been sufficient when all nicks and wire edges have disappeared and the knife, instead of catching and hanging when the edge is drawn lightly across the ball of the thumb, freely enters the moist epidermis. Finally the blade is wiped clean with a soft cloth, great care being taken not to injure the edge.

Some workers prefer to use first a yellow Belgian hone wetted with a moderately thick soap solution, then an Arkansas stone with a thin oil. Hardesty (*Laboratory Guide for Histology*, p. 186) uses only a white Arkansas stone with oil.

Stropping.—A broad firm strop of finest calfskin is best. It should be affixed to a solid back so that it will not spring and thus round off the delicate edge of the knife.

In stropping, the motions are the same as in honing (both sides of blade), only the knife passes *back* foremost and from heel to point. The blade must move lightly over the surface of the strop with very slight pressure on the part of the operator. The stropping is ordinarily considered sufficient when the blade will cut a loose hair freely along every part of the edge. An examination under a low power of the microscope should reveal no nicks in the edge. A good workman strops his knife frequently.

13. **To Remove Pigments and To Bleach Osmic and Chromic Acid Materials,** a 3 per cent solution of peroxide of hydrogen frequently is sufficient. Tissues left too long in this liquid macerate.

Mayer's chlorine method is one of the best for bleaching. To several crystals of chlorate of potash in a glass tube a few drops of hydrochloric acid is added. When the greenish fumes of chlorine appear, add from 5 to 10 c.c. of 50 per cent alcohol. The object, which in the meantime has been standing in 70 per cent alcohol, is transferred to the tube. From 15 minutes to 24 hours are required for bleaching, depending upon the nature of the material. It is well to suspend the object from the mouth of the bottle. Sections on the slide may be bleached in a few minutes. This method is especially recommended for removing natural pigments and for bleaching osmic material.

H. W. Beams finds that one-half of 1 per cent solution of gold chloride, for periods of 12 to 24 hours, is a good bleacher following fixing fluids which contain osmic acid. It is also useful following impregnations in osmic acid where a careful bleaching of the tissue is desired. Also see II, p. 178.

14. **Large Objects May Be Cut in Paraffin** better with a slanting knife than with a square-set one. The block of paraffin must be trimmed to a three-sided prism with its most acute angle farthest from the object. A sliding microtome is used ordinarily and the block of paraffin is so oriented that the knife enters at the sharpest angle of the prism. Each section as cut is removed with a brush.

15. **Metal or Porcelain "L's" Are Frequently Used Instead of Paper Boxes** for molding paraffin blocks. The two L's (Fig. 29) may be placed together on a small glass or metal plate in such a way as to mold blocks of any desired size. Before pouring the melted paraffin in, the inner walls of the metal pieces should be lightly smeared with glycerin so that the block of paraffin will easily separate from them when cool.

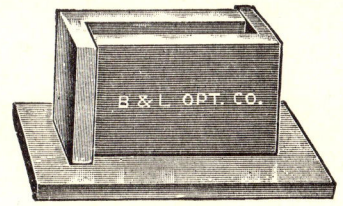

Fig. 29.—Metal L's for molding imbedding-masses.

Many workers, particularly for small objects, imbed in a watch-glass. Glycerin should be smeared over the inside of the glass and then wiped out until almost dry. After the object is oriented and before the paraffin has begun to harden a small strip of paper may be inserted at one edge of the watch-glass to serve as a label. With a little practice, small objects can be imbedded directly on the surface of an ordinary slide without any mold. Baumgartner and Welch (*Stain Technology*, VII [1932], 129) use for molds short sections of rubber tubing set on end on a slide or glass plate.

DIFFICULTIES LIKELY TO BE ENCOUNTERED IN SECTIONING IN PARAFFIN, AND THE PROBABLE REMEDY

1. Crooked Ribbons.—*a*) Usually caused by wedge-shaped sections. Correct by trimming the block of paraffin so that the edge which strikes the knife first and the edge on the opposite side are strictly parallel. See that the block strikes the knife exactly at right angles.

b) The paraffin may be softer at one end of the block than at the other. This can be corrected only by imbedding the object over again in a homogeneous paraffin.

c) Unequal sharpness of the knife-edge may also cause crooked ribbons.

2. The Object Makes a Scratching Noise on the Knife or Cuts with a Gritty Feeling and the sections perhaps crumble and tear out from the paraffin.

a) This is generally caused by too high heating of the object while in the paraffin oven. Not only is such an object worthless, but it endangers the edge of the microtome knife. Correct by limiting the bath in paraffin to the minimum time necessary for a proper penetration of the object and keeping the temperature barely above the melting-point of the paraffin.

b) The fixing reagent has formed crystals (e.g., corrosive sublimate) which have not been thoroughly washed out.

See also 5 (p. 49).

c) The object contains lime. This is often true of pathological tissue.

3. The Sections Wrinkle or Jam Together: the object itself may be compressed before the knife. This is a serious fault because the arrangement of the parts of a tissue is greatly deranged. It may be due to various causes.

a) The microtome knife may be dull. Examine the knife and sharpen it if necessary.

b) The paraffin may be too soft. To remedy this defect employ one or more of the following means: (1) cool the paraffin block in water; (2) cut the sections in a cooler room; (3) cut the sections thicker; (4) reimbed in harder paraffin. See also memorandum 11, p. 50. If sections are not too badly wrinkled they may be flattened out by warming on water as directed in steps 19–21, p. 43.

c) A possible reason is that the tilt of the knife is insufficient (see **step 13, p. 40**).

d) The edge of the knife may be smeared with a layer of paraffin. Clean the edge with a cloth moistened in xylol.

e) The object has not been thoroughly infiltrated with paraffin.

4. The Sections Roll and Refuse To Ribbon.—This is one of the most

The Paraffin Method

exasperating of all defects. If the sections are not tightly curled, they frequently unroll when placed on warm water (step 19, p. 43). Various mechanical devices have been constructed to prevent this evil, but most of them are impractical. Sometimes when a section begins to roll, if the edge is held down by means of a flat-pointed hair brush the curling can be overcome. If a ribbon can once be started, the difficulty is frequently corrected. The sections should be cut rapidly.

a) The commonest cause of rolling is the hardness of the paraffin. This may sometimes be remedied by one or more of the following means: (1) warming the knife with the breath; (2) cutting in a warmer room; (3) placing a lamp or burner near the imbedded object; (4) warming the knife *very carefully* by holding the back on a warm paraffin bath; (5) cutting the sections thinner; (6) reimbedding the object in softer paraffin; (7) dipping the end of the block in melted softer paraffin. Numbers 2 and 5 are the most practical remedies.

b) The tilt of the knife may be too great (step 13, p. 40).

c) The knife may be dull.

5. The Sections Split Longitudinally or Are Crossed by Parallel Scratches.—*a*) Look for a nick in the edge of the knife. Cut in a new place on the knife or sharpen it.

b) A bit of grit may have got into the object or the paraffin, or there may be a hole in the paraffin. Reimbed after carefully cleaning the object in the clearing fluid.

c) Tissues may contain hard substances (lime salts, silica, crystals precipitated from fixing reagents) which have been imperfectly washed out. It is best to take an entirely new piece of tissue in which these defects do not exist.

d) The tilt of the knife may be too great (step 13, p. 40), or the trouble may be due to loose paraffin or dirt on the edge of the knife. Frequent cleaning of the knife is desirable.

e) The object may be too large to cut in paraffin. Try smaller pieces of tissue or use the celloidin method.

f) The paraffin is layered or seamed because of slowness in adding melted paraffin at time of imbedding (step 8, p. 39).

6. The Knife Scrapes or Rings as It Passes Back over the object after having cut a section.

a) This is sometimes caused by a knife with either too great or too little tilt (step 13, p. 40).

b) The object may be too tough or hard to cut in paraffin without springing the edge of the knife (see 7, *b*, p. 50).

c) The blade of the knife may be too thin.

7. The Sections Vary in Thickness: the machine cuts one thick and one thin or misses a section.

a) This may be caused by the imperfect mechanical construction of the machine. Old machines in which the parts are worn are especially liable to this defect. It may be remedied to some extent by tightening up the parts of the machine.

b) The object may be too hard for the knife to cut and, as a consequence, the edge of the knife springs. When tough or hard objects must be cut, use an old microtome knife or a sectioning razor. See if there is not some means of softening such a tissue without obscuring the microscopical structures sought.

c) Either too great or too little tilt may cause the defect (step 13, p. 40).

d) See that the disk bearing the object is securely clamped in the machine and that the knife or other parts are not loose.

8. The Object Crumbles or Drops Out of the Paraffin as Cut.—It has probably been insufficiently penetrated by paraffin. Some of the following precautions may prevent the defect: (1) Leave the object in the paraffin bath longer. (2) See that it is entirely free from the dealcoholizing fluid before placing it into the melted paraffin. Objects which have been immersed in cedar oil are particularly subject to this defect. For this reason xylol should be used to rinse off the cedar oil after dealcoholization. (3) If the object is impervious to paraffin or very friable, as are many ova, some other method must be tried. Consult memoranda 9 and 10; see also the celloidin method (chap. viii) or the combination celloidin-paraffin method (memorandum 8, p. 73).

9. The Ribbon Twists or Curls About or Clings Closely to the Side of the Knife.—This is due to the electrification of sections. It may often be remedied by boiling water in the room, thus increasing the atmospheric humidity. If the fault is excessive, it is best to postpone the cutting until the atmospheric conditions have changed. The difficulty may be minimized by using a drum ribbon-carrier (see step 15, p. 43).

10. The Cut Section Catches On and Clings to the Block as it returns instead of remaining on the knife. Probably the knife is dull or its edge is dirty; the tilt (step 13, p. 40) is insufficient; the paraffin is too soft, or the room temperature is too high. Toughened tissue, following high temperatures or reagents which harden tissues too much, is a common cause.

11. A Simple Cooler for use with the microtome, which facilitates the preparation of thin paraffin sections and is especially useful in a laboratory where the temperature is high, is described by Grave and Glaser in the *Biological Bulletin* for September, 1910. The apparatus "is essentially a hollow truncated pyramid, open at both ends, and suspended in an inverted position from a standard, so adjusted that the lower end of the shoot is at a convenient distance above the knife. At the upper end of the inverted pyramid, and surrounded by it, is a tray whose dimensions are less than those of the base of the shoot. This tray is filled with crushed ice, and from one corner of it a drain leads the water to the escape from the lower end of the air-channel. At that point a rubber tube connects the pipe with a suitable receptacle." Grave and Glaser recommend the following as a convenient size: base, 12.5×8.7 in.; truncated apex, 6.1×2.1 in.; measurements of ice-tray, 8.8×3.3 in.

CHAPTER VI

THE PARAFFIN METHOD: STAINING AND MOUNTING
I. STAINING WITH HEMATOXYLIN

Place enough of the following reagents in tall stender dishes or Coplin staining-jars to cover the slides lengthwise, up beyond the sections affixed to them: xylol, absolute, 95, 70, 50, 35 per cent alcohols respectively, clear water, acid alcohol, and, for washing out the acid alcohol in the case of hematoxylin preparations, a separate jar of 35 per cent alcohol to which a few drops of a 0.1 per cent aqueous solution of bicarbonate of soda has been added. If the tap water is slightly alkaline the alkaline alcohol may be omitted. Arrange these reagents in a row in the order named with the exception of the acid alcohol and its accompanying alkaline alcohol wash of 35 per cent alcohol, which may be placed immediately back of the ordinary 35 per cent alcohol. Put a little vaseline along the upper edges of the jars containing absolute alcohol and xylol and press the cover down tightly to prevent evaporation or the entrance of moisture.

In like manner place in Coplin staining-jars (tall stenders will answer) a supply of Delafield's hematoxylin diluted one-half with distilled water, eosin, Lyons blue, alum-cochineal (or borax-carmine), Congo red, and solutions A and B for the iron-hematoxylin method. Arrange these stains in a row back of the alcohol series.

1. Remove the paraffin from the sections of mussel gill or of intestine (see last lesson) by placing the slides in xylol (turpentine will answer) for 10 or 15 minutes. The process may be hastened by first gently warming the slide until the paraffin begins to melt.

2. Remove the xylol from the sections by transferring the slides to absolute alcohol for 1 minute.

3. Pass the slides through the alcohols (95, 70, 50, and 35 per cent), leaving them for a half-minute in each.

4. Remove to Delafield's hematoxylin for 30 minutes or until stained a pronounced blue.

5. Wash in water for 5 minutes.

6. Dip each slide for from 30 seconds to 5 minutes into the acid alcohol until the sections are of a reddish hue, then rinse them in 35 per

The Paraffin Method

cent alkaline alcohol (or tap water if alkaline) **until the blue color is restored.** If alkaline alcohol is necessary it must be kept very slightly alkaline through the occasional addition of a few drops of a 0.1 per cent solution of bicarbonate of soda (see memorandum 10, p. 59). The alkaline alcohol may be omitted when other than hematoxylin stains are used, as its purpose is merely to restore the blue color of the latter.

7. Pass the slides up through the series of alcohols to 70 per cent, leaving them about half a minute in each alcohol.

8. Pass the slides through 95 per cent alcohol (3 minutes), absolute alcohol (5 minutes), into xylol for 10 minutes or until clear (see memorandum 3, p. 59).

9. Carefully drain off all excess of the clearer from a slide, wipe the under side, and lay it down flat with the sections uppermost. Put a few drops of thin balsam on the sections near one end. Take up a clean cover-glass and, holding it by the edges between the thumb and first finger of one hand, lower it upon the balsam by bringing one end into contact with the slide near the balsam and supporting the other end by means of a needle held in the free hand. Lower the cover slowly so that as the balsam spreads, no air bubbles will be inclosed under the glass. If a slide is tilted a little and allowed to remain in that position small bubbles will frequently work out unaided. They may sometimes be removed by pressing gently above them with the handle of a needle and gradually working them to the edge of the cover-glass. They may often be removed by carefully warming the slide. Keep the slide in a horizontal position until the balsam hardens.

CAUTION.—Do not allow the sections to become dry before adding the balsam and cover.

10. Attach the permanent label. It should contain at least the following data: the number of the record card (3, p. 5); the name of the tissue; the kind of section (plane of section, thickness, etc.); if one of a series, the number of the slide in the series and the number of the first and last section on the slide; the date, and if desired the name or the initials of the preparator. It is well to add the thickness of the cover-glass (see also memorandum 23, p. 63).

It is best to have the label on the left end of the slide, as it will then not be in position to obscure the scale of the detachable mechanical stage so widely used on microscopes today.

NOTE.—Prepare four slides each of the other objects which have been imbedded. Stain and mount one of each kind as you did the intestine, and also one of each kind in the same way, only substitute alum-cochineal (or borax-carmine) for the hematoxylin. The alum-cochineal may require 12 hours or more for staining. Preserve the others for double staining.

As time permits, prepare and section the other tissues which were fixed in alcohol and in Bouin's fluid. After you have had the preliminary practice in double staining, stain and mount these as you prefer.

II. DOUBLE STAINING IN HEMATOXYLIN AND EOSIN

1. Proceed according to the regular schedule with one each of the slides reserved above, and stain in Delafield's hematoxylin.

2. Wash the sections in water, and proceed farther according to the regular schedule to 95 per cent alcohol.

3. Transfer the slide to the eosin stain for 30 to 60 seconds, and after rinsing again in 95 per cent alcohol, place it in absolute alcohol.

4. Clear in xylol and mount in balsam.

NOTE.—The sections should show both the blue stain (in nuclei) and the red stain (in cytoplasm) when examined under the microscope. If either is too dense or too light, make a note of the fact and vary the time accordingly when staining other sections by this method.

III. DOUBLE STAINING IN COCHINEAL (OR BORAX-CARMINE) AND LYONS BLUE

1. Pass the remaining reserved slides through xylol and the alcohols, descending to 35 per cent alcohol.

2. Stain in alum-cochineal (or borax-carmine) for from 6 to 12 hours, or until the sections are well colored. See remarks on alum-cochineal, 34, page 243.

3. Rinse in water or 35 per cent alcohol, and pass the sections up through the alcohols to 95 per cent. If the sections are deeply stained however, remove the excess of stain with acid alcohol (a few seconds).

4. Stain for 10 to 20 seconds in Lyons blue. It is very easy to overstain with this dye.

5. Rinse in 95 per cent alcohol, and transfer the sections to absolute alcohol (5 minutes), clear in xylol, and mount in balsam.

IV. STAINING WITH HEIDENHAIN'S IRON-HEMATOXYLIN

This stain is very valuable in the study of cell division and in determining the finer structure of the cell. The iron-alum acts as a mordant, preparing the tissue for the action of the hematoxylin.

1. Prepare two sets of sections of intestine, testis or ovary, bladder, pancreas, and stomach. The sections should not be over 6 or 7 microns in thickness. Preserve one set for double staining.

The Paraffin Method

2. Pass the other set through xylol, absolute alcohol, 95 per cent alcohol, and thence to water.

3. Transfer from water to the iron-alum, and allow this solution to act for from 30 minutes to 6 hours. The slide may be left in it overnight.

4. Wash thoroughly in water (10 minutes to one hour). If excess alum is carried over into the hematoxylin solution the latter is injured.

5. Stain in the 0.5 per cent hematoxylin 1 hour. If a trace of the iron-alum remains in the sections the hematoxylin will turn black. This, however, does not necessarily impair its power of staining.

6. Rinse in water 5 minutes.

7. Place the sections into iron-alum again (a separate solution from the one used in mordanting), which will now extract the excess of stain. The time required for proper differentiation varies with the kind of tissue and the fixing agent that has been used. From 10 to 30 minutes is usually sufficient, though no definite time limit can be set. Remove the slide from the iron-alum from time to time and inspect it. When the sections become of a dull-grayish hue the decolorization is usually sufficient. If very accurate results are necessary, the slide should be removed from the iron-alum frequently, rinsed in water, and examined under the microscope. When in a dividing cell the chromosomes become sharply defined, the decolorization should be stopped.

8. Wash in running water for 1 hour or in several changes of water for 2 hours. If any of the iron-alum is left in the sections the color will fade later.

9. Wipe off the excess of water, transfer the slide to 95 per cent alcohol, followed by absolute alcohol and xylol.

10. Mount in balsam.

NOTE.—Iron-hematoxylin is perhaps the one most important stain in use today. The student should practice the method until he has mastered it. It is better though not absolutely essential that the stain be "ripe." Some workers use a 4 per cent solution of the iron-alum for mordanting and a 2 per cent solution for destaining. Professor C. E. Allen of the University of Wisconsin botanical department uses successfully a much stronger solution of hematoxylin made as follows: 10 grams of hematoxylin crystals are dissolved by the aid of heat in 800 c.c. of absolute alcohol and then added to 1200 c.c. of distilled water; the solution is allowed to ripen a month before using. Hance (*Science*, LXXVII [1933], 287) has found that if the distilled water in the stain is even slightly acid, muddy-looking preparations result. A trace of sodium bicarbonate added to the hematoxylin solution will correct this.

For demonstration of centrosomes and finer cytological details, the time of staining may require to be lengthened. In my own cytological work I find that 2 to 4 hours in iron-alum followed, after rinsing in water, by hematoxylin for 12 to 24 hours, yields better results than does the shorter method. Also, the practice of destaining with a saturated aqueous solution of picric acid (Hsw-Chuan Tuan, *Stain Technology*, V [1930], 135) instead of iron-alum has met with increasing favor, since it destains the cytoplasm primarily and leaves the chromosomes practically unchanged, thus yielding a well-defined chromosomal image.

V. IRON-HEMATOXYLIN WITH OTHER STAINS

Use the sections which were reserved for this method. The method is identical with the one just outlined, except that between step 8 and step 9 the following directions should be inserted: 8a, transfer the sections from water to Congo red for a minute, or to orange G for 2 hours, then wash them in water and proceed to step 9.

NOTE.—*Before proceeding farther, kill a female cat or rabbit to secure tissues for the celloidin method and to correct failures in the paraffin method. In addition to the tissues specified before, prepare (fix in Zenker's fluid) pieces of tendon, cartilage, spleen, lymph gland, pancreas, and salivary glands. (If the reagents are at hand and time permits, the student, indeed, might advantageously prepare a number of tissues according to the methods indicated in Appendix C). Fix the ovary in Bouin's fluid, and reserve it for the paraffin method for delicate objects. Fix parts of the brain and cord in Zenker's and in formalin as previously indicated, and place bits of muscle in which nerves terminate plentifully (e.g., intercostals) in formalin. Larger pieces (up to 2 cm.) may be used of such tissues as are to be imbedded in celloidin. Bear in mind that the **larger** the tissue the **longer** must it be left in the different reagents. Select the necessary parts of the digestive tract to prepare longitudinal sections in celloidin from esophagus to stomach and from stomach to intestine. As soon as possible begin the preliminary steps in the celloidin method (chap. viii) so that there may be no loss of time. Prepare a piece of intestine for staining in bulk (see vi). It should be placed in the stain after thoroughly washing out the fixing reagent. Preserve parts of it to cut in celloidin. Remove the lower jaw and prepare it for decalcification of teeth (as indicated in chap. xii). Likewise prepare pieces of femur and of tarsal bone for sectioning (chap. xii).*

VI. STAINING IN BULK BEFORE SECTIONING

It is sometimes desirable to stain objects before sectioning. The method is a slow one, and requires stains which penetrate evenly and thoroughly. Various preparations of carmine and cochineal give the best satisfaction, although several hematoxylin stains are also frequently used in this way. It is best to stain immediately after fixing and washing out, before the object has been carried into higher alcohols. In general, it is advisable to section tissues and stain on the slides because the staining can be controlled more effectually. Use the piece of intestine already prepared (see note, p. 53).

The Paraffin Method

1. After fixing in Zenker's fluid and thoroughly washing according to the directions on p. 29, place the tissue in Delafield's hematoxylin for 24 hours.

2. Wash in 35 per cent alcohol for 5 minutes.

3. Decolorize in acid alcohol 20 to 30 minutes, or until the color ceases to come away freely. Restore the bluish-purple tint by treatment with alkaline alcohol. Follow this with 50 and 70 per cent alcohol, 20 minutes each.

4. From this point proceed through 95 per cent alcohol, absolute alcohol, xylol, and imbedding, sectioning, and mounting precisely as in the general paraffin method, except that after the sections have been freed from paraffin in xylol, do *not* mount immediately in balsam, but first transfer the slide back into absolute alcohol, and thoroughly wash it in order to remove the glycerin from the fixative and so prevent cloudiness of the final mount. From alcohol the slide is passed through xylol, or carbol-xylol, and mounted in the usual way.

NOTE.—When certain of the carmines or hematoxylins are used as stains for entire objects, the preparations usually need to be decolorized with acid alcohol. This may be deferred, however, until after the objects are sectioned.

VII. PARAFFIN METHOD FOR DELICATE OBJECTS

To prevent the distortion of delicate objects which are to be sectioned in paraffin, the transition of the material from one reagent to the other must be very gradual and the heat be minimized. Observe the following modifications of the general method and prepare pieces of ovary which have been fixed in Bouin's fluid.

1. Pass the object in the usual manner up through the series of alcohols to absolute. It is sometimes necessary to use a more closely graded series of alcohols if the object be very delicate. See also the "drop" method (memorandum 6, p. 174).

2. From the absolute alcohol pass to a mixture of absolute alcohol two-thirds and chloroform one-third; gradually add more chloroform until at the end of an hour the mixture is at least two-thirds chloroform. (Some workers prefer the anilin and oil of wintergreen method to this chloroform method. See memorandum 7, p. 175.)

3. Transfer to pure chloroform for 30 minutes.

4. Add melted paraffin little by little during the course of an hour or two (24 hours will do no harm), until the chloroform will hold no more in solution.

5. Transfer the object to pure melted paraffin in a small vessel on the paraffin oven for 10 to 10 minutes, changing the paraffin once. Imbed in the usual way.

6. Cut the sections about 7 microns thick. Mount and stain some in Delafield's hematoxylin and eosin and others in iron-hematoxylin and Congo red or orange G, according to the directions already given for these methods.

NOTE.—For very sensitive objects Schultz's dehydrating apparatus (to be obtained from dealers) may be used. It consists of a tube within a tube, each having the lower end covered by an animal membrane. The tubes are suspended in the neck of a much larger bottle which contains 95 per cent alcohol. The object is placed in the inner tube and both tubes are filled with water. When suspended in the alcohol, a very gradual hardening or dehydration of the object takes place as the alcohol slowly diffuses through the membrane. Sometimes it is necessary to use only one tube, and in such a case the hardening proceeds more rapidly. See also memorandum 6, p. 174.

VIII. EUPARAL AS A MOUNTING- AND PRESERVATION-MEDIUM

This reagent, introduced by Gilson, has been highly extolled by various workers as a final mounting-medium, although most still prefer balsam or damar for general work. Two forms, colorless and green, are obtainable from Grübler. The green is used only with hematoxylin stains, which it intensifies. One of the merits of euparal is that delicate tissues may be mounted in it directly from 95 per cent alcohol, thus avoiding the expense and the risks of passage through absolute alcohol and the essential oil. Another value lies in the fact that, because of its lower index of refraction (1.483), unstained or faintly stained elements which are invisible in balsam (index, 1.535) are rendered visible. It thus becomes particularly serviceable in the study of spindle fibers and other such delicate cytological elements. It hardens rapidly, hence preparations can be used within 24 hours. It is well for the beginner to try it along with other methods.

Wright (*Annals of Tropical Medicine and Parasitology*, XXI [1927], 179) recommends euparal highly for mounting blood films stained with various Romanowsky-type stains since it preserves these for years without fading.

MEMORANDA

1. **In Passing from One Liquid to Another,** one corner of the slide bearing sections should first be touched by blotting paper to remove any excess of the

The Paraffin Method

liquid last used. This is especially necessary in transferring from absolute alcohol to xylol, or from 95 per cent to absolute alcohol.

2. **Sections Once Placed in Turpentine or Xylol** for the removal of paraffin must never in any subsequent step be allowed to become dry. Particular care must be taken to prevent sections from drying out after removing them from xylol to mount in balsam because the xylol evaporates rapidly. If the sections become dry the preparation is usually rendered valueless.

3. **Xylol Used for Removing Paraffin** should be kept in a jar separate from that which contains xylol for clearing before mounting, and it should be changed occasionally because it tends to become saturated with paraffin.

4. **Sections Not over 10 Microns Thick** may be plunged directly from 95 per cent alcohol into an aqueous medium and vice versa. If sections are over 10 microns thick it is better to put them through the complete series of alcohols. With thick sections diffusion is less rapid, and too abrupt a change from one fluid to another may produce distortions or wrench the sections loose from the slide.

5. **To Avoid Rubbing Sections off the Slide,** hold the slide with one end toward the light before wiping it and glance obliquely along the surface. The shiny side is the one to wipe.

6. **The Series of Alcohols and Stains** ordinarily may be used a number of times without replenishing. When the alcohols become very much discolored or the stains cloudy, they should be renewed. Alcohols should not be used too often, however, as they soon accumulate particles of dirt which settle upon the sections and render preparations unsightly.

7. **Absolute Alcohol** must be kept free from water. It may be tested from time to time by mixing a few drops with a little turpentine. If the mixture appears milky the alcohol contains a harmful amount of water and should be renewed.

8. **Two Slides Placed Back to Back** can be handled as readily as a single slide in passing through the various liquids. Various kinds of slide-baskets or slide-holders, which enable one to transfer a number of slides through liquids at one time, may be obtained from dealers. Preparation of a large number of slides at one time is conveniently accomplished by using a staining rack such as that shown in Figure 30, and a series of rectangular museum jars measuring about 15×15×9 cm., to hold the necessary reagents.

9. **Gentle Agitation of a Slide** in any liquid facilitates the action of the liquid. Observe this precaution especially with absolute alcohol.

10. **For Washing Sections after Staining in Hematoxylin** tap water is preferable to distilled water because it is usually slightly alkaline. When acid alcohol is used to decolorize sections stained in hematoxylin, the sections may be washed in 35 per cent alcohol rendered alkaline by the addition of a few drops of 0.1 per cent solution of bicarbonate of soda. The alkali neutralizes

the acid and restores the bluish-purple color to the section; it also renders the blue color more permanent. If too much of the soda is added the color will be a hazy disagreeable blue, and the tissue will often not take the counterstain.

11. **To Obtain a More Precise Stain** with Delafield's hematoxylin, it is well to dilute it with three or four times its bulk of distilled water. The sections must be left in this solution a correspondingly longer time. Sections stained in this way may not require treatment with acid alcohol. Most workers, however, prefer to overstain and decolorize.

12. **The Length of Time Required for Staining Different Tissues is** exceedingly variable. Upon removal from the stain after rinsing, if the sections are insufficiently colored, put them back into the stain and examine them from time to time until they are properly stained (30 minutes to 24 hours).

13. **If Objects Refuse To Stain,** it is usually due to one of the following causes: (a) The fixing agent has not been sufficiently washed out. This is a frequent cause of poor staining. (b) The fixation has been poor. The success of a preparation depends largely upon proper fixation in most cases. (c) The stain is at fault. Hematoxylin will not stain properly until ripe (see "Hematoxylin," p. 9). Many stains, especially the anilins, deteriorate and must be replaced. (d) Certain stains will not follow some fixing agents. This can be remedied only by using a different stain or by fixing tissues in a different fluid. The hematoxylins and carmines are applicable after a very large variety of fixing agents. (e) The paraffin has been insufficiently removed from the sections. This may be corrected by dissolving off the cover-glass in xylol and, after thoroughly removing all paraffin, restaining and mounting the sections again in the ordinary way.

14. **Use Only Clean Slides and Covers.**—Always grasp a slide or a cover by its edges to avoid soiling its surface. All cloudiness (seen by looking through the glass toward some dark object) must be removed. For wiping slides and covers, a piece of clean cloth which does not readily form lint should be used. A well-washed linen towel is good, as is also bleached cheesecloth cut into pieces the size of a handkerchief. Slibes may often be cleaned after simply dipping them into alcohol or into alcohol followed by water. A tedious but excellent method is to wash the slide in 95 per cent alcohol, wipe, and smear with Bon Ami as for washing windows. When dry wipe clean with a fine cloth. If this treatment is insufficient, place them for several hours into equal parts of hydrochloric acid and 95 per cent alcohol, keeping them well separated so that the liquid may act on the entire surface of each. Then rinse them in water and place them in 95 per cent alcohol. It is well to keep a stock supply of such slides and cover-glasses in 95 per cent alcohol.

To clean a cover-glass, grasp it by the edges in one hand, cover the thumb and first finger of the other hand with the cleaning-cloth, and rub both surfaces of the glass at the same time. To avoid breaking the cover, keep the

The Paraffin Method

thumb and finger each directly opposite the other. A large cover-glass may be cleaned by rubbing it between two flat blocks which have been wrapped with cleaning-cloths.

To clean slides which have been used, if balsam mounts, warm and place in xylol or turpentine (or, better, trichlorethylene, see mem. 26) to dissolve off the covers. Put the slides and covers into separate glass or porcelain vessels and leave them for a few days in the following cleaning mixture:

Potassium bichromate	10 parts
Hot water	50 parts
Sulphuric acid	50 parts

In making, add the acid very cautiously after the bichromate solution cools. When the slides are freed from balsam, wash them in water, rinse in a dilute solution of caustic soda, again in water, and finally place them in 95 per cent alcohol until needed.

15. If Sections Appear Milky or Hazy under a medium power of the microscope, when finally mounted, the effect is probably due to one of the following causes: (*a*) The clearer is poor and needs replenishing or correcting. (*b*) The absolute alcohol contains water (see memorandum 7, p. 59). (*c*) The cover bore moisture. Passing a cover-glass quickly through a flame before putting it on to the object will remove moisture. (*d*) The acid has not been entirely removed from the sections. (*e*) Too much albumen fixative has been used. (*f*) The glycerin of the albumen fixative has not been removed by passing sections of objects stained in bulk (see vi, 4, p. 57) back into absolute alcohol after removing paraffin from them.

The defect may be remedied frequently by dissolving off the cover in xylol or turpentine, descending through the series of reagents to the point where the fault lies, correcting and ascending again according to the regular method. To remove water, for example, it is only necessary to go back as far as absolute alcohol which has a great affinity for water.

A few drops of n-butyl alcohol added to the jars of absolute alcohol and xylene, respectively, will commonly prevent cloudiness due to water contamination.

16. Dry or Dull-looking Areas under the Cover-Glass indicate that the sections were allowed to get dry after the removal from the clearer, or that insufficient balsam was applied.

17. Delicate Black Pins in sections of tissue fixed with a mercuric fixer are due to deposition of mercury. Remove with iodized alcohol before staining.

18. Balsam Which Exudes from under the Cover may be scraped off with an old knife after it hardens. Remove the last traces by means of a brush or a cloth dipped in turpentine or xylol. Balsam may be removed from the surface of a cover by means of a brush dipped in xylol.

19. If Sections Wash off the Slide the defect is probably due to one of the following causes: (*a*) The slide was soiled or oily. Remedy by cleaning slides thoroughly (see 14, p. 60). (*b*) The albumen fixative is too old. (*c*) The transitions in the alcohol have been too great. This is true sometimes of thick sections. (*d*) The paraffin ribbon was not thoroughly spread when mounted (see p. 44).

Thick sections are more likely to come off the slide than thin ones. To avoid this, the sections may be collodionized by placing them, after the paraffin has been removed, in a thin solution of collodion or celloidin in ether-alcohol ($\frac{3}{4}$ gram in 100 c.c. of ether-alcohol; see 4, p. 8) for a few minutes and then transferring them to 70 per cent alcohol. If carmine dyes are to be used, this method is not satisfactory, as carmine stains collodion.

20. **Flooding Sections with the Dye** by means of a pipette, especially in case of stains which act rapidly (e.g., eosin, acid fuchsin, Lyons blue, picric acid, etc.), is sometimes more convenient than immersing the sections in a jar of the staining fluid. Small bottles with combination rubber stopper and pipette are now provided for this purpose by dealers.

21. **Balsam Mounts in Which the Stain Has Faded** may frequently be restained, either with the original or with other stains. All that is necessary is to dissolve off the cover in xylol (2 to 3 days) and pass the preparation down through the alcohols to the stain in the usual manner. Fading is often caused by insufficient size of cover.

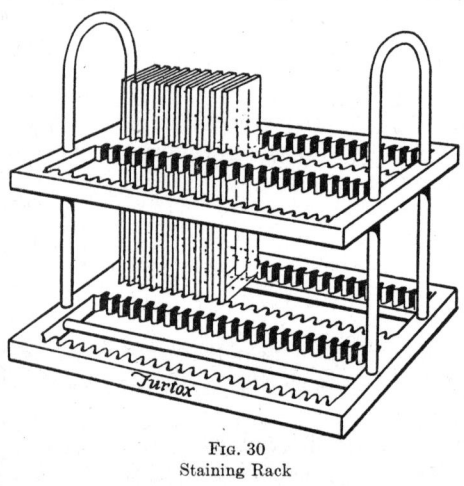

Fig. 30
Staining Rack

22. **Ink for Writing on Glass** (Hubbert, *Journal of Applied Microscopy*, V, 1680).—Mix drop by drop 3 parts of a 13 per cent alcoholic solution of shellac with 5 parts of a 13 per cent aqueous solution of borax. If a precipitate forms, heat the solution until it clears. Add enough methylen blue to color the mass deep blue.

Professor Robert F. Griggs uses common water-glass (an aqueous solution of sodium silicate or potassium silicate) with an ordinary steel pen. After marking, the slide is heated until the water-glass decomposes, leaving behind a rough, sandy surface, which when rubbed away shows the written characters etched on the slide.

"Diamond ink" obtainable from Eimer & Amend is useful for writing on glass. When not in use it is kept sealed with paraffin. See also p. 1, "Carborundum Points."

23. **More Detailed Labeling** than that indicated on p. 50 is sometimes desirable. For well-devised schemes see Richard E. Scammon, "A Method of Recording Embryological Material," *Kansas University Science Bulletin*, IV, No. 5 (March, 1907); also Robert T. Hance, "A System for Recording Cytological Material, Slides and Locations on the Slides," *Transactions of the American Microscopical Society*, XXXV, No. 1 (January, 1916).

24. **For Orientation of Objects in the Imbedding-Mass** see p. 144.

25. **Carbol-Xylol** (to 75 c.c. of xylol add 25 c.c. melted carbolic-acid crystals) instead of xylol alone is recommended where there is much humidity or where absolute alcohol is hard to obtain. Complaint is occasionally made that stains sometimes fade after clearing in carbol-xylol. However, I have preparations more than twenty-five years old cleared in this mixture which are still as bright as the day they were mounted.

26. **Substitution of Trichlorethylene** ($CHCl:CCl_2$) **for xylol** in all histological operations is recommended by Oltman (*Stain Technology*, X [1935], 23). Alleged advantages: non-inflammable; both paraffin and Canada balsam are soluble in it; slides cleared in this solvent, and mounted in balsam dissolved in it, dry very quickly; excellent for removing covers from old balsam mounts.

CHAPTER VII

THE DIOXAN METHOD

Dioxan is diethyl-dioxide or diethylene oxide, $O(CH_2CH_2)_2O$, a colorless liquid, miscible with melted paraffin and also with water in all proportions, and with the usual organic solvents employed in the paraffin imbedding method. This makes it possible to omit the use of such reagents as absolute alcohol and xylol, which cause excessive hardening of certain tissues, and, since the use of alcohols and clearing oils is unnecessary, to simplify the whole technique. When tissue from an aqueous or an alcoholic solution is placed in dioxan, the water or the alcohol is gradually given up. Transfer later of such material to melted paraffin results in the replacement of the dioxan by paraffin. When infiltration is completed, the object is imbedded in paraffin and sectioned as usual. Such difficult objects to prepare without great shrinkage, or to section without crumbling or splitting, as amphibian oviduct, salamander ovaries, mammalian ovaries, pigeon embryos, and decalcified bone are readily prepared by the dioxan method.

The author is indebted to Dr. Harland W. Mossman, not only for first calling his attention to the method and for a full translation of the original papers of Heinz Graupner and Arnold Weissberger, who first presented it (*Zoologischer Anzeiger*, XCVI [1931], 204–6, and *ibid.*, CII [1933], 39–44), but also for a critical appraisal of the technique after using it constantly over a period of some two years.

NOTE.—To get satisfactory results it is necessary to have a good grade of dioxan, particularly one free from water. The reagent can be obtained from Pfaltz & Bauer, 300 Pearl Street, New York City, American representative of Dr. K. Hollborn and Sons (successors to G. Grübler & Co.), or from the Union Carbide and Carbon Corporation, 230 N. Michigan Ave., Chicago, and 30 E. Forty-second Street, New York City. The Eastman Chemical Company of Rochester, N.Y., also supply an excellent, though more expensive, dioxan.

THE METHOD

1. Kill and fix some convenient tissue or object by any of the ordinary fixing methods (Bouin, formol, alcohol-formol-acetic, etc.) and wash it as usual. Transfer it to the vessel containing dioxan and leave

it from 3 to 5 hours, changing the fluid twice. If one has tissues already fixed and stored in such fluids as formol or alcohol, these may be used instead of the freshly fixed tissue. Transfer them directly to the dioxan.

NOTE.—It is all-important to have the dioxan free from water. To insure this, calcium oxide (CaO) may be added to the supply bottle of the reagent and allowed to settle on the bottom. This vessel should be kept tightly covered to prevent absorption of water from the atmosphere and evaporation of the liquid.

2. Transfer the material from dioxan to a mixture of dioxan approximately one-third and melted paraffin two-thirds, and keep at a temperature of about 55° to 60° C. for 2 to 3 hours, then to pure melted paraffin for the same time at the same temperature. It is well to keep the pure paraffin in a dish with a gauze a short distance above the bottom, since dioxan is heavier than the paraffin and any residue of it will thus sink through the gauze to the bottom. With tissues that are easily penetrated, the dioxan-paraffin mixture step may be omitted.

3. Imbed in paraffin, cut serial sections, and mount as in the ordinary paraffin method.

4. If sections are to be stained on the slide, proceed as usual in the paraffin technique; or, transfer from xylol through two changes of dioxan, then to water, followed by the stain (e.g., Delafield's hematoxylin).

5. Wash in water, then destain (as desired) in a mixture of three-fourths dioxan to one-fourth water very slightly acidified with HCl. Follow by a rinse in a mixture of three-fourths dioxan to one-fourth water slightly alkalized with sodium bicarbonate.

6. Counterstain in 0.1 per cent erythrosin in a mixture of 1 part water to 9 parts dioxan, rinse again in the alkalized dioxan, and pass into pure dioxan.

7. From dioxan pass through two changes of xylol and mount in gum damar or in balsam. In transferring from dioxan to xylol, wipe the backs of the slides and drain them well (but do not let the sections become dry), so that the least possible dioxan is carried over into the xylol.

MEMORANDA

1. **Some Properties of Dioxan.**—Dioxan is inflammable and somewhat anesthetic. It has a specific gravity of 1.040 to 1.050 at room temperature, and boils at approximately 101° C. It is an excellent solvent for a wide variety of both the oil-soluble and the alcohol-soluble resins. Such waxes as beeswax

and paraffin, but slightly soluble in it in the cold, are soluble when heated; and practically all vegetable and mineral oils are soluble in all proportions of it. Oil-soluble dyes are soluble in it; water-soluble and alcohol-soluble dyes are practically insoluble, but become freely soluble if 10 per cent of water is added. Dioxan does not dissolve dry nitrocellulose but does so when a small amount of alcohol is added. It is inadvisable for any but a skilled chemist to try to reclaim used dioxan by distillation, as accumulation of peroxides in the distilling apparatus may cause an explosion.

2. **Dioxan Dissolves Picric Acid;** hence, from fixers such as Bouin's, tissues may be transferred directly to dioxan, in which washing out of excess picric acid takes place. Similarly, bone decalcified in trichloracetic acid can be passed directly into dioxan.

3. **Dioxan Dissolves Bichloride of Mercury;** hence materials fixed in such reagents as Zenker's ordinarily require no iodine treatment if they are allowed to stand in one or two changes of dioxan for several hours. If one desires to use iodine, however, it may be applied while the tissues are in dioxan. Iodine crystals should be used instead of the tincture, since the latter causes a precipitate to form.

4. **Potassium Bichromate Is Practically Insoluble in Dioxan;** hence the dioxan technique is useful when one wants to retain the brown stain of the chromate in tissues.

5. **Repeated Use of Dioxan** can be made as long as it is not diluted too much by water or by the fixatives. Water may be removed by adding calcium oxide (CaO); after standing some hours, the solution should be filtered. However, dioxan is not very expensive, so that such reconditioning should not be overdone.

6. **Larger Objects Are Imbedded** in paraffin more successfully, according to W. H. Schaefer, of our laboratory, if when they have been stored in 70 per cent alcohol they are first transferred to 70 per cent alcohol and anilin oil half-and-half, then to pure anilin oil, next to dioxan, then to melted paraffin.

7. **With Hollow Structures Where Gas Bubbles Are Apt To Form,** Mossman finds it necessary to remove the gas while the tissues are still in the fixative or water, by placing them for about 24 hours in a dehydration chamber and exhausting the air with a water suction pump. Dioxan for subsequent use is also placed in the chamber with them. After the tissues have been transferred to the dioxan, he usually places them *in vacuo* again and keeps them so until ready for the paraffin.

8. **To Stop Extraction of a Dye Which Is Insoluble in Pure Dioxan** (e.g., eosin, orange G, anilin blue) but soluble in dioxan diluted 10 per cent with water, one has only to destain to the desired degree in the diluted dioxan and then transfer the preparation to the pure reagent.

9. **In Cytological Work,** Pearl E. Claus, of our laboratory, finds the dioxan technique satisfactory, even after osmic fixation, in such methods as the

Mann-Kopsch, or Sevringhaus for demonstrations of Golgi apparatus, mitochondria, etc. However, she finds that, while sections which have been stained by the Sevringhaus method may be passed up through dioxan to be mounted, it is better to use the ordinary series of alcohols after the Mallory triple stain.

She also finds that, after either Bouin's fluid or the Champy mixture, a preliminary treatment of sections with a 0.125 per cent solution of potassium permanganate for 1 minute, followed by rinsing in distilled water and then immersion for 1 minute in a 1 per cent solution of oxalic acid, sharpens the differentiation of cytoplasmic granules when subsequently stained by either the Mallory or the Sevringhaus triple stains.

10. The Freezing Method (chap. ix) can be used with the dioxan technique.

a) Fix small bits of tissue (3 to 4 mm. in diameter) for 5 to 15 minutes in some such reagent as alcohol-formol-acetic (23, p. 239). Fixation is greatly hastened, without tissue injury, at higher temperatures (not above 55°, in a water bath); e.g., 5 minutes' fixation at 55° C. is equal to 15 minutes at 15° C.

b) Remove to cold dioxan and freeze on the microtome. Long washing-out of the fixing fluid is unnecessary.

c) Transfer the sections to distilled water (3 to 5 minutes) as cut. Keep them flat and spread apart so that the dioxan will be washed out and thus not interfere with subsequent staining. However, sections may be safely stored in dioxan a week or more before staining.

d) Stain in Mayer's hemalum $\frac{1}{2}$ to 1 minute, depending upon the thickness of sections and concentration of stain.

NOTE.—Mayer's hemalum is made by dissolving 1 gram of hematoxylin in 1 liter of distilled water, to which is then added 2 grams of iodate of sodium and 50 grams of alum. This solution acts much like Delafield's hematoxylin but does not keep so well.

e) Rinse sections in distilled water, then pass them through tap water with a drop of ammonia in it, into fresh tap water ($1\frac{1}{2}$ to 2 minutes).

f) Stain in Hollborn's eosin substitute (1 to $1\frac{1}{2}$ minutes), controlling degree of staining under the microscope. With ordinary eosin the color is likely to diffuse if glycerine jelly is used as a mounting medium.

g) Rinse in distilled water and (1) mount in glycerine jelly (p. 109) or (2) transfer to dioxan (1 minute) and mount in balsam.

11. Tertiary Butyl Alcohol has been found by Johansen (*Science*, LXXXII [1935], 253) to have the desirable properties for microtechnique possessed by dioxan.

12. The Poisonous Nature of Dioxan should be called to the attention of users, I am advised by J. B. G. of Flatters and Garnett, Ltd., of Manchester, England. This is probably a very minor hazard in the small quantities in which the reagent is used in biological laboratories, but it should be known.

CHAPTER VIII

THE CELLOIDIN METHOD

Use the tissues which were prepared (p. 56) for this method, including pieces of the brain and spinal cord which were fixed in Zenker's fluid. Reserve a piece of spleen for the freezing method (p. 77).

1. Fixing, washing, and dehydrating are the same as usual (chap. iii). If the object is in 70 per cent alcohol, complete the dehydration by using successively 95 per cent and absolute alcohol. It should remain in the absolute alcohol for at least an hour.

2. From absolute alcohol transfer the object to equal parts of absolute alcohol and ether $1\frac{1}{2}$ hours.

3. Next to thin celloidin (reagent 20, p. 11) for from 24 hours to several days or weeks.

4. Make at least two intergrades between the thin and the thick celloidin. Transfer tissues from thin celloidin to these in succession (using the thinner one first), leaving them from 24 hours to several days in each, according to the size and nature of the tissues.

5. Thence to thick celloidin for from 24 hours to several days. It may be left for weeks.

Thorough dehydration and thorough infiltration are the great essentials for success with the celloidin method.

6. Prepare a vulcanite block (see memorandum 3, p. 71) in such a manner that it will have surface enough to accommodate the object, leaving a small margin, and length enough to be readily clamped into the carrier on the microtome (Fig. 31).

7. Oil one side of a strip of stiff paper by rubbing on a very little vaseline, and wrap it, oiled surface in, about the prepared end of the block in such a way that it will project beyond the end of the block, forming a collar high enough to extend a little beyond the object which is to be placed within it. Tie the paper in place by means of a thread. If the object is small, however, see second paragraph under step 8.

8. Pour a small amount of thick celloidin into the paper cup thus formed and with forceps remove the piece of tissue and place it in celloidin. Add more thick celloidin until the cup is full. By means of needles which have been moistened in ether-alcohol arrange the object so that it will be cut in the desired plane.

Small objects may be mounted very simply on the block by putting on a drop of thick celloidin, orienting the object in it, and adding another drop or two of thick celloidin to cover the object thoroughly.

NOTE.—Instead of being mounted on blocks, objects may, after saturation with thick celloidin, be placed in proper position in a glass dish and covered with thick celloidin. The dish is then loosely covered and placed under a bell-jar so that the ether will gradually evaporate, leaving the mass, in a day or two, of proper consistency for cutting. Each specimen is then cut out in a block of suitable size for sectioning. For fastening to base, see memorandum 5, p. 72.

9. Into a small stender dish put chloroform to the depth of 3 mm. When a film has formed over the exposed surface of the celloidin place it in the chloroform to harden. It need not be submerged. Keep the vessel tightly covered. The object may be left for a day or two, but 1 to 3 hours usually suffices.

10. Transfer the block to 70–83 per cent alcohol, where it may remain indefinitely.

11. Make a careful study of the microtome used for cutting celloidin (Fig. 31).

12. Place the block in the object-carrier of the microtome at the proper level and arrange the microtome knife obliquely, so that it will slice through the object with a long drawing cut for at least half the length of the blade. If the object is oblong it is advantageous to have the long diameter parallel to the edge of the knife.

13. Keep both the knife and the object flooded with 70 per cent alcohol, preferably from an overhanging drop-bottle.

14. Draw the knife through the object with a straight steady pull; avoid pulling down on or lifting the knife-carrier.

15. If the feed is not automatic push the knife back to position always before turning the screw which raises the object. Cut the sections about 15 or 20 microns thick. If they curl they are best unrolled on the surface of the knife with a camel's hair brush just before they are wholly cut free from the block.

16. As the sections are cut, transfer them by means of a small soft brush or a paper spatula to a flat stender or a watch-glass containing 70 per cent alcohol.

17. Transfer some of the sections through 50 and 35 per cent alcohol, 2 minutes each, into alum-cochineal (or borax-carmine) for from 20 to 30 minutes, or until stained (12 to 24 hours).

18. Wash successively in 35, 50, and 70 per cent alcohols, leaving the sections from 2 to 3 minutes in each.

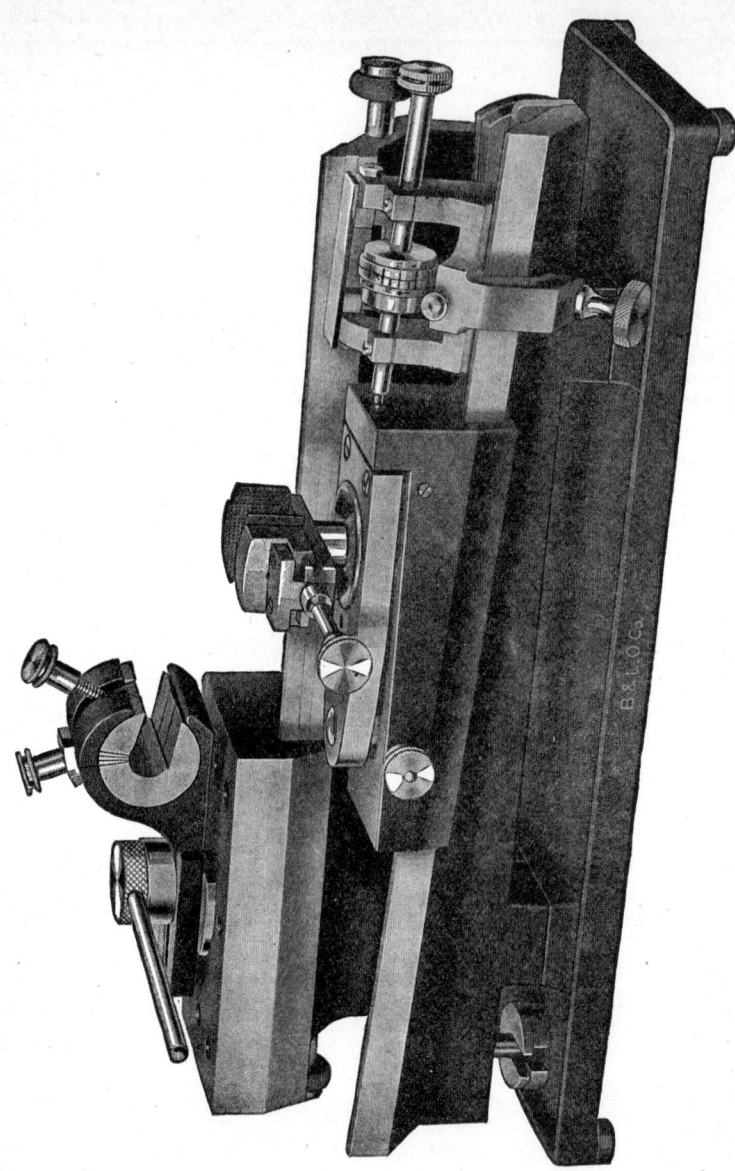

Fig. 31.—Celloidin Microtome (Thomson–Jeffery Model)

The Celloidin Method

19. Transfer the sections to 95 per cent alcohol for 3 to 5 minutes. Absolute alcohol is not to be used with celloidin because it dissolves the celloidin.

20. Clear in cedar oil or beechwood creosote for from 10 to 20 minutes.

21. Mount in balsam (see step 9, p. 53).

STAINING CELLOIDIN SECTIONS IN HEMATOXYLIN AND EOSIN

The objects are killed, fixed, and preserved as usual in 70 per cent alcohol, and sectioned as in the foregoing method.

1. Fifty and 35 per cent alcohol, each 3 to 5 minutes.
2. Delafield's hematoxylin, 20 to 30 minutes, or until stained.
3. Water, 5 minutes.
4. Thirty-five, 50, and 70 per cent alcohol, each 3 to 5 minutes.
5. Acid alcohol, until the celloidin which surrounds the object shows but little of the stain.
6. Seventy per cent alcohol, barely alkaline (see memorandum 10, p. 59), until the red color caused by the acid is replaced by bluish purple.
7. Alcoholic eosin, 30 seconds to 1 minute.
8. Ninety-five per cent alcohol, 2 to 5 minutes. Clear in cedar oil or beechwood creosote and mount in balsam.

NOTE.—As time permits, section other tissues by the celloidin method and stain as above.

MEMORANDA

1. **If Chloroform Is Not at Hand,** 80 per cent alcohol will harden the celloidin, although more slowly.

2. **The Length of Time** that objects should be left in ether-alcohol and the celloidin mixtures depends upon the size and density of the objects. When time permits, it is always best to leave them several days or even weeks in the mixtures of celloidin. For large objects such as the medulla of a large brain this is a necessity. For an embryo of large size months may be required.

3. **Blocks for Celloidin Mounting** may be of white pine, glass, vulcanized fiber, or even a very hard paraffin. Cork should not be used because it is liable to give or bend. The vulcanized fiber is the most satisfactory. It may be purchased from dealers in the form of strips which may easily be sawn to the necessary dimension. It is well to saw several parallel cuts into the upper edge of the block to provide points of attachment for the celloidin. If nothing but a wooden block is available, the end of the block to which the object is to be

attached should first be dipped into ether-alcohol for a minute, then into thick celloidin, and then left to dry so that air bubbles will not work up later out of the wood into the imbedding mass.

4. **Other Clearers** may be substituted for cedar oil or creosote. One which clears from 95 per cent and which does not dissolve celloidin must be chosen. Other good clearers are: (1) origanum oil; (2) a mixture of 3 parts of oil of thyme and 1 part of castor oil; (3) Eycleshymer's clearing fluid, which is a mixture of equal parts of bergamot oil, cedar oil, and anhydrous carbolic acid.

5. **Imbedding a Number of Objects** in one mass is frequently convenient. Fold a stiff paper into a box of the proper size (step 7, p. 39) or use metal L's (Fig. 30). Pour in thick celloidin, put the objects in place, and orient them properly for cutting. Leave a space of about 8 mm. between adjacent objects. Fill the box with thick celloidin and set it in a dish containing a little chloroform, or leave it in 80 per cent alcohol to harden. When ready to proceed, cut the large blocks into smaller ones each containing a piece of tissue. To fasten it to the base, trim the small celloidin block to the proper dimensions, soften for a few minutes in ether-alcohol the side to be attached, then dip it into thick celloidin and apply to the end of a vulcanite block which likewise has been dipped into the ether-alcohol and the thick celloidin. Press the two together and place them in chloroform or 80 per cent alcohol to harden (see also note under step 8, p. 69).

6. **Anilin Dyes** are usually avoided in the celloidin method because they stain the celloidin intensely and are not removed in subsequent treatment. When necessary, however, some (e.g., eosin) may be used. Safranin, for example, may be removed satisfactorily from the celloidin by means of acid alcohol without extracting all the stain from the tissue. If anilin dyes have been used, it is sometimes better to remove the celloidin by treating the sections with absolute alcohol or with ether before the final clearing and mounting (see memorandum 15, p. 75).

7. **Relative Merits of the Paraffin and the Celloidin Methods.**—Celloidin is good for large objects, for brittle or friable objects, and for delicate objects which heat would injure. It does not require removal from the tissues ordinarily, hence it holds delicate structures together permanently. Some tissues are not rendered so hard and so difficult to cut as in paraffin. However, very thin sections cannot be obtained except by great skill. The method, moreover, is extremely slow. The paraffin method is comparatively rapid, serial sections may be cut and mounted with ease, and very thin sections may be obtained. Large objects do not section as satisfactorily as in celloidin, although up to 10 mm. or even considerably greater diameter they cut readily. The rule is to use the paraffin method when you can. However, the advanced student should read what Wetmore has to say about the celloidin method in *Stain Technology*, VII (1932), 37–61.

8. **For Brittle Objects, a Combination of Celloidin and Paraffin Infiltration** sometimes proves successful (see, however, memorandum 9, p. 45). The method is too tedious for ordinary use, although it must sometimes be resorted to with friable or delicate objects such as eggs.

According to Apathy's method, as reported by Kornhauser (*Science*, July 14, 1916), fixed material is dehydrated as usual, finally passing through three changes of absolute alcohol into ether and alcohol, where it is left 5 hours. It is next put into 2 per cent celloidin for 24 hours, then 4 per cent celloidin for 24 hours, and ultimately imbedded in 4 per cent celloidin and hardened in chloroform vapor for 12 hours. The block is then quickly trimmed, leaving a margin beyond the object of a few millimeters on every side, and put into liquid chloroform for 12 hours. It is next transferred to an oil mixture made up by weight instead of volume as follows:

Chloroform	4 parts
Origanum oil	4 parts
Cedar-wood oil	4 parts
Absolute alcohol	1 part
Carbolic-acid crystals	1 part

Anhydrous sodium sulphate should be kept in the bottom of the tube to take up any water brought in, in the celloidin.

The block must remain in this oil mixture until it clears and sinks; this may take from 3 days to a week. It is next washed in three or more changes of benzol to remove oils and alcohol, then infiltrated in paraffin, imbedded, sectioned, and mounted in the usual way. In subsequent handling, slides should not be left for any great length of time in absolute alcohol, as it will dissolve out the celloidin.

9. **To Transfer Celloidin Sections from the Knife,** it is an excellent plan to use a paper spatula; a bit of postal card held in the cleft end of a small stick answers very well. Press the paper down evenly on the section and then slide it off the edge of the knife. The section adheres to the paper. In carrying loose sections from one fluid to another an ordinary section-lifter may be used, or a glass rod around which the section is allowed to curl answers very well.

10. **Objects Stained in Bulk May Be Cleared While Yet in the Block,** then sectioned, and mounted without passing back into the alcohols. After the block of celloidin has hardened sufficiently in chloroform it is transferred directly to the clearer (cedar oil, or a mixture of oil of thyme 3 parts and castor oil 1 part). In cutting objects thus cleared the knife must be flooded with the clearer instead of alcohol. Do not allow the sections to become dry. If it is desired to use this method for a celloidin block which has already been preserved in 70 to 83 per cent alcohol, the block must pass through 95 per cent alcohol (1 to 2 hours) before it is placed in the clearer.

11. **Celloidin, Collodion, Parlodion,** etc., are trade names for tri-nitro-cel-

lulose (also called pyroxylin or soluble guncotton). Thin and thick solutions are employed and the method is in every respect similar to the celloidin method.

Imbedding with low viscosity nitrocellulose (Hercules Powder Co., R.S., 0.5 sec.) instead of celloidin is strongly recommended by Davenport and Swank (*Anatomical Record*, LVIII [1934], Sup., 58). To quote them:

"The embedding solution was made as follows: 100 gm. of nitrocellulose (containing 30 per cent absolute alcohol) were macerated in 130 cc. of 95 per cent alcohol and 160 cc. of anhydrous ether added. Some fine grit was present but could be eliminated by allowing the solution to stand a week or two and decanting before using. Greasing glass embedding dishes with a thin film of mineral oil prevented sticking. Sections as thin as 4 microns (blocks cut dry) were obtained. The dehydration of sections with absolute alcohol had to be avoided because the nitrocellulose is more soluble than celloidin in absolute. Clearing oils or butyl alcohol were substituted for absolute between 95 per cent and xylene."

12. **Fixing Celloidin Sections to the Slide** is accomplished (1) by covering the sections, when mounted in proper order, with a strip of tissue paper, which is then bound fast by wrapping thread around it. Lee (*Microtomist's Vade-Mecum*, 9th ed., p. 119) recommends (2) the albumen method for celloidin sections as well as for paraffin. (3) If the sections on the slide are carefully flooded with 95 per cent alcohol two or three times, this drained off and followed by a small amount of ether-alcohol or ether fumes until the edges of the sections begin to soften perceptibly (10 to 20 seconds), the sections will generally adhere to the slide sufficiently when the celloidin becomes hard again upon exposure to the air (30 seconds) after the ether-alcohol has been drained off; they must then be immersed in 95 per cent alcohol before any further steps are taken.

13. **For Serial Sections in Celloidin** some one of the so-called *plating* or *sheet* methods will be found most satisfactory. That of Linstaedt (*Anatomical Record*, November, 1912) is among the best. It is a modification of the celloidin sheet method suggested by Huber for paraffin sections and of Weigert's method for serial sections in celloidin.

Plates of glass, thoroughly clean and of suitable size (e.g., 5×7 inches), are coated on one side with the following solution and allowed to dry:

Saccharose	3 grams
Dextrin	3 grams
Distilled water	100 c.c.

To which as a preservative is added a bit of thymol.

When the sugary layer is thoroughly dry, coat it with a 4 per cent solution of celluloid in acetone. A number of such sheets may be made up and kept dried if desired.

As sections are cut, place them in the desired order on the celluloid sheets, moistening from time to time to prevent drying out. Frilling may be prevented by brief treatment with absolute alcohol. When the plate is filled with sections, blot with a smooth-surfaced toilet-paper, then, in order to fix the sections to the celluloid, spray by means of an atomizer with a 1 to 2 per cent solution of celloidin in ether and alcohol. When it is partially dry immerse the plate in 70 per cent alcohol, then in water. The water will dissolve the sugary solution and the celluloid sheet bearing the sections will float off. The sheet may be preserved indefinitely in 70 or 80 per cent alcohol, or stained, cleared, and mounted at any time by any of the methods suitable for ordinary celloidin sections. Strips may be cut and mounted serially on properly numbered slides.

For another "sheet" method see Böhm, Davidoff, and Huber, *A Text-Book of Histology*. Also see Carothers, in *Science*, April 13, 1928, p. 400.

14. **Gilson's Rapid Celloidin Process** (Lee, *The Microtomist's Vade-Mecum*, 9th ed., p. 106) is valuable under some circumstances because of the great saving of time. After dehydration the object is saturated with ether and finally placed into a test-tube containing thin celloidin. The lower end of the tube is then dipped into melted paraffin and allowed to remain there until the celloidin solution has boiled down to about one-third of its original volume. The mass is then mounted in the ordinary way, hardened for an hour or more in chloroform, and cleared in cedar oil. Sections are cut as directed under memorandum 10 above.

15. **Celloidin May Be Removed from Sections** when necessary by passing them through absolute alcohol into ether and alcohol or oil of cloves for from 5 to 10 minutes, then back through absolute to ordinary alcohol.

16. **For Orientation of Objects in the Celloidin Mass** see p. 144.

17. **The "Dry-cutting" Celloidin Method** may be used for objects which are not too bulky. Dehydrate from 95 per cent alcohol with cedar oil, using first equal parts of cedar oil and 95 per cent alcohol (add a little absolute if mixture is foggy), followed by pure cedar oil. Instead of the "thin" and "thick" celloidins, use only one celloidin, a little thinner than the standard "thick," replacing the cedar oil directly with it. Place specimens in a corked, large-mouthed bottle in an incubator at 37° C. Two to three days should suffice for penetration of any ordinary-sized block. In any event, leave the bottle in the incubator until the celloidin has become very thick, then imbed in *new* thick celloidin.

Immerse the block, after preliminary hardening in air a few minutes, in chloroform and leave several hours. Transfer to chloroform and cedar oil equal parts for several hours, then to pure cedar oil. The block may be left in cedar oil in tightly stoppered bottle indefinitely.

If not too large, blocks thus prepared may be cut "dry" on the ordinary paraffin microtome, with straight knife. Dr. Harvey M. Smith says in a personal note:

"In my experience this method is far preferable to the standard technique, especially for eyes. Sections are removed from the knife and transferred to 95 per cent alcohol, where they instantly straighten out, and clear quickly. They are then transferred through 83 per cent to 70 per cent alcohol, in which they are stored.

"For bulky specimens the standard absolute alcohol-ether dehydration, thin and thick celloidin, cutting in alcohol, seems preferable. However the incubator idea for hastening penetration is applicable to large as well as to smaller specimens.

"Celloidin sections should be much overstained in hematoxylin, the excess washed out in 35 per cent of alcohol plus 0.5 per cent hydrochloric acid. Bluing after the acid is best done in tap water. Aqueous eosin is preferable to alcoholic, because then a graded series of passages through alcohol allows the operator to watch the section and obtain the desired shade of eosin.

"Dehydration from 95 per cent alcohol is best performed in origanum oil. I have tried them all, and have found nothing that comes near equaling it. The undesirable browning of the oil after it has been exposed to air can be completely overcome by subsequently washing out the origanum oil in xylol, and finally mounting the section from the latter. The browning does not injure the clearing power in the least."

18. **A Hot Celloidin Technique for Animal Tissues** is given in detail by Walls in *Stain Technology*, VII (1932), 135–45.

CHAPTER IX
THE FREEZING METHOD

1. Use a piece of spleen which has been properly fixed and later preserved in 70 per cent alcohol. Transfer it through 50 and 35 per cent alcohol successively to water, and wash it for 12 hours in running water.

2. Place it into a gum and syrup mass for 24 hours (a saturated solution of loaf sugar in 30 c.c. of distilled water, added to 50 c.c. of gum mucilage. Prepare a supply of gum mucilage by dissolving 60 grams of best gum acacia in 80 c.c. of distilled water).

3. Examine the freezing microtome carefully (Figs. 32 and 33).

4. Remove the gum and syrup mixture from the outside of the tissue with a cloth, put a little gum mucilage (not gum and syrup) on the freezing disk of the microtome, and place the tissue in it in such a way that longitudinal sections through the hilum may be cut. Surround the object with gum mucilage and set the freezing apparatus to going. If carbon dioxide is used, open the valve very cautiously, and let only a small quantity of the gas escape.

NOTE.—Carbon dioxide is commonly used for charging soda water and beer. It may be purchased in iron cylinders containing about 20 pounds of the liquefied gas. The cylinder, when empty, is exchanged for a charged one, so that the purchaser pays only for the contents. The freezing attachment shown in Fig. 33 has a hard-rubber ring between the plate to which the object is frozen and the rest of the apparatus. This localizes the cold at the specimen and thus saves time and gas. To operate, the valve at the freezing chamber should first be closed and the valve at the cylinder be slightly opened to admit the gas into the tube; then by opening and closing the small valve at the chamber three or four times in quick succession, the tissue is frozen without waste of gas or the inconvenience caused by the connections freezing up.

Small tubes of compressed carbon dioxide, sufficient for one or two freezings, may sometimes be obtained from stores carrying automobile supplies. These can be utilized in operations where immediate diagnosis by means of sections is required.

5. As soon as the gum is frozen, continue to add more until the tissue is completely covered and frozen.

6. If a non-automatic microtome is used work the microtome screw with one hand and plane off sections (15 to 20 microns thick) with the

other. The well-sharpened blade of a carpenter's plane is the best instrument for cutting. It must be frequently stropped. If an automatic microtome is available the proper type of knife is provided with the machine.

NOTE.—The blade for cutting by hand should be mounted in a short, broad handle, which may be grasped easily and firmly with one hand. In cutting, the bevel edge of the knife should set squarely on the glass ways of the microtome so that the handle of the knife is inclined toward the operator at about an angle of 45 degrees from the perpendicular. The hand guiding the knife should be firmly supported against the chest while pressing the cutting edge steadily against the glass ways of the microtome. The cutting stroke is made by bending the body forward from the waist and thus forcing the blade squarely across the surface of the tissue.

The knife must be kept cold to prevent sections from sticking to

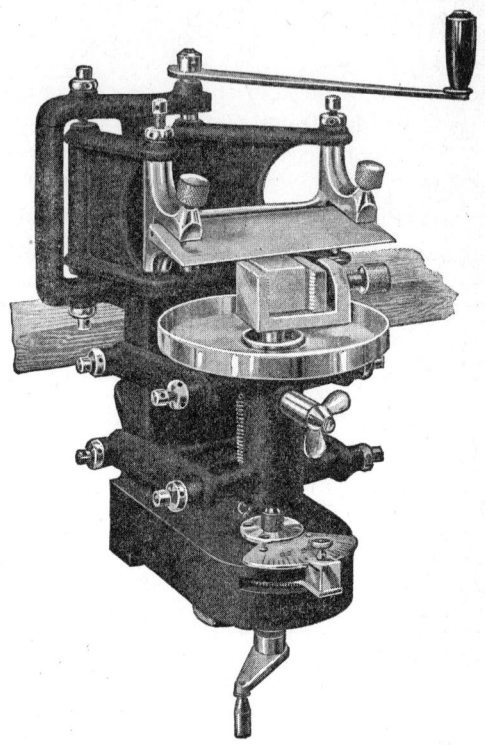

FIG. 32.—Freezing Microtome

it. If the sections fly off or roll, the tissue is probably frozen too hard. The same defect may arise if there is insufficient syrup in the gum with which the tissue has been saturated. To correct, let the tissue thaw a little, and if it is still at fault, soak it again in a mixture which contains a greater proportion of syrup. Work rapidly, so as to cut sections in quick succession. Several sections may be allowed to collect on the blade before they need be removed.

7. Transfer the sections to distilled water. The water should be changed several times to dissolve out the gum. Reserve a few sections in water for later use (step 11, p. 79).

8. Immerse a few of the sections for 10 to 30 minutes, or until stained, in Delafield's hematoxylin, then wash them in several changes of tap water.

9. Transfer the sections through the successive grades of alcohol (decolorizing with acid alcohol if necessary) up to absolute alcohol, leaving them 2 minutes in each, after which remove them to xylol for

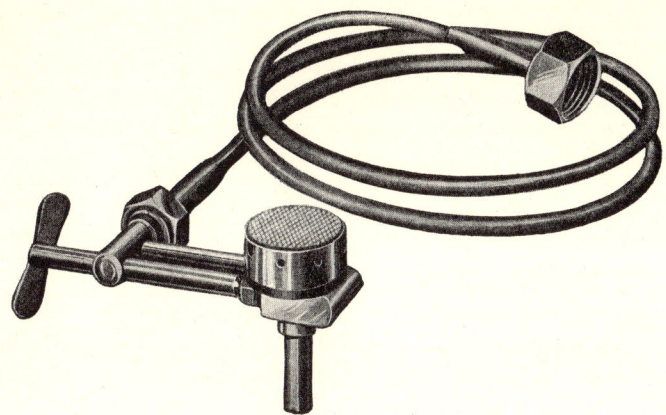

Fig. 33.—Freezing Attachment for CO_2

5 minutes or until clear. If desired, stain with eosin (30 to 60 seconds) after 70 per cent alcohol.

10. Remove one or two of the best to a slide, drain off the excess of xylol, add a few drops of balsam and a cover-glass of suitable size, and label.

11. Remove the sections reserved in step 7 (p. **78**) to a test-tube containing a small amount of water, and shake the test-tube vigorously for a minute or two. This removes the lymphocytes from the sections, and exposes the reticular connective tissue so that it may be examined. Dehydrate the sections and mount in balsam.

MEMORANDA

1. **Fresh Tissues Are Frequently Sectioned** by the freezing method. The tissue may be transferred directly to the disk of a microtome without previous imbedding, and sectioned after freezing. This affords a ready means of rapidly determining the nature of a given tissue, and is very serviceable, especially to the pathologist. The principal objection is that crystals of ice form in the cells and distort them badly. This is avoided when syrup and gum are used for imbedding.

2. The Fresh Tissue Method Used at the Mayo Clinic is as follows: (1) Bits of fresh tissue (alive, or at least not kept in the ice-chest longer than two hours), not more than 2×10×10 mm. in size, are frozen in saturated dextrine solution on a good microtome, and cut in sections 5 to 15 microns thick. (2) Sections are removed from knife with tip of finger and allowed to thaw thereon, thus avoiding later development of air bubbles. (3) They are then unrolled with a camel's hair brush in 1 per cent NaCl solution, stained 10 to 20 seconds in Unna's polychrome methylene blue (Grübler's), and washed out again momentarily in 1 per cent NaCl solution. (4) Sections are mounted in Bruin's glucose medium, formula:

Glucose	240 c.c.
Distilled water	840 c.c.
Spirits of camphor	60 c.c.
Glycerin	60 c.c.

Sections are handled from the first NaCl solution through the remaining operations with a small glass-rod lifter; they are kept constantly moving while in the stain. Permanent sections may be made similarly from formalin-fixed material which has been thoroughly washed in water.

3. **Fixed Tissues** may be sectioned by the freezing method if they are first washed in running water for some hours. Even tissue fixed in formalin—perhaps the best fixing fluid for tissues which are to be frozen—is better for a washing of at least 30 minutes before being frozen.

4. **Sections May Be Preserved** in alcohol in the usual way after being cut by the freezing method. All trace of gum should be washed out and the sections passed through the grades of alcohol to 83 per cent, where they may remain indefinitely.

5. **Sections of Fresh Tissue May Be Fixed** and washed out after cutting if desired. This requires but little time, and the sections will take stain much more satisfactorily after having been subjected to a fixing reagent.

6. **Objects Which Alcohol Would Injure** may be sectioned by the freezing method and mounted in aqueous media.

7. **Ether or Rhigolene Is Sometimes Used for Freezing**, although the method is more expensive and less satisfactory on the whole than the carbon-dioxide method. A special freezing attachment for such liquids may be obtained from microtome makers.

8. **Organs or Parts Varying Greatly in Density** may sometimes be cut more successfully by the freezing than by any other infiltration method.

9. **To Fix Frozen Sections to the Slide** after treatment with absolute alcohol, flow a little three-fourths of 1 per cent solution of celloidin in ether and alcohol over the sections and drain off at once. After a few seconds of exposure to the air, place in 80 per cent alcohol for a minute. The film of celloidin should be very thin. If it turns white upon immersing in the alcohol, the solution of celloidin was too thick; it should be thinned by adding more ether and alcohol

10. **A Rapid Method for Tissue Diagnosis** applicable to both fresh and formalin-fixed frozen sections is presented in detail by Geschichter, Walker, Hjort, and Moulton in *Stain Technology*, VI (1931), 3. It permits of balsam mounting and differential staining pictures as satisfactory as those obtained with fixed tissues.

CHAPTER X

METALLIC SUBSTANCES FOR COLOR DIFFERENTIATION

I. A GOLGI METHOD FOR NERVE CELLS AND THEIR RAMIFICATIONS

The Golgi chrom-silver method is one widely used for the demonstration of nerve cells together with their various processes. There are many modifications of the method, all of which are more or less inconstant in their results. In a successful preparation the various cells and nerve processes are not equally blackened, a fact which allows of discrimination between the different elements. Sometimes the ganglion cells and fibers remain unstained while the neuroglia cells are impregnated, or occasionally other elements than nervous tissue (e.g., blood vessels) are affected.

The following method is applied to material preserved in 10 per cent formalin and is a so-called "rapid method."

1. From the brain and spinal cord, which have previously (see note, p. 56) been subdivided and placed in at least 10 times their volume of 10 per cent formalin (3 days to an indefinite time), cut out small pieces 4 to 5 mm. thick from the region desired for study and transfer them to a vessel containing from 15 to 20 times their volume of a 3.5 per cent aqueous solution of potassium bichromate. They should remain in this solution for 2 days. Renew the fluid at the end of 20 hours. Keep the different pieces of tissue in separate vessels so as to avoid confusion.

2. For impregnation, transfer the tissues to a silver-nitrate solution made as follows:

    Silver nitrate (crystals).................... 1.5 grams
    Distilled water............................. 200 c.c.
    Concentrated formic acid................... 1 drop

3. Rock the tissues gently in a small amount of this fluid until the brown precipitate of silver chromate ceases to appear, then transfer them into from 20 to 40 times their bulk of fresh silver-nitrate solution and leave them *in the dark* for 3 days. Change the fluid after the first 20 hours.

4. Transfer a few of the brown pieces of tissue to 95 per cent alcohol for half an hour, renewing it once or twice during this time. **Leave**

the rest of the tissue in the silver nitrate solution for future use in case the first attempt proves unsuccessful.

5. Remove the pieces from 95 per cent to absolute alcohol for 20 minutes, changing the latter once. Then transfer them to ether-alcohol for 20 minutes.

6. Imbed in celloidin without waiting for infiltration to occur (thin celloidin 30 minutes, thick celloidin 10 minutes). Mount directly on a block and harden in chloroform for 20 minutes.

7. From chloroform transfer directly to the clearing fluid (e.g., cedar oil), and as soon as clear (30 to 60 minutes) cut sections 50 to 100 microns thick, but keep the knife flooded with the clearing fluid instead of alcohol. Cut sections of cortex so that they will be perpendicular to the surface of the brain.

8. When the sections are thoroughly cleared, transfer them to a slide flooded with the clearing fluid, select such as prove desirable upon microscopic inspection, and discard the remainder.

9. Replace the oil with xylol, then remove the xylol by pressing upon the sections with blotting paper. Add enough *thick* Canada balsam to cover the sections.

CAUTION.—Do not put on a cover-glass; moisture must evaporate from the section. If this is prevented, the metal deposits break up and the sections become worthless.

10. Keep the preparations level and put them away in a dry place free from dust. If the balsam runs off the sections, more balsam must be added at once. Do not attempt to examine under a high power until the balsam is thoroughly hardened.

MEMORANDA

1. **A Fuller Account** of the Golgi Methods will be found in Hardesty's *Neurological Technique* (pp. 55–61); in Lee's *Microtomist's Vade-Mecum* (9th ed.) (pp. 600–624); or in McClung's *Microscopical Technique* (pp. 344–58).

2. **An Osmium-Bichromate Mixture** is frequently used instead of formalin for fixing fresh tissues. To 85 parts of a 3.5 per cent solution of potassium bichromate add 15 parts of a 1 per cent solution of osmic acid. Small pieces (4 to 6 mm. thick) of fresh tissue are placed in 40 times their volume of this mixture and kept in the dark for from 12 to 24 hours. This fixing fluid is then replaced by a 3.5 per cent solution of potassium bichromate, as in the case of material fixed in formalin (see above). From this point the method is identical with the one given above.

3. **The Determination of the Elements That Will Be Impregnated** appears to depend upon the length of time the tissue is left in the 3.5 per cent solution

of potassium bichromate. Hardesty gives the following lengths of time for different structures: neuroglia, 2 to 3 days; cortical cells, 3 to 4 days; Purkinje cells, spinal cord, peripheral ganglion cells, 4 to 5 days; nerve fibers of the spinal cord, 5 to 7 days. Axones are impregnated ordinarily only in so far as they are not medullated.

4. Mounting the Sections upon a Cover-Glass is preferred by some workers. The cover-slip is then fastened over the opening of a perforated slide with the section downward.

5. For Permanently Mounting Golgi Preparations under a Cover-Glass Huber recommends the following method: The sections are removed from xylol to the slide and the xylol then removed by pressing blotting paper over the sections. A large drop of xylol-balsam is then quickly applied and the slide is carefully heated over a flame from 3 to 5 minutes. A large cover-glass is warmed and put in place before the balsam cools.

6. The Cox Modification of Golgi's corrosive-sublimate method is widely used. It is likely to impregnate nearly all of the cells in the section. This may prove to be disadvantageous rather than otherwise, however, where cells are numerous and close together. Small pieces of nervous tissue are placed for from 1 month in summer to 2 or 3 months in winter in the following solution:

Potassium bichromate, 5 per cent solution....... 20 parts
Corrosive sublimate, 5 per cent solution......... 20 parts
Distilled water............................30 to 40 parts
Simple chromate of potassium, 5 per cent solution 16 parts

The later treatment is the same as for ordinary Golgi preparations.

7. Tracheae of Insects, Bile Capillaries, and Gland Ducts may also be studied by the Golgi chrom-silver method. A bit of the wing muscle of a bumble bee is a good object in which to demonstrate the finer ramifications of tracheae.

II. OTHER SILVER-NITRATE METHODS

a) For Nerves (after Hardesty)

1. The fresh nerve, or, better, a spinal nerve root, may be obtained from a frog which has just been killed. Without stretching the nerve, carefully insert beneath it the end of a strip of postal card or similar card which has been trimmed to the width of 50 mm. The nerve when cut off at each side of the card will adhere to it and remain straight and at approximately normal tension.

2. Clip off the end of the card bearing the nerve into a clean vial which contains 0.75 per cent aqueous solution of silver nitrate. Place the vial in the dark for from 12 to 24 hours.

3. Transfer the nerve to pure glycerin on a slide and tease the fibers apart thoroughly under the dissecting microscope.

Metallic Substances for Color Differentiation 85

4. Add a cover-glass and expose the fibers to sunlight until they become brown (30 minutes).

5. To make the preparation permanent, take off the cover and remove the glycerin by means of filter paper, add a few drops of warm glycerin-jelly (p. 109), put on a clean cover-glass, and press it down. Wipe away the exuded jelly, and when the preparation has cooled seal the cover with asphaltum (see steps 5 and 6, p. 108).

The preparation should show the "cross of Ranvier" and the "lines of Fromman."

b) For the Cornea

1. Quickly rub a piece of silver nitrate over the cornea of an eye which has been removed from a recently killed frog.

2. Slice off the cornea and place in distilled water. Brush the surface with a camel's hair brush to remove the epithelium (conjunctiva).

3. Expose to the action of sunlight or strong daylight until the tissue turns brown.

4. Wash in distilled water and mount in glycerin, or mount in balsam, after proper dehydration.

If the preparation is successful the cells should be strongly outlined by the precipitated silver. If desired, after washing, the nuclei of such cells may be stained in hematoxylin according to the usual method.

MEMORANDA

1. **Fresh Membranes** are also commonly treated with silver nitrate to outline cells. The membrane should be stretched over some smooth surface, or, better, by means of two small vulcanite rings which fit into the other in such a way as to stretch bits of membrane like a drum-head, and hold them fast. Such stretched membrane is first washed with distilled water, then agitated in a 1:300 silver-nitrate solution, in direct sunlight, until it darkens. It is then washed in distilled water, removed from the rings, and mounted in glycerin, glycerin-jelly (p. 109), or, after dehydration, in balsam. If preferred, after washing it may be stained in hematoxylin to bring out the nuclei, and then mounted.

2. **Cajal's Method for Neurofibrils** is widely used. Small pieces of nervous tissue are fixed in formalin for 6 hours, washed in water 4 hours, and transferred to 40 per cent alcohol for 6 hours. They are next kept for 24 hours in 40 per cent alcohol to which ammonia has been added in the proportion of 5 drops of ammonium hydrate to 50 c.c. of the alcohol. The tissues are then placed in an incubator in a 1.5 per cent silver-nitrate solution and kept for 5 days at a temperature of 38° C. Next they are placed in a mixture of 100 parts of water,

15 parts of formalin, and 1 part of pyrogallic acid or hydrochinon for 24 hours, after which they are ready to be passed through graded alcohols into paraffin or celloidin and sectioned in the usual way.

An excellent application of the Cajal method to serial sections has been devised by Malone (*Anatomical Record*, IX [1915], 791), who also describes how to obtain satisfactory Cajal preparations from sections previously stained by the Nissl method.

3. **The Pyridine-Silver Method,** a modification of the Cajal method, devised as a differential stain for non-medullated nerve fibers, has come into wide use in American laboratories in the study of various other problems. It is often used in the preparation of sections of spinal ganglia, sympathetic ganglia, and spinal cord and is the most reliable of the silver stains. Relatively large pieces of tissue can be successfully stained. Ranson's technique (*American Journal of Anatomy*, XII [1911], 69) is as follows: "The nerve or ganglion is placed in 100 per cent alcohol, with 1 per cent ammonia for 48 hours (95 per cent alcohol with 5 per cent ammonia will give much the same results, but seems more likely to bring out the neurilemma nuclei). The pieces are then washed for from $\frac{1}{2}$ to 3 minutes (according to their size) in distilled water and transferred to pyridine for 24 hours, after which they are washed in many changes of distilled water for 24 hours. They are then placed in the dark for 3 days in a 2 per cent aqueous solution of silver nitrate at 35° C., then rinsed in distilled water and placed for 1 to 2 days in a 4 per cent solution of pyrogallic acid in 5 per cent formalin. Sections are made in paraffin, and after mounting are ready for examination."

For use of the method in staining and sectioning the entire head of a small animal or embryo, after decalcification, see Huber and Guild, *Anatomical Record*, VII (1913), 253 and 331. For a discussion of the method with bibliography, see Ranson, *Review of Neurology and Psychiatry*, November, 1914.

4. **Klein's Silver-Line Method** for demonstration of the fibrillar system of ciliates is given on p. 294.

5. **Davenport's Silver Impregnation Method for Nerve Fibers in Celloidin Sections** (*Anatomical Record*, XLIV [1929], 2) is a very useful one for nerve fibers (axis cylinders) of the central system, since it can be applied to tissues which have already been sectioned in celloidin. Davenport used formalin-fixed material (cat and man) stored in 80 per cent alcohol. The procedure is as follows:

a) Slowly stir 15 c.c. of concentrated nitric acid into 85 c.c. of 95 per cent alcohol. Immerse the sections (from 80 per cent alcohol) in this solution for 2 hours at 37° to 38° C., then wash them thoroughly in three or four changes of 80 per cent alcohol.

b) Dissolve 10 grams of silver nitrate in 10 c.c. of distilled water and add 90 c.c. of 95 per cent alcohol. Immerse the sections for 12 to 24 hours in the dark at 37° to 38° C. in this solution, then rinse them in alcohol (preferably absolute; certainly not less than 95 per cent).

c) Develop for from 3 to 5 minutes (avoid bright light) in a mixture of pyrogallic acid, 3 grams; neutral formalin, 5. cc.; and 95 per cent alcohol, 95 c.c. The intensity of the stain can be controlled by varying the time of development. For very dark sections soak for 1 to 15 minutes in alcohol before developing and develop only 1 or 2 minutes; pass very light sections through the alcohol rapidly and develop 10 or more minutes.

d) Rinse in 50 per cent alcohol, then fix until chocolate brown in a 10 per cent aqueous solution of sodium thiosulphate.

e) Wash thoroughly in five or six changes of distilled water, dehydrate, and mount. Precipitates should be brushed off the sections with a camel's hair brush or a cotton swab just before removing from xylol.

Gold toning by any standard method is optional after rinsing in 50 per cent alcohol following step *d*. A good toning solution is made by dissolving 1 gram of gold chloride (yellow) in 500 c.c. of distilled water. Overstained sections can often be saved by gold toning.

Davenport's technique has been used very successfully by James O. Foley in studying axons in peripheral nerves and the normal vagus and normal glossopharyngeal of the cat, as follows:

a) Peripheral nerves.—Bouin fixation; paraffin sectioning; sections silvered by Davenport method, reduced, toned by gold chloride and counterstained by Heidenhain's (*Zeitschrift für wissenschaftliche Mikroskopie*, XXIII [1915], 361–72) modification of Mallory's triple stain. See note below.

b) Normal vagus nerve of cat (below nodose ganglion).—Formalin-acetic acid fixation; paraffin sectioning; staining as in *a*.

c) Normal glossopharyngeal nerve of cat (below petrous ganglion).—Technique as in *b*, only light green is substituted for the anilin blue in Mallory's triple stain.

NOTE.—Foley's variants (*Stain Technology*, January, 1936; 1938) of Heidenhain's modification of the Mallory triple stain are as follows:

Solution 1: Azocarmine Mixture.—Azocarmine (Grübler), 1 gram; glacial acetic acid (Merck, c.p.), 4.0 c.c.; distilled water, 100 c.c.

Solution 2: Anilin-Blue–Orange-G Mixture.—Anilin blue (Grübler), 0.5 gram; Orange G (Coleman & Bell Co., certified 82 per cent), 2.0 grams; glacial acetic acid (Merck, c.p.), 8.0 c.c.; distilled water, 100 c.c.

Solution 3: Light-Green–Orange-G Mixture.—Light green (Eberbachs & Sons Co.), 0.5 grams; Orange G (Coleman & Bell Co.), 2.0 grams; glacial acetic acid (Merck, c.p.), 8.0 c.c.; distilled water, 100 c.c.

Solution 1 is used full strength; both solutions 2 and 3 are diluted with 40 c.c. of distilled water before using.

III. GOLD-CHLORIDE METHOD FOR NERVE-ENDINGS

1. Trace some of the motor nerves of a reptile or mammal to where they enter the muscles (intercostals are best), and clip out small pieces of the muscle. Use material that has been preserved in 10 per cent formalin (see note, p. 56).

2. Place the bits of muscle in 10 or 12 times their volume of a 10 per cent solution of formic acid in distilled water and leave them for from 30 to 40 minutes.

3. Transfer the tissue into from 8 to 10 times its volume of a 1 per cent solution of gold chloride in distilled water for from 40 to 30 minutes. Avoid direct sunlight. The muscle should become yellow in color.

4. Remove the tissue without washing it to about 25 volumes of a 2 per cent formic-acid solution and keep it in the dark until it assumes a purple color (24 to 48 hours). When the fibers appear reddish violet in color the reduction has gone far enough; if they show a decidedly bluish tinge the process has gone too far.

5. Wash the tissue in several changes of distilled water for an hour and transfer a small piece to a slide. Tease the fibers apart very carefully under a dissecting lens. Great care must be exercised to avoid tearing the nerve fiber from its endings. Examine from time to time under a low power of the compound microscope, and when a nerve fiber with its termination is found, carefully separate it as much as possible from the other fibers.

6. Add glycerin-jelly and a cover-glass. Seal in the ordinary way (p. 108).

NOTE.—Tissues may be dehydrated in the ordinary way and mounted in balsam or imbedded in paraffin or celloidin and sectioned.

CHAPTER XI

ISOLATION OF HISTOLOGICAL ELEMENTS MINUTE DISSECTIONS

I. ISOLATION

A. Dissociation by Means of Formaldehyde; ciliated and columnar epithelium.—1. Kill a frog and secure the hinder part of the roof of the mouth, bits of the brain, and a small piece of the intestine. Slit open the latter. Leave the objects for 24 hours in a dissociating fluid made by adding 0.5 c.c. of formalin to 250 c.c. of normal saline solution.

2. Scrape the roof of the mouth after removal from the fluid and mount the ciliated cells thus obtained on a slide. Similarly remove some columnar epithelium from the internal surface of the stomach and mount on another slide.

3. Add a cover-glass and examine. If the cells cling together in clumps, separate them by drumming gently upon the cover-glass with the handle of a needle.

4. Stain by placing a drop of alum-cochineal on the slide just at the edge of the cover and applying a bit of filter paper to the opposite edge of the cover. The filter paper absorbs the fluid from under the cover and the stain replaces it. Keep the preparations under a bell-jar or other cover to prevent evaporation of the staining fluid.

5. After a few hours replace the stain by glycerin in a similar manner.

6. If a permanent preparation is desired, the cover-glass must be sealed (p. 108), or, after staining, the tissue must be dehydrated and mounted in balsam in the usual manner.

B. Isolation of Muscle Fibers by Maceration and Teasing.—1. Place small fragments of voluntary muscle, of the root of the tongue and of the heart muscle of the frog into separate vials containing MacCallum's macerating fluid (reagent 9, p. 9). After 2 days pour off the fluid, fill the vials about half full of water, and separate the fascicles by shaking each vial. Further isolate the fibers by teasing.

Teasing.—In teasing, the important thing to remember is that the elements of the tissue are to be separated, not broken up. Both patience and sharp clean needles are indispensable. The process is best carried on under the lens of a dissecting microscope or a binocular dissector, although it may be done without such aid. A background which enables the tissue to be seen distinctly should be selected, black for colorless or white for colored objects. Black-and-white porcelain slabs are made for this purpose and are very convenient. A good dissecting microscope has attached beneath the stage a reversible plate one side of which is black, the other white. Use a small piece of tissue and begin teasing at one end of it.

2. With the aid of a dissecting microscope carefully tease out in water a number of fibers. Use a small piece and, beginning at one end, with both needles separate the piece along its entire length into two; likewise further subdivide these until the ultimate fibers are isolated.

3. Transfer some of the fibers through the alcohols and xylol and mount in balsam. Stain others in alum-cochineal for some hours and mount in glycerin as above.

C. **Maceration by Means of Hertwig's Fluid** (Hydra, Testis).—1. The solution consists of:

0.05 per cent aqueous solution of osmic acid...	1 part
0.2 per cent acetic acid......................	1 part

Prepare the ingredients for this mixture by diluting the stock solution (1 per cent) in each case with distilled water. Make a separate 0.1 per cent solution of acetic acid also.

2. Treat a hydra with the osmic and acetic acid mixture for 3 minutes and then transfer it to the 0.1 per cent solution of acetic acid. Wash in several changes of this fluid to remove all osmic acid and let the hydra remain in the acetic acid for 12 hours.

3. Wash in water, stain in alum-cochineal or in acid carmine (reagent 44, p. 248) and mount in glycerin as above. If the cells are not sufficiently separated, gently tap on the cover-glass.

4. Submit small bits of the testis of some animal to the same treatment. Stain with methyl green (reagent 67, p. 257) or acid carmine (reagent 44, p. 248).

D. **Mall's Differential Method for Reticulum.**—1. Cut sections of fresh spleen or lymph gland 40 to 80 microns thick by the freezing method and digest for 24 hours in the following solution:

Pancreatin (Park, Davis & Co.)..............	5 grams
Bicarbonate of soda.........................	10 grams
Water.......................................	100 c.c.

2. Wash thoroughly in water, then, in order to remove cellular débris, shake for some minutes in a test-tube half full of water. Spread out on a slide and allow to dry.

3. Apply a few drops of a 3.5 per cent solution of picric acid in 11 per cent alcohol and allow it to dry on the preparation.

4. Stain for about half an hour in a 10 per cent solution of acid fuchsin in 35 per cent alcohol.

5. Wash in the picric-acid solution (step 3) for a moment, then pass through alcohol and xylol and mount in balsam.

II. MINUTE DISSECTIONS

A. Alimentary Canal and Nervous System of Insects.—1. Carefully dissect out the alimentary canal and the central nervous system of a cockroach with the aid of the dissecting microscope or lens. Wash each by gently flooding it with distilled water from a pipette, and then cover it with Bouin's fluid or corrosive sublimate (reagent 17, p. 237) for 30 minutes.

2. Wash in several changes of water during the course of half an hour and stain for 40 minutes or more in borax-carmine.

3. Wash in 50 per cent alcohol and decolorize in 70 per cent acid alcohol until the objects become bright scarlet in color.

4. Wash in 95 per cent alcohol for 5 minutes and then transfer to absolute alcohol for 5 minutes, xylol or turpentine 10 minutes, and mount in balsam. Apply cover and label.

B. Gizzard of Cricket or Katydid.—Pull off the head of a cricket or katydid. The gizzard usually remains attached to the head part. Cut it open lengthwise, wash out the contents and mount as above, but omit the staining. The inside should be turned uppermost.

C. Sting of Wasp or Bee.—1. Place a wasp or bee in water, cover to keep out dust, and let it stand for two or three days until the smell becomes unpleasant.

2. Wash in clear water and squeeze the abdomen gently until the **sting** protrudes. With forceps pull it out carefully. The poison gland and **duct** should come away with it.

3. Place the parts removed on a slide and under a lens draw the sting **out** of its sheath by means of a small needle which should be drawn over the **outer** surface of the sheath from the base to the apex of the sting.

4. Stain and follow out the same subsequent treatment as for II, A, above, or mount without staining. It is advisable to compress the object between two slides as soon as the acid alcohol is washed out. The slides should be tied together and left in 95 per cent alcohol several hours. Then proceed in the ordinary way.

D. Salivary Gland of Cockroach or Cricket.—Let the animal soak in water as for preparation of sting. When sufficiently decayed pull off the head carefully with forceps. The esophagus, the salivary glands, and crop usually come along with it. Stain and mount as for sting. For preparation of fresh salivary gland use the salivary gland of a *Chironomous* larva.

E. Mouth Parts of Insect.—1. Place the head of a bee or cockroach in 95 per cent alcohol for 2 or 3 hours. Transfer to absolute alcohol for 30 minutes, and then to cedar oil for 30 minutes to an hour.

2. Remove the head to a slide and in a drop of the oil dissect out the mouth parts. Transfer them to a clean slide, remove the excess of oil, and arrange them in their relative positions in sufficient balsam to hold them in place, then set the slide aside in a place free from dust until the balsam hardens enough to keep the parts from shifting. Make any necessary rearrangement. Add more balsam and a cover.

MEMORANDA

1. **The Cover-Glass May Be Supported** by means of small wax feet, bits of broken cover-glass, or fine glass threads when the tissue is too bulky to allow the cover-glass to fit down closely to the slide.

2. **A General Rule for Dissociating Tissues** is to use small pieces of the tissue and not a very great amount of the fluid.

3. **For Minute Dissections** clove oil is often a convenient medium. It tends to form very convex drops, clears well, and renders the object brittle; any or all of which properties may be useful in such dissections.

4. **The Fixation of Pieces of Macerated Tissue** (e.g., macerated epithelium in 0.5 to 1 per cent osmic acid for an hour or so often proves advantageous.

5. **Congo Glycerin** is recommended by Gage as especially good for isolated preparations, particularly nerve cells. It is made by dissolving $\frac{1}{2}$ gram of Congo red in glycerin. It acts both as a stain and as a mounting-medium. The preparation may be sealed (p. 108) if desired.

6. **To Dissect Small Insects** Melvin Doner imbeds them in parawax in the concavity of a hollow-ground slide and dissects under a binocular dissecting microscope, using the finely drawn points of glass tubes with the tips so broken as to give a sharp cutting edge. Freshly molted individuals are to be preferred. In imbedding, the parawax is melted around the insect's body with a warm needle, and is then permitted to set so that the body is firmly enveloped and carries a thin film of parawax over its exposed surface. The concavity of the slide is then filled with normal salt solution, completely submerging the object. A 250-watt bulb is used for illumination and to warm the parawax to such an extent that after the incision is made the two sides of the insect's body will be held apart by the wax, permitting dissection of the viscera. The fatty tissues are first carefully dissected out to reveal the under-

lying structures. It is well to make several changes of the salt solution to remove the loosened fatty tissue and to prevent undue warming of the wax. Where differentiation by staining is desirable, the salt solution is replaced for a few minutes by a weak solution of gentian violet, and then destained to the proper degree with weak alcohol. The viscera may then be restained, if desired, with a permanent dye such as alum cochineal, hemalum, or basic fuchsin. Light-green is especially useful for differentiating the chorion of the developing eggs in the ovarioles. Permanent mounts of organs or organ-systems *in situ* in the body cavity may be made by removing the undesired structures, freeing the insect from the parawax with xylol, and mounting in balsam. Isolated organ-systems may be run up *in situ* through the alcohols in the usual way and hardened, then removed from the body, cleared in cedar oil, and mounted in balsam. It is well to support the cover-slip by small bits of glass.

CHAPTER XII

TOOTH, BONE, AND OTHER HARD OBJECTS

Sectioning Decalcified Tooth.—1. Kill a cat and remove the lower jaw (p. 56). With a fine saw cut out about a quarter of an inch of the bone bearing a tooth (e.g., canine), remove as much of the surrounding tissue as possible, and place the object in Zenker's fluid for 1 or 2 days. Wash thoroughly in water and place in alcohol for at least 24 hours. Transfer to nitric-acid decalcifying fluid (reagent 10, p. 9). Use a relatively large quantity of the fluid and change it each day until the tooth is decalcified (2 to 6 days). It is sufficiently soft to cut when a needle can be thrust into it easily. Use this test sparingly, however, as it injures the tissue.

2. Wash it in repeated changes of 70 per cent alcohol until all trace of the acid is removed.

3. Transfer the object through 50 and 35 per cent alcohol successively to running water and wash for 24 hours.

4. Cut sections by means of the freezing microtome as directed under that method (p. 77). If a freezing microtome is not available use the celloidin method.

5. After dissolving out all of the gum from the sections in distilled water, stain in alum-cochineal and Lyon's blue (see method, p. 54). Dehydrate. Remove one or two of the best sections (through the center of the tooth) to a slide, clear, and mount in the usual way in balsam.

6. Stain other sections in 1 per cent osmic acid for 24 hours and mount in glycerin-jelly. When the jelly has hardened, seal the cover with asphaltum, and when this is dry add a thinner coat. If preferred, dehydrate and mount in balsam instead of glycerin-jelly.

Sectioning Decalcified Bone.—Saw out a short piece from the femur of a cat (p. 56). Prepare transverse sections by decalcifying and sectioning as directed in memorandum 5. Do not destroy the periosteum. Prepare likewise longitudinal sections of a tarsal bone.

Sectioning Bone by Grinding.—1. With a fine saw cut a thin transverse section of the femur of a cat. Let it macerate in water until quite clean, then dry it carefully.

Tooth, Bone, and Other Hard Objects

2. Grind the disk of bone between two hones, keeping the hones parallel in order to avoid wedge-shaped sections. The section is not thin enough until fine print can readily be distinguished through it.

3. Wash the section thoroughly in water, transfer it to absolute alcohol for 10 minutes, then to pure ether for half an hour.

4. After removal from the ether, clamp it between two slides by means of a string or a rubber band and let it dry thoroughly.

5. Place some xylol-balsam in the center of a slide and heat it for a few minutes to drive off the xylol, then press the section of bone down firmly into it and put on the cover-glass. The air in the spaces of the bone makes them stand out black. The balsam should not be thin enough to enter these spaces.

MEMORANDA

1. **Failure To Stain Properly** is due ordinarily to insufficient washing out of the acid.

2. **Teeth and Other Hard Objects** may be prepared by grinding in the same way as bone.

3. **For Other Decalcifying Fluids** than nitric acid, see Appendix B, v.

4. **Decalcification by the Fixing Fluid** often results where acid is present. C. J. Hamre finds that 10 days' immersion in Bouin's fluid produces excellent decalcification of the bones of the bat without injury to surrounding tissues.

5. **A Technique for the Decalcification of Bone** which has proved very satisfactory in the hands of S. J. Martin is the following: (*a*) Fix the bony tissue in Bouin's fluid and wash out in 70 per cent alcohol. (*b*) Place it in the decalcifying solution (reagent 10, p. 9) the volume of which is at least thirty times that of the tissue; shake the reagent several times a day to insure uniform and thorough decalcification. (*c*) Change the reagent daily for the following number of days, depending upon the kind of tissue:

Guinea-pig (pubic symphysis)	18–21 days
Rat (femur, scapula)	12–15 days
Frog (femur)	8–10 days
Bat (pubic symphysis)	7–9 days
Rabbit (femur)	16–18 days
Fish (vertebra)	7–9 days

(*d*) When decalcification is complete remove all acid by repeated washings in 70 per cent alcohol. (*e*) When the tissue is acid-free remove it to 83 per cent alcohol, 10 to 15 minutes; 95 per cent alcohol, 15 to 20 minutes; absolute alcohol 1 to 1½ hours; cedarwood oil, until clear (not more than 5 hours). To minimize hardening the decalcified tissue should not remain in the alcohol longer than necessary. (*f*) Wash in xylol for 1 to 3 minutes (but do not substitute xylol for cedarwood oil in clearing). (*g*) Place in paraffin melting at 42 to 47 degrees for about an hour, then in paraffin melting at 50 to 52 degrees for 4 or 5 hours (infiltrate for the minimum time, and keep down the temperature). (*h*) Imbed in paraffin of 52 degrees melting-point. Sections can be cut as thin as 10 microns. For cutting serial sections, a "bone" knife (one with a small bevel and thick edge) is desirable.

CHAPTER XIII

INJECTION OF BLOOD AND LYMPH VESSELS

Red Injection Mass.—1. Rub up 4 grams of carmine thoroughly with 8 c.c. of distilled water in a mortar and add ammonium hydrate drop by drop until a transparent red color results.

2. After quickly washing it to remove dust, etc., soak 10 grams of best French gelatin in distilled water until it is swollen and soft (18 hours), then remove it to a porcelain evaporating-dish and melt it at a temperature of about 45° C.

3. While the gelatin is yet fluid, slowly add the coloring matter, stirring constantly until a homogeneous mixture is obtained.

4. Before the mass cools add also some 25 per cent acetic-acid solution drop by drop, stirring thoroughly until the mass becomes slightly opaque and the odor of ammonia gives place to a faint acid smell. Watch for this change closely, for a few drops too much of the acid will spoil the entire mass by precipitating the carmine. If the ammonia is not completely neutralized, on the other hand, the coloring matter will diffuse through the walls of the injected vessels and stain the surrounding tissues. Walker (*American Journal of Anatomy* [1905], p. 74) makes results more certain by mixing 1 part of the laboratory ammonia with 4 parts of distilled water, and then determining the exact amount of the laboratory acetic acid which will neutralize it. Knowing this, it is easy to determine the total amount of acetic acid which must be added for the amount of ammonia which has been used in any quantity of the gelatin mass. Just before using, the mass should be heated and strained through clean flannel wrung out of hot water.

With a large animal it is advisable to keep animal and apparatus submerged in warm normal saline during the operation of injection, but with a small animal this is unnecessary if the operator works rapidly.

Blue Injection Mass.—Prepare a gelatin mass as directed above. To the warm mass add sufficient quantity of saturated aqueous solution of Berlin blue to give the desired blue color. If the blue does not dissolve, add a little oxalic acid to the mixture. The blue mass need not be made for the present practical exercise unless the student wishes to undertake a double injection as indicated in memorandum 2, p. 100.

Yellow Injection Mass.—Prepare a gelatin vehicle consisting of 1 part of gelatin to 4 parts of distilled water. Take equal volumes of the gelatin mass, a cold, saturated solution of bichromate of potassium, and a cold, saturated solution of lead acetate. Add the bichromate solution to the gelatin and heat almost to boiling; then add slowly, while stirring, the solution of lead acetate.

INJECTING WITH A SYRINGE: SINGLE INJECTION

A common method of injection, and one which proves satisfactory in many instances, is by means of a metal or glass syringe. Although not as desirable in the main as the method of continuous air pressure, many good injections may be made by means of the syringe. The apparatus consists of a syringe fitted with a stop-cock in the nozzle, and a separate tube, known as the cannula, which fits on to the end of the nozzle. The syringes are made in different sizes, and each is provided with an assortment of cannulae to fit vessels of different caliber.

1. Provide yourself with several strong threads about four inches in length for ligating blood vessels. Have the red injection mass melted and heated to about 50° C. Also have ready some hot water to warm the syringe.

2. Kill a cat or a rabbit by means of chloroform or illuminating gas. The latter acts more rapidly and causes less struggle on the part of the animal. Work rapidly so that the entire animal may be injected while yet warm. Stretch it out in a dissecting pan or tie it out on to a board, or, better, keep it immersed in a vessel of warm normal saline solution.

3. Slit the skin along the ventral surface of the body to the middle of the neck and reflect it to the right and left side. Pin it back out of the way.

4. Snip a small hole through the body wall just posterior to the ensiform cartilage. Insert the index finger of the left hand to guide the scissors and prevent injury to the underlying organs, and cut the costernal cartilages of the right side up to the first rib. In like manner cut the cartilages of the left side up to the first rib.

5. Ligate the sternum tightly as close to the first ribs as possible to prevent leakage from cut blood vessels.

6. Cut off the apex of the heart and expose the ventricles. The left ventricle is seen as a round opening, the right as a slit.

7. With a sponge wrung out of warm water rapidly absorb the blood from the thorax.

Injection of Blood and Lymph Vessels

8. Choose the largest cannula that the aorta will admit and thrust it through the left ventricle into the aorta.

9. With a pair of fine-pointed forceps (preferably with curved points) pick up one end of a thread for ligating and carefully work it through under the aorta (do not mistake the vena cava superior for the aorta). Tie the thread around the aorta over the cannula, making a double or surgeon's knot. Draw it tightly on the cannula so that the latter will be held firmly in place. Run another thread through under the aorta and have it in readiness to ligate the aorta when the cannula is withdrawn.

10. Warm the syringe by sucking hot water into it repeatedly, then fill it and the cannula with the warm injecting fluid.

11. Force out a little of the fluid from the syringe to expel all air, and connect it carefully with the cannula.

12. Force the injecting mass into the blood vessels by a slow steady pressure. Begin with a very low pressure, so that the large vessels will be thoroughly filled before the mass enters the capillaries. The pressure should be gradually increased. Avoid sudden increase of pressure or too strong pressure, for either may cause a rupture of the blood vessels and consequent extravasation. From 8 to 10 minutes is about the time required to make a good injection of the cat.

13. Examine the intestines and the gums from time to time and also the inside of the thigh (from which the skin has been reflected); they should be deeply colored by the mass before the injection is complete. If the mass begins early to flow from the right ventricle, the ventricle should be ligated. In any event, it is well to tie the ventricle a few minutes before completion of the injection, to insure filling of all blood vessels.

NOTE.—If the gums remain uncolored, the cannula has probably been forced past the arteries which lead to the head. In such a case, complete the injection of the trunk and then, if injected tissue from the head region is desired, cut obliquely into one side of the innominate artery, tie a cannula in place, and inject toward the head as in the case of the aorta.

14. When the injection is complete, shut the stop-cock, ligate the aorta, or clamp it with pressure forceps beyond the end of the cannula and then remove the latter.

15. Place the animal in cold water or cold alcohol for half an hour, then remove pieces of liver, spleen, pancreas, stomach, intestine, salivary glands, kidneys, and voluntary muscles and harden in strong alcohol or in 10 per cent formalin.

16. When sufficiently hardened transfer the objects to ether-alcohol and proceed to imbed and cut in celloidin according to the method already given. Make longitudinal sections of the kidney parallel to its flat surface. Cut transverse sections of liver, stomach, and intestine, longitudinal ones of the muscle, and sections passing longitudinally through the hilum of the salivary glands and spleen. The sections should not be under 30 microns thick. Mount some unstained; stain others in diluted Delafield's hematoxylin or in hemalum.

MEMORANDA

1. **Apparatus for Continuous Air-Pressure Injections** is now provided in many laboratories. If a regular cylinder for air pressure is not present, how-

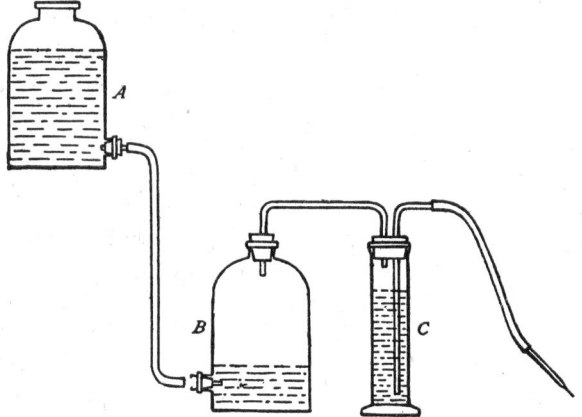

Fig. 34.—Apparatus for Continuous Air-Pressure Injection (after B. G. Smith)

ever, anyone with a little ingenuity can readily fit up a suitable apparatus. A carboy or large-mouthed bottle which can be tightly corked will answer as a chamber for compressed air, a water tap, or a tank of water elevated to the height of 7 or 8 feet will provide sufficient pressure. By making the proper connections by means of rubber and glass tubing a steady stream of compressed air may finally be conducted to a flask containing the injection mass; the flask works in the same way as an ordinary wash-bottle (Fig. 34). All corks and fittings must be tightly secured with wire or strong cord. If desired, by adding an extra perforation to the cork in the air chamber, a mercury manometer may be added to register the amount of air pressure. If a metal cannula is not at hand a glass one may be made as indicated under memorandum 9, p. 103. In lieu of a stop-cock, use a pinch-cock on the rubber delivery tube.

2. **A Double Injection of the Vascular System** may be made by first in-

jecting the blue mass until it is seen to flow from the right ventricle, then detaching the tube which conveys the blue mass, and slipping over the end of the cannula a tube conveying a red mass. This second mass should be in a bottle or flask connected with the pressure bottle by means of an additional tube through the cork of the latter, or the two flasks containing the colored masses may be connected with the tube from the pressure bottle by means of a Y-tube. Each must be provided with a pinch-cock or clamp to hold back its contents while the other is in operation. If a syringe is used, it is better to have a second syringe for the second mass, although one will answer if it is rinsed out with hot water before being filled with the second mass. The second mass should have a quantity of very finely pulverized cornstarch mixed with it, so that when it reaches the capillaries they will become completely plugged. Walker uses the red gelatin mass first, then follows with a gelatin colored with ultramarine blue. The granules of the latter are too large to enter the capillaries, hence a double injection with veins red and arteries blue is obtained.

It should be borne in mind that the larger veins cannot be injected in a direction contrary to their flow because of the valves they contain.

3. **The Lungs, Liver, and Kidneys** are readily injected through their larger blood vessels with two masses, and afford very instructive material when thus prepared. A triple injection of the liver may be made by injecting the hepatic artery and the hepatic and portal veins. The third mass may be colored with China ink. Whitman (*Methods in Microscopical Anatomy and Embryology*) recommends first injecting the hepatic artery and afterward the two veins. The blood should be washed out of the organ to be injected, with warm salt solution.

4. **To Inject Lymphatics,** the puncture method is commonly employed. For example, an aqueous solution of Berlin blue is drawn into a hypodermic syringe, the sharp point of the cannula is thrust into the tissue, and the syringe emptied by slight, steady pressure. For practice, thrust the cannula into the pad of a cat's foot, and force in some of the injection mass. If the leg is rubbed upward, the fluid will flow along the lymph channels and into the glands of the groin. Instead of this haphazard method, however, much better results will be insured by the use of the needle-and-clamp device of W. S. Miller (*Johns Hopkins Hospital Bulletin*, XVI, No. 173 [1905]).

5. **Micro-Injection of Embryonic Vessels** has been much resorted to in recent years for the study of early stages of lymph and blood vessels, and some very delicate and effective methods have been devised. India ink is the medium ordinarily used. Embryos of medium size are generally injected with a hypodermic syringe, but smaller embryos require a more delicate procedure in which glass tubes with the finest possible capillary points are used.

Anyone who has seen the beautifully injected specimens of Dr. H. McE. Knower will concede the success of his method. The general scheme of his apparatus is shown in Fig. 35, and the accompanying legend is self-explanatory.

102 *Animal Micrology*

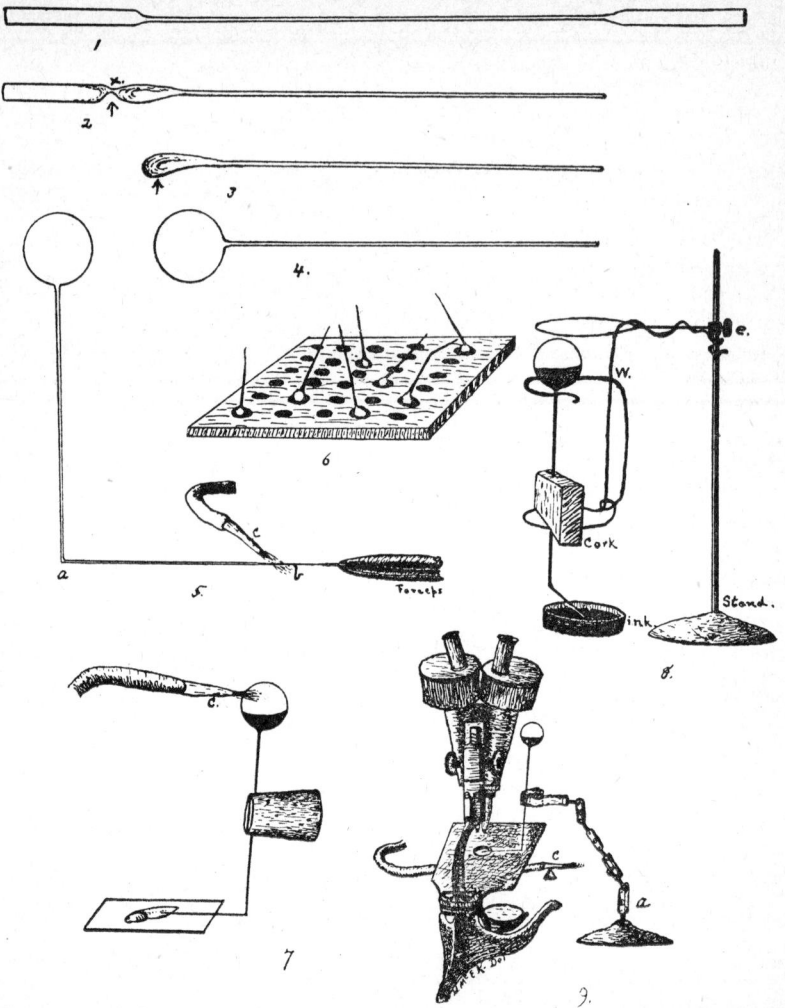

Fig. 35.—Apparatus for the Injection of Small Embryos, etc., under the Microscope (after Knower).

    1. Glass tube drawn out after moderate heating near middle. Tube should be turned while heating. The capillary tube will usually be three or four times the length of that in the figure.
    2. Superfluous tube burned off at x by a fine blow-pipe flame. Tube should be turned while heating.
    3. Enlarged end, molten and heated around the end (note arrow) while being turned.
    4. Bulb blown as soon as tube in Fig. 3 is molten. The end is usually removed from the flame, when hot enough, and blown quickly, but without too great pressure, through the capillary tube. Size usually about twice that of figure.
    5. The bend at a is given before the fine tip b is drawn out over the gentle heat of c, or near a hot iron. Reduced size.
    6. Board with auger holes to hold bulbs. Size much reduced.
    7. Method of applying heat and introducing tip with free hand, and simple cork holder. Size reduced.
    8. Wire support w, for holding bulb while filling. Reduced size.
    9. Method of using jointed holder a, with binocular. The small gas-jet c lies in front ready for heating the bulb. Size greatly reduced.

He expresses the essence of his method as follows: "If a gentle warmth is applied to a glass bulb blown on the end of a capillary tube, while the fine point of the tube is held beneath the surface of some fluid, such as India ink, air will be driven out of the bulb and ink will run up to replace it as the bulb cools. When the system has come to equilibrium, the point of the tube is inserted into the desired blood vessel under a dissecting microscope (binocular if possible), the tube being carried on a holder to avoid warming it or the bulb. India ink is now injected, as desired, by warming the bulb when ready." The method is explained in minute detail in the *Anatomical Record* for August, 1908 (Vol. II, No. 5), together with applications to fish, amphibia, reptiles, birds, and mammals.

Instead of using a glass bulb, as does Knower, Heuser inserts a simply made resistance coil of fine German-silver wire into an ordinary salt-mouth bottle. The ends of the wire extend out through the cork for electrical connection. When a current is passed through the wire the air surrounding it in the bottle, becoming heated, expands and affords a steady and prolonged pressure.

Dr. Emily Ray Gregory uses direct pressure from a good grade of De Vilbiss' atomizer bulb which lies on the floor. The bulb, kept from rolling by a crocheted net cover, is operated by the foot. Pressure is transmitted through a $\frac{3}{16}$-inch red-rubber tube to the short glass injection tube which has the outer end drawn into a capillary tip at right angles to itself. The method is given in detail in the August number of the *Anatomical Record*, 1916 (Vol. XI, No. 1)

For other suggestions regarding micro-injections see Hoyer, *Zeitschrift für wissenschaftliche Mikroskopie*, Band XXV (1908); Evans, *American Journal of Anatomy*, IX (1910); Sabin, *Contributions to Embryology*, III, No. 7 (1915), Carnegie Institution of Washington; Brown, *Anatomical Record*, XXIV (1922) 295; Reagan, *University of California Publications in Zoölogy* (1926); and McClung, *Microscopical Technique*, (1929).

6. **To Keep Gelatin Injection Masses** let them congeal, then cover the surface with 95 per cent alcohol, and leave in a well-stoppered vessel until needed.

7. **Injection through the Femoral Artery** is frequently practiced, and is preferred to injection through the aorta by some workers. An oblique cut is made in one side of the artery and the cannula inserted pointing toward the heart. Others prefer to cut into the dorsal aorta and inject both anteriorly and posteriorly.

8. **The Injecting Syringe** must work without jerking or catching along the wall of the barrel. It should always be carefully cleaned after using. If the piston does not fit the barrel tightly enough it should be wrapped with gauze.

9. **Glass Cannulae** may be made by grasping the ends of a short piece of soft glass tubing and heating the middle in a flame until the glass becomes soft, which is indicated by the yellow color of the flame. The tubing should be constantly rotated, so that all sides heat equally. When the glass becomes

soft, **draw the tube out steadily** until the diameter of the soft portion becomes as small as desired. When the glass has cooled, the tube should be cut with a file at the proper place to make two cannulae of it.

10. **If the Blue Color Fades** in the gelatin mass in the tissues, it may frequently be restored by treating the tissue or section with oil of cloves or turpentine.

11. **A Cold Fluid Gelatin Mass** has been used successfully by Tandler (see abstract by A. M. C. in *Journal of Applied Microscopy*, V, 1625). To prepare the mass, dissolve 5 grams of finest gelatin in 100 c.c. of tepid distilled water. Color to the desired shade with Berlin blue, and then add slowly 5 to 6 grams of potassium iodide. The mass remains fluid at ordinary temperatures, but when injected objects are placed in 5 per cent formalin it sets completely and is thereafter unaffected by reagents. The minutest vessels are injected, and sections may be stained in the usual ways. Subjection to strong acids, such as sulphuric or hydrochloric, does not affect the mass; hence it may be used for injecting specimens that are to be decalcified afterward. To preserve the fresh mass, add a few crystals of thymol and keep in a stoppered bottle.

12. **Corrosion of Injected Vessels or Cavities** is sometimes practiced. A mass must be employed which will not be attacked by the reagent used for destroying the surrounding tissues. One of the best masses consists of white wax 5 parts and rosin 6 parts, melted together at a temperature of about 75° C. For fine vessels increase the proportions of wax, for larger ones add more rosin. Vermilion, Prussian blue, or chromate of lead may be used for coloring. The part to be injected should be placed in warm water and the mass injected at a temperature of from 50° to 60° C. The injected part is left in cold water for from 1 to 2 hours, and is then corroded in pure hydrochloric acid for from 6 to 48 hours, according to the resistance of the tissue. Finally, wash the preparation thoroughly in running water. For bibliography and more detailed directions see *Technique des injections*, by Hermann Joris, Université Libre de Bruxelles, 1903. See also Hinman, Morrison, and Brown, *Journal of the American Medical Association*, LXXXI, No. 3 (1923), 177–84.

13. **Wood's Metal** is one of the commonest injection media used for corrosion preparations. It is a fusible alloy consisting of 1 or 2 parts of cadmium, 2 parts of tin, 4 of lead, with 7 or 8 parts of bismuth. It melts at from 66° to 71° C.

14. **Celluloid Dissolved in Acetone** was found preferable by Flint (*American Journal of Anatomy*, VI [1906]), in his study of lung development, to celloidin or Wood's metal for corrosion preparations. He injected from aspiration bottles into the lungs through the trachea, and used hydrochloric acid for corrosion.

15. **Dehydrated Nitrocellulose for Corrosion Preparations** is highly recommended by Batson (*Science*, LXXXI, No. 2108 [May 24, 1935]). He uses "dehydrated nitrocellulose, RS $\frac{1}{2}$ second, viscosity 3.2" (obtainable

from Hercules Powder Co., Wilmington, Del.), dissolving 1,000 grams of it in 1,000 c.c. of acetone (technical grade.) For color, artists' oil pigments worked up in a small quantity of dioxan may be used. Acetone soluble dyes are preferable; they may be dissolved in the acetone at the beginning or added later. English vermilion is particularly suitable. For maceration, concentrated technical hydrochloric acid is used, or sometimes, 5 parts of this acid to 1 part of water. If bone is to be retained, the preparation is macerated in water at body temperature.

16. **Air Injection of Minute Vessels** was found to be an indispensable method by Locy (*American Journal of Anatomy*, XIX [May, 1916], 3) in his work on the lung and air passages of the chick: "In stages subsequent to 96 hours, the lungs and air sacs were dissected out of the previously fixed and hardened specimens, then cleared in cedar oil, after which the organs were placed in a mixture of 1 part cedar oil and 2 parts chloroform. On becoming permeated with this fluid, the preparation was removed from the mixture and placed on a filter paper until the chloroform might evaporate. The evaporation of the chloroform served to draw the cedar oil from the lumina of the various branches of the bronchial tree into the lung tissue and to fill the spaces thus made with air. When this preparation was replaced in pure cedar oil, the difference between the refractive index of the imprisoned air and the surrounding medium gave the lung tubes the appearance of being filled with a metallic cast. Thus minute air passages that could not be injected by other means were made clear. The finer details would disappear after a few minutes as the cedar oil percolated into them, but the same specimen, if carefully manipulated, can be treated repeatedly without apparent injury, and a complete picture could finally be obtained."

For later stages Locy also used celloidin and Wood's metal injections followed by corrosions. He obtained some beautiful Wood's metal casts of the adult lung. The lungs of the freshly killed fowl were distended under pressure with 80 per cent alcohol until the air sacs were fully expanded. The entire bird was then immersed in alcohol for 24 hours or more before metallic injection was attempted.

17. **An Excellent Injection Mass** for other than histological purposes as used in our own laboratories is made as follows (Wagner):

Water	100 c.c.
Glycerin	20 c.c.
Strong formalin	20 c.c.
Cornstarch, powdered	75 grams

Mix by gradually adding water and glycerin to the starch, rubbing out all lumps. For yellow color add 10 grams of chrome yellow; for green, 10 grams of chrome green; for red, 10 grams of vermilion. Strain through cheesecloth and add the formalin. If the mass is too thick to strain, add the formalin first. Cornstarch is vastly superior to laundry starch and the colors recommended diffuse less into tissues than carmine or Berlin blue.

18. **For Clearing of Injected Organs or Embryos "in Toto,"** see memoranda 17, 18, and 20 (pp. 117–19).

CHAPTER XIV

OBJECTS OF GENERAL INTEREST: CELL-MAKING, FLUID MOUNTS, "IN TOTO" PREPARATIONS, DRY MOUNTS, OPAQUE MOUNTS

When objects of considerable thickness are to be mounted, it is sometimes necessary to resort to cells which will contain the object and support the cover-glass. Fluid mounts and aqueous media must occasionally be used for delicate objects which would be injuriously affected by alcohol, or which are unsuitable for mounting in balsam. When such mounts are used, whether in a cell or not, the cover-glass must ordinarily be sealed with a cement if the preparation is to be permanent. In all cases where it is at all practicable, balsam mounts are to be preferred for permanent preparations. Glycerin is a convenient mounting-medium for many objects, especially for temporary mounts. It is often used where such media as balsam would render the preparation too transparent; it is much more favorable, moreover, to the preservation of color than are resinous media. For making cells and sealing circular covers, a turntable (Fig. 36) is desirable, although the work may be done by following a guide ring drawn on paper and placed under the slide.

I. TURNING CELLS

Prepare 12 or 15 slides as follows: 1. Place a slide on a turntable and adjust it so that its center lies over the center of the turntable.

2. Dip a small camels' hair pencil into asphaltum (asphalt varnish) but do not take up enough of the fluid to drop (see also memorandum 15, p. 117).

3. Choose a guide ring on the turntable which is of slightly smaller diameter than the cover-glass to be used, whirl the table and hold the pencil lightly over the guide ring. The ring which has been spun should be even. If it is not, practice turning rings until satisfactory ones are made. If the asphaltum is old it is probably too thick to make suitable rings.

4. The slide must be set aside to dry before it can be used for mounting. A gentle heat will aid in drying.

5. To some of the cells add successive coats of the varnish as the previous one dries, so that you will have cells of varying depth.

II. MOUNTING IN GLYCERIN

A. Water Mites and Transparent Larvae.—1. Kill several small, colored water mites or transparent larvae of insects by means of chloroform (a few drops in water) and place them for half an hour (two or three hours for larger objects) into a mixture of water and glycerin equal parts, after which transfer them to pure glycerin.

2. Apply a thin coat of asphaltum to the upper edge of a cell which is of sufficient depth to accommodate the object.

3. Breathe into the cell to moisten it so that the glycerin will adhere throughout and prevent the formation of air bubbles.

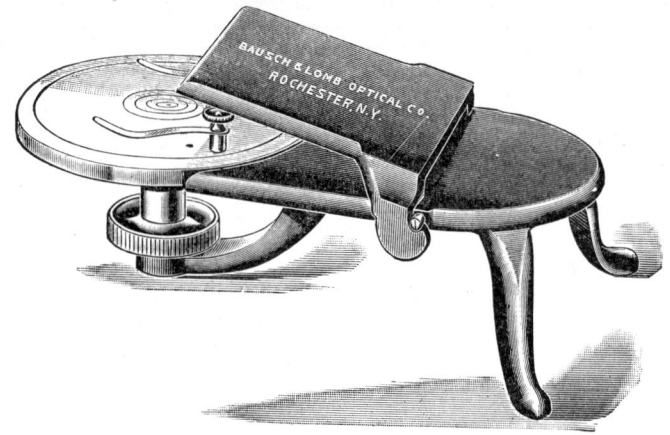

FIG. 36.—Turntable

4. Fill the cell flush with glycerin and put the object into it, carefully spreading out all parts.

5. Breathe on the lower surface of a clean cover-glass, put one edge down on the edge of the cell, and then gradually lower the cover so as to avoid bubbles of air. When in place, press the cover down gently with the handle of a needle and see that it adheres all around. Wash off the exuded glycerin and carefully wipe the slide with a cloth.

6. Turn a comparatively broad ring of asphaltum around the edge of the cover to seal it, and when this is dry add a thinner coat. Label and put away in a horizontal position until dry.

CAUTION.—It is *indispensable* that the edges of the cover-glass be perfectly dry before attempting to seal the preparation; otherwise the cement will not adhere.

B. **Killing and Mounting Hydra.**—1. With a dipping-tube (memorandum 12, p. 116) remove a hydra to a warm watch-glass and leave it in only a few drops of water. Have ready some hot Bouin's fluid or corrosive acetic, and when the hydra sends out its tentacles and expands its body, apply the reagent by suddenly squirting it into the watch-glass so that it sweeps over the hydra from aboral to oral extremity and carries the tentacles out straight. Then fill the watch-glass with the hot fluid. If quantities are to be fixed use Wu's technique (p. 299).

2. After 10 minutes pour off the fixing fluid and wash the animal thoroughly in 70 per cent alcohol.

3. Replace the alcohol with alum-cochineal or dilute hematoxylin and stain for from 5 minutes to several hours.

4. Remove the stain with a pipette and replace it with a mixture of equal parts of glycerin and water for half an hour, followed by pure glycerin. Proceed farther as in the preceding exercise.

NOTE.—After removal from the stain, if necessary, decolorize in acidulated water or alcohol (0.5 per cent hydrochloric acid); then, wash out the acid thoroughly in tap water.

Hydra may also be dehydrated, cleared, and mounted in balsam (see also "Hydra," p. 297).

III. MOUNTING IN GLYCERIN-JELLY

Glycerin-jelly is frequently preferable to pure glycerin for mounting because it is a solid at ordinary temperatures. A good formula for making it is as follows:

Water	400 c.c.
Gelatin	65 grams
Glycerin, carbolic-acid crystals, and egg white, as directed below.	

Let the gelatin soak in the water overnight, then dissolve with gentle heat. Add the whites of two eggs and heat (but do not boil) for half an hour. If an autoclave is available heating is best done in it. Heat for 15 minutes at a pressure of 15 to 20 pounds, then turn off steam and allow to cool without opening exhaust valve. The gelatin will boil if pressure is reduced too rapidly, and boiling transforms it into metagelatin, which will not harden at room temperature. The egg albumen gradually precipitates and carries down all fine particles of dust, etc., so that the gelatin is left perfectly clear.

Filter through filter-paper on a hot filter, add an equal volume of glycerin and 2 grams of carbolic acid. If a hot-water funnel is not available, invert a metal container hood-fashion over the funnel, let the warm gases from a low flame rise into this cover, and thus heat the funnel. Use only clean gelatin of the best quality. Warm for 10 or 15 minutes, stirring continually until the mixture is homogeneous.

A. Small Crustacea.—1. By means of a dipping-tube isolate such small creatures as Cyclops, Daphnia, or Cypris.

2. Kill by warming slowly in a drop of water on a slide.

3. Place them in a cell of proper depth, draw off all water with a pipette, and gently warm the slide.

4. Place the bottle of glycerin-jelly into a vessel containing warm water until the jelly becomes liquid, but do not let it get any warmer.

5. Fill the cell flush with the warm jelly and arrange the objects in suitable positions.

6. Breathe upon the lower surface of a clean cover-glass and put it in place in the usual way.

7. Wash away any trace of the jelly from the outside of the cell and when the slide is dry run a ring of asphalt-cement around the edge of the cover. After this dries, varnish with more cement. It is not an absolute necessity to seal glycerin-jelly mounts, but the writer has always found it a wise precaution.

B. Muscle of Insect.—1. Cut off the head of an insect and bisect the trunk so as to expose the interior. Observe two kinds of muscular tissue, that of grayish color belonging to the legs, the yellowish to the wings.

2. Take a shred of muscle and on a dry slide carefully separate pieces of muscle fiber and stretch them out, while keeping them moist by breathing on them.

3. Mount in glycerin-jelly as directed in the previous exercise (see also p. 278).

IV. MOUNTING IN BALSAM

A. Flatworms.—1. Obtain specimens of Planaria from the under surface of flat rocks in the edge of streams (see "Planaria," p. 299).

2. Place the animal in a little tepid water. Watch until it is extended full length, then flood it quickly with corrosive sublimate to which 1 to 3 per cent of acetic acid has been added. (For Wu's method see p. 299). The animal may be removed after 30 minutes or an hour

and washed thoroughly in 50 per cent alcohol to which a little tincture of iodine has been added.

3. Stain for 24 hours in alum-cochineal or in Delafield's hematoxylin diluted one-half with water.

4. Wash in water followed by 35 and 50 per cent alcohol each 15 minutes.

5. Decolorize in acid alcohol until the color ceases to come away freely (10 to 30 minutes).

6. Wash out the acid in 70 per cent alcohol, using the alkaline alcohol if hematoxylin was used in staining.

7. Flatten the animal by compressing it between two slides by means of a rubber band, and place it for 24 hours in 95 per cent alcohol.

8. Transfer to absolute alcohol for 1 hour, and to xylol until clear.

9. Mount in balsam in a thin cell or without a cell at pleasure. If on examination the separate organs of the animal are not seen distinctly, it probably has not been compressed sufficiently. This difficulty may sometimes be avoided in a measure by letting a cover-glass rest upon the live planarian to flatten it out slightly, and then running the fixing fluid under the cover. Specimens which have been in the laboratory for some weeks or months make better preparations than those fresh from the stream.

B. **Mosquito, Gnat, or Aphid.**—1. Kill a mosquito with cyanide or chloroform and place it in cedar oil or turpentine for an hour.

2. Remove, and place it on its back on filter-paper. Carefully spread the legs of the insect, put a drop of thick balsam on a slide, invert the slide, and bring the balsam in contact with the thorax of the mosquito. Spread the wings and the legs of the insect and gently press it down into the balsam.

3. Add thinner balsam, see that the proboscis and antennae are floated out properly, then add more balsam, and put on a cover-glass.

V. OPAQUE MOUNTS

Some objects are mounted to be viewed by reflected instead of transmitted light. They may be mounted in the ordinary way, and when they are examined as opaque objects, the light from the mirror should be turned away and, if neceessary, a strip of dark paper placed under the slide to shut off all light from below.

A. **Beetles.**—Choose a shallow cell for mounting the wing cases and legs of one of the Curculionidae, preferably *Curculio imperalis*, the South American diamond beetle.

1. Soak the part in cedar oil or turpentine for half an hour, then place it in the cell in the proper position, the outer side of the case toward the observer.

2. Fill up the cell with balsam and add the cover.

B. Wings of Moths or Butterflies.—Prepare parts of the wings of moths or butterflies as in A. The wing of the clothes moth makes a good opaque mount.

C. Head of a Fly.—1. Secure the specimen (preferably one having colored eyes, as one of the gadflies) and choose a cell of the proper size for it. The cell should be of such a depth that the cover will rest lightly upon the object and retain it in the center of the cell. The head should present the front view when mounted.

2. Spin a very thin coat of asphaltum on to the dry edge of the cell so that the cover will adhere.

3. Soak the head of the fly for a couple of hours in equal parts of glycerin and water.

4. Moisten the cell by breathing into it, fill it with glycerin, and transfer the object to it.

5. Breathe on the cover-glass and apply it very carefully to avoid air bubbles. When the cover settles into place, press it down gently to make it adhere to the cement.

6. Set it aside to harden. When hard, seal on the turntable with asphaltum followed by a second coat when the first is dry.

D. Foreleg of Dytiscus, the Great Water Beetle.—1. Detach the foreleg of a male, and soak it in 10 per cent potash solution (see reagent 93, p. 264) for a day or two.

2. Wash it in water, run it up to 95 per cent alcohol, and leave it there for 24 hours.

3. Pass it through absolute alcohol and clear in cedar oil, turpentine, or xylol.

4. Lay the leg, disk side uppermost, in a drop of balsam on a slide, add another drop of balsam, and carefully cover with a clean cover-glass. Place a small weight (e.g., half of a bullet) on top of the cover to hold it down until the balsam hardens.

VI. DRY MOUNTS

A. Scales.—Prepare a very shallow cell and let it dry. Thoroughly dry the scales from a moth's wing by gently heating them on a slide over a flame. Place the scales in a cell, warm the slide until the cell wall becomes sticky, put on the cover and press it down until it adheres all around, and finally seal as in previous exercises.

B. Eggs of Butterflies, Small Feathers, Antennae of Insects, etc., may be mounted as dry objects. Care must be taken to have them *perfectly* dry, or they will in time cloud the cover with moisture from within.

MEMORANDA

1. Small or Soft Insects or Their Larvae may frequently be mounted directly in glycerin, or they may be dehydrated and mounted in balsam. A method often used is to kill them in strong carbolic acid and mount them directly in balsam. The carbolic acid both dehydrates and clears. It is better, however, to clear the preparation further by immersion in cedar oil or xylol before adding the balsam. For dissection of small insects, see page 92, memorandum 6.

Melvin Doner submits the following techniques as specially valuable with certain insects and allied forms:

a) For permanent mounts of ticks, mites, mosquito larvae, etc., in extended condition.

Berlese's fluid:

Distilled water	20 c.c.
Chloral hydrate	160 grams
Gum arabic	15 grams
Glucose syrup	10 grams
Glacial acetic acid	5 grams

Dissolve the ingredients in the order named, on a water filter bath. Using No. 5 Whatman filter-paper, filter with suction pump. For mosquito larvae, ring slides with celluloid dissolved in amyl acetate, and finish with one or two coats of asphaltum.

b) For endoparasites, hymenopterous and dipterous larvae, scale insects and their nymphs, etc.

De Faure's fluid:

Gum arabic	30 grams
Chloral hydrate	50 grams
Glycerin	20 c.c.
Chlorhydrate of cocaine	0.5 grams
Distilled water	50 c.c.

c) For ticks, such as the American dog tick.

Zebrowski's mixture:

Iodine (sat. in 95 per cent alcohol)	20 c.c.
Eosin (sat. in 95 per cent alcohol)	10 c.c.
Delafield's hematoxylin	10 c.c.
Picric acid (sat. in 95 per cent alcohol)	5 c.c.
Alum cochineal	10 c.c.
Glacial acetic acid	5 c.c.
Alcohol, 50 per cent	90 c.c.

Leave the ticks in this solution for 10 days. Destain some in acid alcohol; others in alkaline alcohol. Mount in balsam. Different ticks react differently.

2. **Insects Having Hard Shells** may first be soaked in 10 per cent potash to soften them and render them transparent if they are to be examined by transmitted light. The softer parts of insects so treated are destroyed and only the external parts remain. Such insects may be mounted in glycerin or glycerin-jelly, or they may be dehydrated, cleared, and mounted in balsam. If heavily pigmented they should be bleached in chlorine vapor or hydrogen peroxide before dehydration.

For softening chitin, hypochlorite of soda is sometimes effective. It is recommended that insects (larval, pupal, or adult) be dropped into boiling hypochlorite of soda which has been diluted with four volumes of water. The heat fixes the soft parts. Let stand for 24 hours or more, according to size of specimen; the chitin of many forms will become soft, transparent, and permeable to stains. Better staining will result if, after removal from the hypochlorite, the preparation is further fixed in some of the standard picro-formol solutions. Instead of depending on softeners, however, it is wisest, when possible, to fix insect larvae just after they have molted (or pupae just before emergence), before the chitin hardens.

3. **Delicate Insects** which are too frail to withstand much handling may be placed at once in cedar oil or turpentine and after an hour mounted in balsam (see "Mosquito," p. 111).

4. **Wings, Legs, Antennae, Mouth-Parts, etc., of Such Forms as Flies and Bees,** which have been preserved in alcohol, should be completely dehydrated, cleared, and mounted in balsam in cells of the proper depth.

5. **Transparent and Soft Insects** may be stained in alum-cochineal or hematoxylin in the ordinary way and mounted as whole objects, if desired. They will stain better if they have been fixed previously in some corrosive-sublimate mixture and then washed properly (see reagent 17, p. 237). To stain, follow the method outlined in IV, A, p. 102.

6. **To Demonstrate Insect Spiracles and Tracheal Systems** caterpillars of the sphinx moth (e.g., the tomato or the tobacco "worm") are excellent. Cut out the spiracles, together with a small area of the body-wall, dehydrate in absolute alcohol, clear, and mount in balsam. To see the tracheal systems to advantage soak the caterpillar for 24 hours in a 20 per cent sodium hydroxide solution; then carefully tease out the tubules and their finer branches.

7. **To Center an Object in a Cell** (the head of an insect, for example), thread a fine needle with a hair and run it through the object. Remove the needle and imbed the ends of the hair in the cement on opposite sides of the cell. When the cover-glass is put in place the object may be adjusted by pulling the hair. After the slide is finished and dry, the ends of the hair should be cut off at the edge of the cell.

Another method which will frequently answer for an object to be mounted in balsam is to place the object (after clearing) in the center of the cell, coat it with balsam, adjust it properly, and then set the slide away in a place free from dust till the balsam thickens. Finally fill the cell with balsam and add the cover.

8. **The Radula or Lingual Ribbon of the Snail or Slug** should be dissected out and soaked for a day or two in a 10 per cent solution of potash. If the animal is a small one, cut off the head including the buccal mass and soak it in a solution of potash until the soft tissues are destroyed and only the radula remains. From the potash the radula is transferred to water and washed for some hours. With a strip of paper on each side to prevent crushing it, it should be placed between two slides, and the slides bound together by means of string or rubber bands. While held in this position, dehydrate and clear it. Finally remove one slide and the paper and mount the object in balsam on the other slide. A shallow cell may be used if desired.

9. **Flukes and Tapeworms** are prepared in the same manner as Planaria (p. 110). The time of immersion in the various fluids should be lengthened in proportion as the object is larger than the planarian. See also pp. 300, 301.

For *in toto* staining, Mayer's paracarmine and Mayer's hemalum are highly recommended by nearly all specialists on these forms. The animals should be much overstained and then very rapidly and completely destained in strongly acidulated (2 to 4 per cent HCl) 70 per cent alcohol (Cort, *Transactions of the American Microscopical Society*, XXXIV, No. 4 [October, 1916]).

To prevent the curling up of flatworms which are to be infiltrated for sectioning, Peaslee binds them by wrappings of thread to a bit of bristol board which is not removed until the animal is to be imbedded.

10. **Nematodes** frequently turn black after mounting in balsam; even when successfully so mounted they may be too transparent to show details of structure. An adaptation of Diehl's method for botanical objects (*Science*, March 8, 1929) has given excellent results in our laboratory, in the hands of J. B. Goldsmith, for glycerin-jelly mounts of certain flatworms, tapeworm eggs, and nematodes, all of which display better optical properties in aqueous media than in balsam. A drop of liquid glycerin-jelly is placed in the center of a large cover-glass (22 mm.) and the object oriented in it; a smaller cover-glass (12 mm.) is dropped on top and the jelly allowed to solidify. Then sufficient Canada balsam is placed on top to cover the smaller cover-glass and the exposed under surface of the large one and to surround completely the edges of the mount with a ring of balsam. The whole is covered gently with a slide, and the slide is then inverted revealing the mount immediately under the larger cover-glass and surrounded by a protecting balsam seal. If a nematode such as ascaris is to be sectioned, its cuticle may be made more permeable by soaking for several hours in 5 per cent potassium hydroxide. See also p. 300.

11. **Spirogyra, Protococcus,** Volvox, Desmids, etc., may be mounted in a cell in the following copper solution:

Acetate of copper	1	gram
Camphor water	240	c.c.
Glycerin	240	c.c.
Glacial acetic acid	0.3	c.c.
Corrosive sublimate, saturated aqueous solution	0.1	c.c.

Mix thoroughly, filter, and keep in a glass-stoppered bottle. The green color of the plant may frequently be preserved for some time in this medium. The specimen is washed in water, transferred to the cell, then the solution is added. The cell is covered and sealed in the usual way.

Keefe (*Science*, October 1, 1926) successfully employs the following formula for many green plants:

50 per cent alcohol	90	c.c.
Commercial formalin	5	c.c.
Glycerin	2.5	c.c.
Glacial acetic acid	2.5	c.c.
Copper chloride	10	grams
Uranium nitrate	1.5	grams

For blue-green algae he substitutes 10 grams of copper acetate for the copper chloride and uranium nitrate.

Herbert N. Clarke has shown me algae mounted directly in dilute water-glass which are still green after four years.

12. **A Dipping-Tube** is a simple glass tube. To operate it, hold the tip of the forefinger over the upper end and dip the lower end into the water until it comes just above the object desired; lift the finger and let the air out of the tube, and the water will rush in at the lower end carrying the object with it. Replace the finger over the top of the tube and remove it; the water will remain in it as long as the finger is held firmly over the upper end. When the finger is removed the water and the object pass out. The object may sometimes be more readily discharged if the tube is rotated. A pipette made of a large-bore glass tube and an atomizer bulb is also very serviceable.

13. **To Keep Water from Evaporating from a Cell Too Freely,** use a round cell and cover it with a square cover-glass. Apply a brush wet with water to the slide beneath one of the projecting corners of the cover from time to time. Capillary attraction will draw in the water and will keep the cell full. If a continuous supply of fresh water is necessary, one end of a loosely twisted cotton thread may be laid along one side of the cover and the other end of the thread immersed in a small vessel of water which stands within half or three-quarters of an inch of the cell. A reservoir made from the bottom of a shell vial or homeopathic vial answers very well; it may be cemented to the slide.

Protozoa and other small forms may be kept alive on a slide for a number of hours by simply mounting them in water under a cover in a cell of blotting paper which has been saturated with water. For aquaria for studying microscopic organisms, Walton (*Ohio State University Bulletin*, XIX, No. 5 [1915]) used ringlike pieces of lens paper cut somewhat smaller than the cover-glass. Such aquaria keep for several hours. They may be made more permanent by letting them stand 15 to 30 minutes in order to allow the outside water to evaporate, and then running paraffin oil around the margin of the cover-glass.

14. **Deep Cells** are made frequently by cutting out rings of paper, lead, or block-tin with gun punches and cementing them to the slide. Glass and hard-rubber rings of various sizes may be purchased from dealers. To support cover-glasses Barker uses circular cloth patches with a hole in the center. These may be bought of a stationer.

15. **Filtered Shellac** is recommended by McClung as excellent both for making and for sealing cells. It may be colored with Bismarck brown and similar dyes. Barker uses any good quality of enamel paint. Clarke uses ordinary duco and finds it excellent.

16. **For a Method of Preserving Fine Dissections** for microscopic study as opaque objects, see memorandum 15, p. 152.

17. **The Clearing of Total Specimens** as developed by Spalteholz (*Ueber das Durchsichtigmachen von menschlichen und tierischen Präparaten*, 2d ed., 1914, S. Hirzel, Leipzig), whereby relatively large anatomical and embryological preparations can be made transparent, is an extremely useful method, particularly with injected objects. The following account of the method together with modifications introduced by herself is taken from Miss Sabin's article on *Contributions to Embryology*, III, No. 7, Carnegie Institution of Washington, 1915: "In general the essentials of the method are, first, fixation in formalin; second, a thorough bleaching of the tissues with hydrogen peroxide to remove the hemoglobin and other pigments; third, dehydration; and fourth, clearing the specimens in an oil which has the same index of refraction as the tissues. As applied to embryonic tissues, the method, developed by Professor Spalteholz, to whom I am very much indebted, is as follows: The specimens which have been injected with India-ink are fixed for 24 to 48 hours in 5 and 10 per cent formalin. Commercial formalin is slightly acid, which is an advantage for the India-ink injections, since the ink diffuses in an alkaline solution. Specimens which have been injected with silver nitrate are ruined by fixation in formalin, because the silver salt is changed to a white precipitate which obscures the vessels. If injections of bone are desired, the formalin may be made slightly alkaline and the diffusion of the ink prevented as much as possible by tying off all vessels before fixation. For large fetuses, which are to be cleared *in toto*, Dr. P. G. Shipley has found that the subsequent bleaching is made easier by washing the specimen in running water before fixation, thus removing much of the hemoglobin. After fixation, the specimens are washed in

running tap water from 12 to 24 hours, followed by distilled water to remove the formalin. The bleaching is done in hydrogen peroxide. Spalteholz adds a few drops of ammonia to precipitate the barium salts. This is not necessary with barium-free oxide. For adult tissues, Spalteholz uses undiluted peroxide; for the embryonic tissues about 2 to 3 per cent is the best strength. The small embryos with ink injections take about 20 minutes to bleach; for the silver specimens, 2 or 3 minutes suffice, and they must be watched constantly and the bleaching stopped before the silver is affected. Following the bleaching, the specimens must be washed thoroughly in running water and in distilled water. The dehydration may be begun with 50 per cent alcohol and the percentage increased successively by five points or less. After two changes of a good grade of absolute alcohol, the specimens are passed through changes of benzene into the synthetic oil of wintergreen. The small amount of benzene which is carried over evaporates quickly, and the few bubbles which develop in the bleaching process can be removed with needles. The oil of wintergreen should be entirely colorless, but both the specimens and the oil will gradually become brown with age. This is especially true of the silver-nitrate specimens, but they will keep for six months or a year in oil. They can be returned to alcohol for storage and recleared when desired, or they may be made permanent in balsam. The advantage of keeping the total specimens in oil rather than in balsam is that they can be dissected. On the other hand, they are made more permanent in balsam. The oil of wintergreen makes the tissues tough, so that it is possible to obtain minute dissections of the injected specimens.

"The Spalteholz method as applied to embryos can be very much simplified by changing the fixative. For mammalian embryos the best fixative is Carnoy's mixture. This is absolute alcohol 60 parts, chloroform 30 parts, and glacial acetic acid 10 parts. In this mixture the acid is sufficiently strong to bleach the hemoglobin so that the peroxide is unnecessary. The penetrating power of the fixative is very great, which is of importance, since no injected specimen can be cut into until it is thoroughly fixed. The relations of the tissues are well maintained and the swelling due to the acetic acid tends to counteract the shrinkage that always takes place in the oil of wintergreen. The fixative does not affect any of the injection fluids. The process after fixation in the Carnoy's mixture is simple; specimens remain in the fixative from 2 to 12 hours and are then placed directly into 70 per cent alcohol, dehydrated in graded alcohols, and cleared as before. The specimens can then be studied *in toto*, or dissected or imbedded in paraffin and sectioned. They should be imbedded through a mixture of the oil of wintergreen and paraffin. They do not become brittle in the oil, so that they may be sectioned after staying in the oil for many weeks. The shrinkage in the oil, however, seems to increase on long standing. The advantages of the fixation in Carnoy's mixture are that the specimens are even clearer than after bleaching with peroxide, there are no bubbles formed to damage the tissues, the time of the procedure is shortened and the fixation is much better should it be desired to section the specimens after studying the vessels in whole embryos. Specimens which are strongly pigmented, however, must be bleached with hydrogen peroxide before they can be cleared."

For other methods of clearing *in toto* preparations see memoranda 18 and 20.

18. The Potash Clearing Method (or Modifications of It) for "in Toto" Preparations is, according to Thurlow C. Nelson, one of the best as well as one of the simplest methods for the demonstration of skeletal and cartilaginous structures. Unless nervous or other structures are to be stained, the animal should be put into a 1 per cent potash solution immediately after killing. Twenty-four hours in this medium should be sufficient to clear the overlying tissues to such an extent that the skeletal elements are clearly visible. After removal to glycerin the specimen will keep indefinitely. Lipman (*Stain Technology*, X [1935], 61) finds that the skeletons of embryos fixed in alcohol and cleared in KOH will stain selectively with Alizarin Red S (1 part Alizarin to 10,000 parts of a 2 per cent solution of KOH). After the desired intensity of color is attained, the clearing and mounting process is completed in increasing concentrations of glycerin. See also p. 157.

19. For Staining Nerve Tissue "in Toto," Nelson combines the Charles Sihler hematoxylin stain with the potash method. Three solutions are required:

1

Potassium hydroxide, 1 per cent aqueous solution

2

Glacial acetic acid.	1 part
Glycerin.	1 part
Chloral hydrate, 1 per cent solution.	6 parts

3

Glycerin.	1 part
Ehrlich's acid hematoxylin.	1 part
Chloral hydrate, 1 per cent solution.	6 parts

A minnow, for example, is killed in 95 per cent alcohol and left for 48 hours. After the viscera are removed, it is next transferred to the potash solution for from 1 to 3 days. When transparent, the specimen is put into solution No. 2 for 72 hours, then into solution No. 3 for a week. It is then destained in solution No. 2 for 18 hours and cleared in glycerin. The nervous tissue should show up dark purple in the semi-transparent muscular tissue of the body.

20. A Benzaldehyde Clearing and Fixing Method for "in Toto" Preparations has also been devised by Nelson as follows:

Small objects, such as eggs and small fish, may be put into the aldehyde directly as it acts as a dehydrating agent to a slight extent, being miscible with 30 parts of water. Larger objects are dehydrated by running up through the alcohols and then clearing in the benzaldehyde. If a specimen of the freshwater mussel, for example, *Anodon*, is treated with 50 per cent alcohol for 2 hours, 70 per cent for 5 hours, 80 per cent for 15 hours, and 95 per cent for 2 hours, and is then put into benzaldehyde, the mantle begins to clear at once, the whole preparation being almost transparent in 12 hours. As benzaldehyde is very unstable, being oxidized to form benzoic acid, it must be removed before the tissue is exposed to the air for any long period. Likewise vessels containing it must be kept tightly stoppered. Objects may be kept in the fluid permanently by sealing the container in such a way as to exclude all air. This method of clearing finds its chief use in clearing up the entire bodies of small animals so as to show injected circulatory systems and such structures. Small objects may be mounted in balsam from benzaldehyde if preferred.

To infiltrate with paraffin or to stain *in toto*, the object should first be passed through a mixture of 6 parts of xylol to 1 part of absolute alcohol. A satisfactory stain may be prepared by dissolving methylen-blue crystals in 2 c.c. of absolute alcohol and adding 6 c.c. of benzene.

21. **Farrant's Solution** is a popular aqueous mounting-medium. Twenty-five grams of dry gum arabic is soaked for several days in 25 c.c. of a saturated aqueous solution of arsenious acid, and then mixed carefully with 25 c.c. of glycerin. Air bubbles are likely to be troublesome in this medium.

22. **Anatomical Preparations** such as bisected eyes, etc., may be preserved in stender dishes or the like in the glycerin-jelly medium given on p. 109.

23. **Clear Corn Syrup** has been found very useful in our own laboratory for mounts of such objects as fish scales and the like. It is clearer than glycerine jelly. Mix thoroughly 50 parts of water, 45 parts of white corn syrup (Karo), and 5 parts of formalin. Let the mixture stand several days for the escape of air bubbles. The object to be mounted is fixed, stained, washed in water, and put into glycerine in an open vessel. After the water has evaporated, the object is placed on a slide in the smallest possible amount of glycerine. Enough corn syrup and formalin mixture is added to fill the space under a cover-glass. Add the cover-glass, let dry 10 days or longer, if necessary, and seal with Duco enamel.

CHAPTER XV

BLOOD

I. EXAMINATION OF FRESH BLOOD

a) General.—1. Thoroughly clean a slide and cover, bathe one thumb in ether-alcohol (reagent 4, p. 8), sterilize a blood lancet or a sharp needle by heating it in a flame, and then prick the back of the thumb with it. It is well first to swing the arm vigorously back and forth a few times to drive blood into the fingers, then beginning at the base of the thumb, to wind a handkerchief or other cloth around the thumb spirally toward the tip so that a very slight prick will yield abundant blood.

2. Place a small drop of the resulting blood on a slide and quickly put on a cover-glass. To prevent evaporation, the edges of the cover may be surrounded by olive oil or vaseline.

Living corpuscles may also be studied in a drop of normal saline or in Ringer's solution (90 and 91, p. 264).

b) Effects of reagents.—When it is desired to study the effects of reagents on fresh blood (e.g., distilled water, 1 per cent tannic acid, etc.), a drop of fresh blood is placed on a slide, the cover is put on, and then the blood is "irrigated" with the reagent. That is, a drop of the reagent is placed at the edge of the cover to be drawn under by capillary action. The process may be hastened by applying the edge of a bit of blotting paper to the opposite edge of the cover.

c) To demonstrate blood-platelets.—Place a small drop of a 1 per cent solution of methyl violet (reagent 68, p. 257) in normal salt solution, on the back of a thumb which has been cleaned by washing it in ether-alcohol. With a sterilized needle prick the thumb through the stain and mount a drop of the blood which exudes. Examine it under a high power. Both platelets and white corpuscles are stained.

d) Stained preparation of fibrin.—Mount a drop of blood on a slide as in *a*. Place it in a moist chamber for from 20 to 30 minutes to coagulate. Loosen the cover with a few drops of water and then thoroughly irrigate the preparation with water. Drain off the water, blot the preparation with blotting paper, and add immediately a drop of a 1 per cent aqueous solution of eosin (reagent 49, p. 249). Remove this after 3 minutes, rinse the preparation in water, then treat it 3 minutes with a 1 per cent aqueous solution of methyl violet (reagent 68, p. 257). Rinse the preparation in water, let it dry, and finally mount in balsam.

e) Crystals of the blood.—

1. **Hemoglobin Crystals.**—Allow a drop of blood to dry on the slide with-

out covering it. Long rhombic prisms of a red color crystallize out. The blood of a rat is best for demonstration. A more certain method is as follows: To 5 c.c. of blood in a test-tube add a few drops of ether and shake the mixture vigorously until the blood becomes laky. Place a drop or two of the laked blood on a slide and allow it to dry in the cold.

2. **Hematoidin Crystals:** reddish-yellow crystals (rhombic plates).—They can be obtained from old blood extravasations (e.g., cerebral hemorrhage, corpora lutea, etc.) by teasing. Mount in Canada balsam.

3. **Hemin or Teichmann's Crystals.**—To a small drop of blood on a slide or a bit of cloth which has been previously saturated with blood, add a few crystals of common salt. Heat over a flame until the mixture has become dry, leaving a reddish brown residue. Apply a cover-glass and flood the preparation with as much acetic acid as will remain in place under the cover. Heat the preparation until the acetic acid boils. After the acid has evaporated, the preparation may be made permanent by adding Canada balsam. The crystals are very small, narrow rhombic plates of dark-brown color. They vary in size and may lie singly, across one another, or in stellate groups.

The presence of these crystals is positive evidence of the presence of blood, hence their demonstration is of great importance in stains or fluid suspected of containing blood.

II. COVER-GLASS PREPARATIONS

a) Dry preparations (Ehrlich's method)—1. In this method the preparation is "fixed" by means of heat. Under one end of a copper bar or copper triangle (Fig. 24) place a flame. After 15 or 20 minutes a given point on the bar will have a practically constant temperature. Thoroughly clean the bar, run a stream of water along the top of it toward the flame, and locate the point farthest from the flame at which the water boils. The blood smears when prepared are to be placed film side up in a row across the bar about three-fourths of an inch nearer the flame than the point at which the water just boiled. This will subject them to a temperature of about 120° C.

2. Thoroughly clean and dry two cover-glasses, touch one to a *small* drop of perfectly fresh blood as it comes from the finger or lobe of the ear and instantly drop it on to the second cover. The blood should spread in a thin film between the covers; if it does not, it has begun to coagulate and the preparation will be inferior. Rapidly separate the covers by sliding them apart, wave them in the air a minute to dry the films, then place them down with the smear side uppermost. *Do not press the covers together* to spread the blood because this ruins the corpuscles. If the red corpuscles are to retain their shape the film of blood must be *extremely* and *uniformly* thin. Practice until you have prepared such a film.

Some workers prefer to make smears on slides instead of on cover-slips. A small drop of blood placed near one end of a perfectly clean slide is spread by putting the end of another slide up against it so that it barely touches and then pulling it away rapidly along the first slide so that the blood follows in a thin film. Crushing is thus avoided.

NOTE.—In the clinical examination of blood great care must be exercised to have it absolutely fresh; furthermore, the cover-glasses should be handled with forceps instead of by means of the fingers. It is recommended that one pair of the forceps be Coronet or spring forceps of some kind (Fig. 39). The lobe of the ear is perhaps the best region from which to obtain the blood. The needle with which the puncture is made should always be sterilized. Wipe away the first drop of blood that appears. The drop finally chosen should be one that has appeared *immediately* after the spot has been wiped and it should be but little larger than a pin-head. The whole operation cannot be performed too rapidly. To shorten the time it is well to have an assistant to prick and manipulate the ear while the operator attends to the preparation of the film.

3. When several satisfactory films have been prepared, place them on the heated bar, as indicated in step 1. Cover them to keep out dust and leave them for from 30 to 60 minutes.

4. Remove the covers and stain the preparations 15 or 20 minutes with Ehrlich's triple stain (reagent 48, p. 249) by flooding the film with the stain by means of a pipette. Rinse off the surplus stain with water, blot the film with blotting paper, and dry it by holding it with the edge downward high above the flame. When dry, mount in balsam on a slide.

NOTE.—Instead of heating the preparation, much the same results may be obtained by subjecting films (prepared as in step 2) to ether and alcohol (p. 8) for from 1 to 12 hours, drying them again in the air, and then staining as above.

b) Rapid method.—1. Prepare a film as above (*a*, 2), but before it has dried treat it for 30 minutes with a saturated aqueous solution of corrosive sublimate (reagent 17, p. 237).

2. Wash the preparation throughly in water or in 50 per cent alcohol.

3. Stain for 10 minutes in Delafield's or Ehrlich's hematoxylin (reagent 59, or 60, pp. 252, 253), rinse in 70 per cent alcohol, and stain for 20 seconds in eosin (0.5 per cent solution in 70 or 95 per cent alcohol).

4. Rinse in 95 per cent and in absolute alcohol each for 2 minutes, pass through xylol, and mount in balsam.

After rinsing following staining, some workers simply blot the preparation with blotting paper, dry it in the air, and mount it in balsam.

NOTE.—The foregoing "dry" method, while still popular in some laboratories, is rapidly being superseded by "wet" methods in which the film is made from fresh blood as above and put directly into the fixing liquid or, at most, after drying in the air without heat for not over 30 seconds. Almost any fixing agent is applicable to blood although osmic acid probably gives the most faithful fixation. Two drops of blood mixed with 5 c.c. of a 1.5 per cent solution of osmic acid and left for from 2 to 24 hours is a good method for fixing blood in bulk.

III. ENUMERATION OF BLOOD CORPUSCLES

The instrument used is the hemacytometer (Figs. 37 and 38). It consists of a special slide for counting and two graduated pipettes (one for red, one for white corpuscles) for diluting and measuring the blood. There are various forms of hemacytometers on the market. The newer counting chambers made of one piece of glass are preferable to those cemented to the slide because there is no cement to become dissolved; also they are easier to clean. Of the numerous types of rulings the Neubauer ruling in the improved form has proved most practical in my own laboratory. The counting chamber, the cover, and the

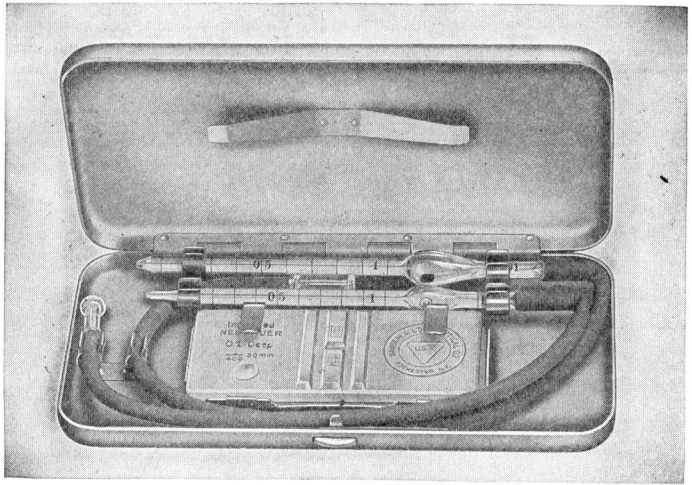

FIG. 37.—Bausch & Lomb Improved Neubauer Hemacytometer, Including Mixing Pipettes for Red and for White Corpuscles, Respectively

mixing pipettes should conform to the specifications established by the United States Bureau of Standards.

Obtain a drop of blood from the lobe of the ear or from the finger. Fill the smaller pipette (with the 101 mark just above the chamber containing bead) of the hemacytometer to the mark 1 by careful suction. Keep the point of the pipette downward. The tip of the tongue placed firmly over the hole in the mouthpiece will keep the blood from dropping back. If the blood is drawn beyond the 1 mark, blow it out immediately, clean the tube, and repeat the operation.

Wipe the blood from the outside of the pipette quickly and draw in sufficient Toisson's solution to make the level of the combined liquids stand precisely at the mark 101. It is well to snap the side of the pipette with the finger once or twice to prevent the glass bead from adhering to the wall and causing an air bubble beneath it. Close the ends of the pipette with thumb and middle finger and mix the blood thoroughly with the solution by shaking the tube for

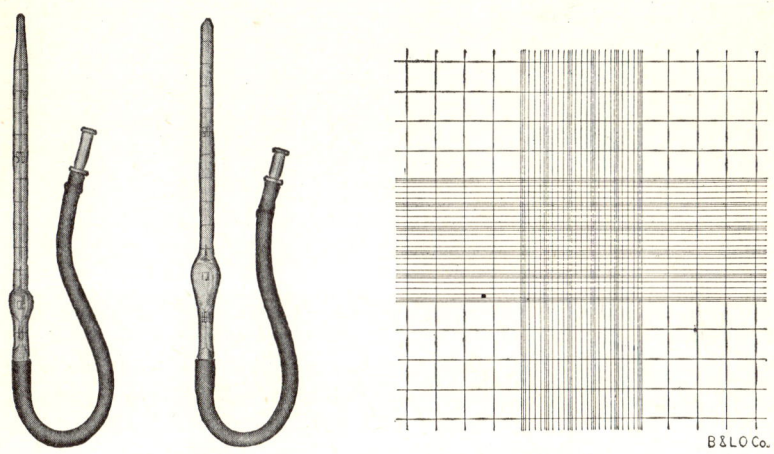

Fig. 38.—Thoma Mixing Pipettes and Improved Double Neubauer Ruling for Counting Chamber

3 minutes. The blood is thus diluted 100 times. The graduation on some tubes differs from that of the Thoma. If a different tube is used this fact must be kept in mind in calculating the dilution.

Toisson's solution:

Sodium sulphate	8.0	grams
Sodium chloride	1.0	gram
Neutral glycerin	30	c.c.
Methyl violet, 5b	0.025	gram
Distilled water	160	c.c.

Blow out a drop of the liquid to remove the unmixed solution remaining in the capillary tube. Have the counting-disk and cover-glass perfectly clean and grease-free. Allow a drop of the diluted blood to flow on to the disk and place the cover-glass over the drop. The cell of the disk must be entirely filled by the drop of blood. Press slightly until Newton's colored rings (rainbow effect) can be seen where the glass surfaces come in contact. Let the corpuscles settle a minute or two before beginning to count. Examine the field under a low-power objective to see that the corpuscles are evenly distributed. If they are not, the blood was not thoroughly mixed and the whole operation should be repeated after thoroughly cleansing the pipette.

Examine the preparation under a high power of the microscope, and count the number of red corpuscles in 20 to 40 small squares (the greater the number of squares counted the greater the accuracy). Of those corpuscles which happen to lie on the boundary line, count the ones that lie only in the upper and on the left sides of each square. Take the average number in a square and calculate the number of corpuscles in a cubic millimeter of blood.

The depth of the entire cell is 0.1 mm., the area of each small square is $\frac{1}{400}$ sq. mm., consequently the volume of blood in each square column is $\frac{1}{4000}$ cu. mm., or 1 cu. mm. of diluted blood would contain 4,000 times the average number in a square. One cubic millimeter of undiluted blood contains 100 times as many, or 400,000 times the number in one square. What result do you obtain? For accuracy, three separate counts should be made and the average taken.

After finishing the count, clean the pipette by successively drawing into and expelling from it water, alcohol, and finally ether. Do not blow through it, but cause the ether to evaporate by sucking air through the tube. For counting the white corpuscles use the large pipette and dilute the blood 10 times with one-third of 1 per cent glacial acetic acid. The acid destroys the red corpuscles, and thus the white corpuscles are more readily seen. Proceed in the same manner as for red corpuscles.

Use water instead of alcohol for cleaning the counting-slide if the chamber is cemented to the slide, as alcohol dissolves the cement.

For other methods of manipulation and other types of counting chambers see chap. vii by Raphael Isaacs in McClung's *Microscopical Technique*.

IV. OBSERVATION OF THE BLOOD CURRENT

a) Circulation in the web of a frog's foot.—Wind a long strip of cheesecloth around a frog stretched out upon a narrow piece of thin board, leaving one hind foot exposed. Soak the cloth in water in order to keep the animal's skin moist. Pin the extended foot in such a way that the web between the toes is stretched over a notch or hole in the end of the board. Examine under the microscope. If the preparation is favorable, leucocytes may perhaps be seen penetrating the walls of the vessel (*diapedesis*) and passing into the surrounding tissues.

b) Circulation in the mesentery. Inflammation.—Immobilize a frog (the male is better) by injecting a few drops of a 1 per cent solution of curare into one of the dorsal lymph sacs. Curare paralyzes the nerve-endings. After waiting 20 minutes for the curare to be absorbed into the circulation, cut open the abdominal wall for a short distance along the left side and draw out several loops of the intestine. Pin out a favorable area of mesentery over a cork ring, and, after covering it with a cover-glass, examine under the microscope. Keep the parts moistened with normal salt solution. Such a preparation is especially favorable for studying the migrations of leucocytes through the walls of the vessels. Do not have the mesentery stretched too tightly or the circulation will cease. After a time the phenomena of inflammation may readily be observed. It is hastened if some irritant (e.g., a drop of creosote) is applied to the mesentery.

MEMORANDA

1. **For Demonstration of the Different Granules of Leucocytes, etc.,** see memorandum 5 and p. 267, under the general topic of blood. For differential reactions of the various forms of leucocytes to a dye or dyes see Sabin; *Johns Hopkins University Hospital Bulletin*, No. 34 (1923, p. 277; and *Carnegie Inst. Embryol.*, No. 16 (1925), p. 165.

2. **To Study Blood in Sections,** ligate a small vessel in two places to keep in the corpuscles, then remove the piece so prepared and fix it in Bouin's fluid or osmic acid. Imbed in paraffin and cut thin sections. Stain material fixed in Bouin or corrosive-sublimate reagents by the hematoxylin-eosin method (p. 54) or with the Ehrlich-Biondi stain (47, p. 248). The blood fixed in osmic acid may be stained by the saffranin-gentian violet method (80, p. 261).

3. **Ameboid Movements in Leucocytes** may readily be observed in blood (preferably amphibian) which has been mounted on a slide in very slightly warmed normal saline. Place a hair under the cover-glass and seal the edges of the latter with vaseline or melted paraffin. For continuous study of the white corpuscles of warm-blooded animals a warm stage of some kind is necessary to keep the temperature of the blood near the temperature of the body.

For classroom demonstration the living leucocytes of lamellibranchs (Ostrea, Venus, Anodonta) are especially serviceable (Breder and Nigrelli, *Science*, LXXVIII [1933], 128). After removing one valve, a capillary pipette is thrust into the heart and a quantity of blood withdrawn. When placed on a slide, the leucocytes in a drop of the fluid contract and clump together, but after some 5 or 10 minutes they begin to thrust out pseudopods and move apart. They will remain active for several hours if the preparation is sealed (e.g., with paraffin). The leucocytes in a few drops of blood placed in a dish of the same water from which the animals came will live for several days. Such leucocytes show ameboid movement, ingestion of food, and other activities that are commonly demonstrated in free-living amebae. Operation is not imperative, for, if these mollusks are placed in water that is allowed to warm slightly, they will emit quantities of leucocytes. Samples taken in this way, however, commonly show contamination.

4. **Ingestion by Leucocytes.**—Rub up sufficient India ink in a few drops of normal saline to make a grayish fluid. With fine scissors make an incision into one of the dorsal lymph sacs of a chloroformed frog (parallel to and close beside the urostyle). Introduce a capillary pipette into the wound and obtain a small drop of lymph. Mix it on a slide with a drop or two of the prepared ink.

After placing a hair across the field, put on a cover-glass and seal the edges with vaseline or melted paraffin. Under a high power of the microscope the cells may be seen engulfing the colored particles.

Gage (*The Microscope*) recommends a mixture of lamp-black, 2 grams; sodium chloride, 1 gram; gum Arabic, 1 gram; distilled water, 100 c.c. Mix thoroughly in a mortar and filter through one layer of gauze and one of lens paper. When injected into an animal the leucocytes will ingest the particles of carbon.

5. Wright's Stain for Blood.—This is a modification of Leishman's Romanowsky stain. To prepare the stain make a 0.5 per cent solution of sodium bicarbonate in distilled water and add to it 1 per cent of methylen blue (B.X., or "medicinally pure"). Subject the mixture to live steam in an ordinary steam sterilizer (e.g., Arnold; not a pressure sterilizer or a water-bath) for one hour. The container should be of such size that the liquid forms a layer not more than 6 cm. deep. When the mixture is cool, filter to remove any precipitate. To each 100 c.c. of the filtered mixture add, with constant stirring, 500 c.c. of a 0.1 per cent aqueous solution of "yellow, water-soluble" eosin. Collect the resulting precipitate on a filter, dry it thoroughly, and, rubbing up in a porcelain dish or mortar if necessary, make a 5 per cent solution in pure methylic alcohol. To prevent the alcohol from evaporating, keep the bottle containing the solution tightly stoppered. Should precipitation occur, filter the stain and add a small quantity of methyl alcohol. The stain is best made up and allowed to "ripen" one month before using. Use undiluted. The prepared stain may be obtained from dealers.

Mallory and Wright in their *Pathological Technique*, p. 383, give the following summary of the method for staining blood films:

1. Make films of the blood, spread thinly, and allow them to dry in the air.

2. Cover the preparation with a measured quantity of the staining fluid for one minute.

3. Add to the staining fluid on the preparation the same quantity of distilled water as there was of the stain. Allow this mixture to remain on the preparation for two or three minutes, according to the intensity of the staining desired. Eosinophilic granules are best brought out by briefer staining.

4. Wash in water, preferably in distilled water, until the film has a pinkish tint in its thinner or better-spread portions and the red corpuscles acquire a yellow or pink color.

5. Dry between filter-paper and mount in balsam. The preparations retain their colors as long as any preparations stained with anilin dyes. Fresh films stain better than those which are several hours old. See, however, comment under "Euparal," p. 58.

Erythrocytes when stained by Wright's stain should appear orange or pink in color (with deep-blue nuclei, when **nucleated**); *lymphocytes* should show purplish-blue nuclei and cytoplasm of robin's-egg blue with occasional dark blue or purplish granules; *polynuclear neutrophilic leucocytes* should have blue or dark lilac-colored nuclei with cytoplasmic granules of reddish-lilac color; *eosinophilic leucocytes* should show blue or dark lilac-colored nuclei, and blue cytoplasm with granules the color of eosin; *large mononuclear leucocytes* should show blue or dark lilac-colored nuclei with cytoplasm pale blue in one form and blue with dark lilac or deep purple-colored granules in the other; *mast-cells* should exhibit irregular-shaped nuclei stained purplish or dark blue in bluish cytoplasm in which numerous coarse spherical granules of variable size, dark purple to black in color, are imbedded; *myelocytes* have dark blue or dark lilac-colored nuclei and blue cytoplasm containing numerous dark-lilac or reddish-lilac-colored granules; *blood-platelets* are stained blue.

6. **For Malarial Parasites** Wright's stain (memorandum 5) is excellent. It yields the so-called Romanowsky stain; the color of the chromatin varies from lilac to very dark red, while the body of the parasite stains blue. A full account of the method will be found in Mallory and Wright's *Pathological Technique*.

7. **Ehrlich's Triple Stain** for blood is given on p. 249.

8. **Giesma's Stain** is a popular one for blood. Methylen azure (instead of methylen blue) is used in combination with eosin. It is difficult to prepare, hence is best obtained from dealers. To prepare from Grübler's powders, dissolve 3 grams of Azur II-eosin and 8 decigrams of Azur II in 125 grams of glycerin and 375 c.c. of methyl alcohol.

For air-dried films fix in alcohol three minutes; dry with blotting paper; stain 20 minutes with a dilution of 1 drop of the stock stain to 1 c.c. of water; rinse in tap water, blot, and dry. Mount in neutral balsam or preserve unmounted.

For wet films fix for 12 to 24 hours in a mixture of 95 per cent alcohol 1 part, saturated aqueous solution of corrosive sublimate 2 parts; wash in water, then treat with a mixture of 3 parts Lugol's solution (p. 135), 2 parts iodide of potassium, and 100 parts of water; wash, and apply 0.5 per cent solution of sodium thiosulphate for 10 minutes; wash, and stain in diluted stain (as for air-dried film) for 2 to 12 hours, changing stain for fresh at the end of 30 minutes; pass through mixtures of acetone and xylol (5 xylol to 50 acetone, then 30 to 50, and 50 to 50) to pure xylol; mount in cedar oil or neutral balsam. This process is also applicable to sections.

9. **Neutral Balsam** is particularly necessary for blood mounts. Two or three fragments of marble kept in the balsam bottle will neutralize the acidity.

CHAPTER XVI
BACTERIA

No attempt is made here to give even an elementary account of bacteriological technique. Only such phases of the work as are concerned with the immediate microscopical examination of bacteria are touched upon, and these chiefly to afford some practice in this kind of manipulation. For special technique, identification, or descriptions of apparatus and accessories, the student is referred to standard textbooks.

For approved up-to-date methods the reader should consult publications by the Committee on Bacterial Technique, of the Society of American Bacteriologists, Geneva, New York. See particularly *Manual of Methods for Pure Culture Study of Bacteria*, a loose-leaf publication which permits of frequent and independent revision of its various sections.

BACTERIAL EXAMINATION

Bacteria when prepared for microscopical examination are in the form of
A. Cover-glass preparations,
B. Bacteria in tissues (section method), or
C. Hanging-drop preparations.

A. Cover-Glass Preparations

I. *Killing and fixing.*—

1. **From Fluid Media** (e.g., bouillon, milk, water, saliva, blood, pus, etc.).—Sterilize a platinum wire loop by heating it red hot in a flame. When cool, touch the loop to the culture and spread the adherent bacteria in a thin film over the surface of a cover-glass which has been sterilized in a flame. After the

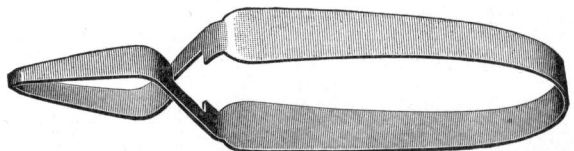

Fig. 39.—Cornet's Cover-Glass Forceps

film has dried in the air, kill and fix the bacteria to the cover by passing it three times, film side uppermost, through the apex of a flame. Each time should not exceed half a second. Prepare several films from a given material. Coronet or similar forceps (Fig. 39) should be used for handling such films, because the cover-glass can be left in them through the entire operation of fixing and staining.

If a platinum loop is not at hand a second cover-glass may be used to spread the smear. The first cover-glass is held in a pair of cover-glass forceps and the second cover-glass is dropped on to it. The glasses are then rapidly drawn apart with a sliding motion by means of forceps. The glasses should not be pressed tightly together. Proficiency in making such preparations is gained only after considerable practice. The chief secret in making a good preparation is to get the films extremely thin and evenly distributed.

2. **From Solid Media** (gelatin, agar, meat, potato, animal tissues and organs, etc.).—The procedure is the same as for 1, except that a drop of sterilized water or bouillon is put on the cover-glass to facilitate the spreading of the bacteria in a film over the cover.

II. *Staining and mounting.—*

1. Gentian violet (memorandum 3, *a*, p. 134), 5 minutes. The cover-glass is left in the forceps, film-side up, and the film flooded with the staining fluid.

2. Rinse in water.

3. Lugol's solution (memorandum 3, *f*, p. 135) until the color becomes black (2 to 3 minutes).

4. Ninety-five per cent alcohol until the violet color has almost completely disappeared.

5. Rinse in water and examine by placing the cover-glass film-side downward on a slide. Only a thin film of water should remain between the slide and the cover. Remove surplus water by means of blotting paper. If a prolonged examination is to be made, water lost by evaporation must be replaced by occasionally placing a small drop of water at the edge of the cover. In ordinary work the final inspection is frequently made at this stage. If a permanent preparation is desired, however, proceed with the following steps:

6. If the bacteria are well stained, a counterstain of Bismarck brown (memorandum 3, *d*, solution 2, p. 134) may be added (5 to 10 seconds). This step may be omitted.

7. Absolute alcohol, 10 to 15 seconds.

8. Xylol.

9. Xylol-balsam.

NOTE.—In staining, if the cover-glass is warmed over a flame some 15 or 20 seconds until the stain steams, the action of the stain is usually more intense and more rapid. Boiling, however, must be avoided.

B. Bacteria in Tissues

Tissues may be fixed and hardened (e.g., Zenker's fluid, Appendix B, p. 232, reagent 6; or Gilson's, reagent 20; or formalin, reagent 22) in the ordinary way and sections made by the usual methods. Where practicable, paraffin sections are preferable to celloidin sections, because the celloidin tends to hold the stain and thus obscure the bacteria. Sections should be fixed to the slide (paraffin by albumen fixative, celloidin by ether vapor).

Bacteria which do not stain by the Gram method (memorandum 3, *f*, p. 135) or the tubercle-bacillus method (memorandum 3, *e*, p. 135) are difficult to demonstrate, because it is hard to stain them so as to differentiate them from the tissues in which they lie; furthermore, most of them easily lose whatever stain they may have taken up. Löffler's alkaline methylen blue (memorandum 3, *b*, p. 134) is a useful stain for such organisms.

Methylen-Blue Stain for Bacteria in Tissues.—1. Stain sections (paraffin) 30 minutes to 24 hours.

2. Acetic acid (1 to 1,000 of water), 10 to 20 seconds.
3. Rinse in absolute alcohol 20 to 30 seconds.
4. Xylol.
5. Xylol-balsam.

With celloidin sections substitute 95 per cent alcohol for absolute (step 3), then treat with creosote or cedar oil until sections are clear. Mount in xylol-balsam.

Anilin gentian violet, methyl blue, methyl violet, or fuchsin (memorandum 3, *a*, p. 134), also carbol-fuchsin (memorandum 3, *c*, p. 134) may be used in the same way.

Gram's Method for Bacteria in Tissues (Weigert's modification).—1. Stain sections (any kind) in lithium carmine 2 to 5 minutes.

Lithium Carmine (Orth's):
Carmine .... 2.5 to 5 grams
Carbonate of lithium, saturated aqueous solution .... 100 c.c.
Thymol .... a crystal or two
Filter

2. Anilin gentian violet, 5 to 20 minutes (celloidin sections should first be dehydrated in 95 per cent alcohol and affixed to the slide with ether vapor).
3. Rinse in normal saline.
4. Lugol's solution (memorandum 3, *f*, p. 135), 1 to 2 minutes.
5. Rinse in water.
6. Blot sections with filter-paper to remove as much water as possible.
7. Anilin oil, several changes. The oil dehydrates, and at the same time decolorizes the celloidin.
8. Xylol, several changes.
9. Xylol-balsam.

C. Hanging-Drop Preparations

1. A slide with a concave center is used (Fig. 40). With a fine-pointed brush paint a narrow strip

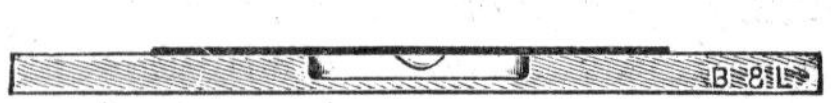

Fig. 40.—Depression Slide

of vaseline around the margin of the concavity. The vaseline makes the cover-glass stick to the slide and also prevents evaporation.

2. Place a small drop of the fluid containing bacteria in the center of the cover-glass. If the bacteria to be examined are on a solid medium, the "drop" should be made by mixing a small portion of the growth with a drop of bouillon, normal saline, or serum. Place the cover-glass, drop downward, over the depression in the slide and press it down well into the vaseline.

3. Use only a small opening in the diaphragm when examining the bacteria, in order to get as much contrast by refraction as possible. Focus first with a medium-power dry objective on the edge of the drop, then employ the oil immersion. Such unstained organisms are frequently difficult to find and there is great danger of breaking the cover-glass with the objective.

Hanging-drop preparations are used mainly in determining the motility of bacteria, or in the study of spore formation. For the latter purpose the slide and cover-glass must be carefully sterilized and the sealing with vaseline complete. The preparation may then be placed on a warm stage or in an incubator and examined from time to time. Instead of vaseline, pyroxylin cement may be used for sealing the cover-slip in place.

MEMORANDA

1. **The Main Points To Be Observed in the Microscopical Examination of Bacteria** are as follows: (1) form of the individual, whether spherical (*coccus*), spiral (*spirillum*), or rodlike (*bacillus*) with end square, pointed, or rounded; (2) uniformity in size; (3) the arrangements of individuals whether single (*micrococci, etc.*), in pairs (e.g., *diplococci*), in chains (e.g., *streptococci*), groups of four (e.g., *tetracocci*), cubical groups of eight or more (*sarcinae*), or small grape-like bunches of various-sized cocci (*staphylococci*); (4) presence or absence of cell wall, gelatinous capsule, etc.; (5) motility in living forms (do not confuse with Brownian movement); (6) reaction to stains; (7) presence of spores which are recognizable as bright, highly refractive rounded bodies.

2. **Material for the Demonstration of Bacteria** (coccus, bacillus, spirillum, and beggiatoa forms) will be found in abundance in foul water, especially when contaminated with sewage. By scraping the inside of the cheek such forms as Leptothrix may often be found. Make a cover-glass preparation; kill and fix in the flame in the ordinary way; stain in methyl violet, gentian violet, or fuchsin (basic) and, if desired, counterstain lightly with Bismarck brown; examine in water or dehydrate in absolute alcohol, clear in xylol, and mount in balsam.

To demonstrate bacteria in tissues, a mouse may be inoculated with anthrax, and paraffin sections of the spleen prepared. Stain by the gentian-violet method.

3. Some of the Most Important Stains for Bacteria are as follows:

a) *Anilin water solution of gentian violet* (Koch-Ehrlich's).—

Crystal (Gentian) violet, saturated in 95 per cent alcohol	10 c.c.
Anilin water (see reagent 36, p. 244)	100 c.c.

After shaking, the mixture should be set aside for 24 hours and filtered because of the precipitation which takes place soon after making. Solutions of fuchsin (basic) and methyl blue are made in the same way. These solutions begin to decompose after about 10 days and must then be freshly prepared. They yield good results with many species of bacteria. The gentian violet, particularly, is widely used in connection with Gram's method (see *f*).

b) *Alkaline methylen blue* (Löffler's).—

Methylen blue, saturated solution (about 0.3 grams) in 95 per cent alcohol	30 c.c.
Caustic potash, aqueous solution (1:10,000)	100 c.c.

This stain keeps well and is one of the most widely used of the general stains. It is especially serviceable in staining the bacillus of diphtheria or of glanders. With methylen blues of American origin, Conn (*Stain Technology*, IV [1929], 27) recommends the following formula as preferable to Löffler's:

Methylen blue, medicinal (90 per cent dye content)	0.3 gram
Ethyl alcohol, 95 per cent	30.0 c.c.

Dissolve and dilute in 100 c.c. of distilled water.

c) *Carbol-fuchsin* (Ziehl-Neelson's).—

Basic fuchsin, saturated solution (about 0.3 gram) in 95 per cent alcohol	10 c.c.
Carbolic acid, 5 per cent aqueous solution	90 c.c.

This stain keeps well, stains powerfully, and can be used on many forms of bacteria.

d) *Neisser's method for diagnosis of diphtheria.*—

Solution I:

Methylen blue (Grübler's)	1 gram
Alcohol, 96 per cent	20 c.c.
Distilled water (add after the methylen blue has dissolved in the alcohol)	950 c.c.
Glacial acetic acid	50 c.c.

Solution II:

Bismarck brown	1 gram
Distilled water (should be boiling when the Bismarck brown is added)	500 c.c.

Cover-glass preparations are stained for from 2 to 3 seconds in Solution I, rinsed in distilled water, placed in Solution II for from 3 to 5 seconds, rinsed again in water, and examined in the ordinary way. The bacteria of virulent diphtheria should appear as pale brown rods, some of which show at one or both ends bluish-black oval bodies of greater diameter than the rod. Such dark bodies will not be seen in the pseudo-diphtheria bacilli.

The bacilli must have been grown for from 12 to 18 hours on Löffler's blood-serum which is a mixture of glucose boullion 1 part and beef-blood serum 3 parts. The mixture is run into test-tubes and coagulated at 100° C.; the tube should be tilted to one side to give a slanting surface for culture purposes. The formula for glucose bouillon is as follows: dry glucose, 10 grams; Liebig's extract of beef, 3 grams; peptone, 10 grams; sodium chloride, 5 grams; water, 1,000 c.c.

e) Gabbet's solution for demonstrating tubercle bacilli.—

Methylen blue	1 to 2 grams
Distilled water	75 c.c.
Concentrated sulphuric acid	25 c.c.

The acid decolorizes, while the methylen blue serves as a contrast stain. The solution acts rapidly. A modification of the method to be commended is first to stain the preparation with carbol-fuchsin (see *c*) by warming the stain on the slide until it steams, rinsing in water, and then proceeding with the methylen-blue solution. Smegma and leprosy bacilli, and the treponema of syphilis are also stained by this method. Tubercle bacilli are also stained by Gram's method (see *f*). To examine sputum for tubercle bacilli, the sputum is carefully inspected for small yellowish-white cheesy masses varying in size from the diameter of a pin-head to that of a small pea. Very thin smear preparations (see A, p. 131) are made from such masses.

f) Gram's method; Hucker modification.—

Lugol's iodine solution:

Iodine crystals	1 gram
Iodide of potassium	2 grams
Distilled water	300 c.c.

Stain:

Crystal violet (85 per cent dye content)	4 grams
Alcohol, 95 per cent	20 c.c.

Mix with 0.8-gram ammonium oxalate dissolved in 80 c.c. of distilled water.

Counterstain:

Safranin (saturated solution in 95 per cent alcohol)	10 c.c.
Distilled water	100 c.c.

The preparations are first stained in the violet solution 1 minute, washed in water, and then immersed in iodine solution for 1 minute. Next, wash in water and blot dry; decolorize in 95 per cent alcohol for 30 seconds with gentle agitation; cover with counterstain for 10 seconds; rinse in water, dry and examine, or, if a permanent preparation is desired, rinse in absolute alcohol, transfer to xylol, and mount in balsam. If the preparations are from cultures, it should be borne in mind that the method works well only when applied to bacteria from actively growing cultures; old cultures seldom yield satisfactory results.

Pathogenic Bacteria Stained by Gram's Method	Pathogenic Bacteria Decolorized by Gram's Method
Bacillus aerogenes capsulatus	Bacillus of bubonic plague
Bacillus of anthrax	Bacillus of chancroid
Bacillus diphtheriae	Bacillus coli communis
Bacillus of malignant edema	Bacillus of dysentery
Bacillus of tetanus	Bacillus of glanders
Bacillus tuberculosis	Bacillus of influenza
Micrococcus tetragenus	Bacillus mucosus capsulatus
Pneumococcus	Bacillus proteus
Staphylococcus pyogenes aureus	Bacillus pyocyaneus
Staphylococcus pyogenes albus	Bacillus of typhoid
Streptococcus pyogenes	Diplococcus intra cellularis meningitidis
Streptococcus capsulatus	Gonococcus
	Spirillum of Asiatic cholera

4. Staining Spores (Abbott's method).—Prepare a cover-glass smear in the usual way. Apply the stain (e.g., methylen blue) and hold the cover-glass over a flame until the liquid steams. Repeat the heating several times, but do not boil continuously. Rinse the cover-glass in water and then decolorize the preparation in a 0.3 per cent solution of hydrochloric acid in 95 per cent alcohol until all color visible to the naked eye has disappeared. Wash in water. If a counterstain is desired, stain for from 8 to 10 seconds in anilin-fuchsin solution. Rinse in water and mount in the usual way. The spores are stained blue.

5. Staining Flagella (Bunge's modification of Löffler's method).—The locomotor organs of motile bacteria are long, hairlike prolongations (1 to many) termed flagella. Special methods of staining are necessary for their demonstration.

Make thin cover-glass smears of an 18-hour culture which contains motile forms. Dry and fix in the ordinary way.

The mordant.—
    Ferric chloride, aqueous solution (1:20)....... 25 c.c.
    Alum, saturated aqueous solution........... 75 c.c.
Shake well and add
    Fuchsin (basic), saturated aqueous solution.... 10 c.c.

Filter and allow to stand for some time before using. Treat the smear for 5 minutes with this preparation, gently warming by holding it high above a flame. The fluid must not boil. Rinse in water, then stain faintly with carbol-fuchsin. Repeat the process until a successful result is obtained. Mount in the usual way.

CHAPTER XVII

SOME EMBRYOLOGICAL METHODS; SECTIONS AND "IN TOTO" MOUNTS OF FROG AND CHICK; AMPHIBIA; FISH; MAMMALS; OTHER FORMS

THE FROG

Frog eggs and tadpoles are best fixed in Smith's modification of Tellyesnicky's fluid (5, p. 232). Eggs in early cleavage stages should, after such fixation, be preserved in 3 per cent formalin, but later stages and tadpoles are better preserved in 70 to 80 per cent alcohol. Before eggs can be sectioned, the thick albuminous coats which surround them must be removed (4, p. 139). In addition to the ordinary cleavage and yolk-plug stages, I find 3, 5, 7, and 9 mm. tadpoles, both as whole mounts and sectioned, the most useful stages for a course in embryology. Older stages are also necessary for the study of external features of later development.

Allen (see mem. 9, p. 140) finds that cleavage furrows show with greater distinctness if the eggs are bleached. The fixed and hardened specimens are placed in ordinary commercial hydrogen peroxide for a week or more until the pigmented area is of a light-brown color. If formalin-hardened material is used, the formalin should be washed out before the objects are placed in the peroxide, otherwise the tissue will be distorted by the rapid liberation of oxygen. Tadpoles may be bleached white by this method and mounted entire (see *B*, p. 138).

A. Section Method

1. Select several 7 mm. tadpoles which have been fixed in Smith's modification of Tellyesnicky's fiuid (5, p. 232)—if preserved in formalin, wash for at least 3 hours in water—and stain for from 12 to 24 hours in alum-cochineal.

2. Run the stained specimens up through the grades of alcohol into absolute alcohol, leaving it not longer than 1 hour in the latter.

3. Transfer to absolute alcohol and chloroform, equal parts, for an hour, then to pure chloroform. After an hour add melted paraffin to the chloroform from time to time until the latter contains all the paraffin it will hold in solution. Leave the objects in this mixture for at least 24 hours.

4. Transfer to melted paraffin (melting-point about 48° C.) and keep for 2 or 3 hours at a temperature just high enough to liquefy the paraffin. Imbed after reading step 5.

5. Prepare at least three sets of sections, one set in each of the three different planes of the body. The sections should be cut some 20 or 30 microns thick. Read carefully the directions under memorandum 1, p. 143, before imbedding, so that the sections will be properly oriented. Read also caution under step 10 on p. 143.

6. Carefully following directions under memorandum 1, p. 143, mount the sections with albumen fixative and albuminized water as usual (steps 17–21, p. 43).

7. Dissolve out the paraffin from the sections in the usual way after the latter are *thoroughly* dry. If not completely dry and tightly stuck to the slide some of the sections will float off.

8. Pass the slides back into absolute alcohol for a few minutes, then into fresh xylol until clear. Add balsam and the cover-slip.

B. Whole Mounts (after B. M. Allen, mem. 9, p. 140)

1. Select 5 to 7 mm. tadpoles which have been fixed in Tellyesnicky's fluid (5, p. 232) and bleached in hydrogen peroxide for one week, or until white.

2. Stain for 12 to 24 hours in alum-cochineal diluted with distilled water until the stain shows but a faint tinge of color.

3. Run the tadpoles up through the increasing grades of alcohol into cedar oil, creosote, or synthetic oil of wintergreen and leave until clear.

4. Using Canada balsam or damar which has been heated in an oven for some days until it will harden immediately upon cooling, prepare several slides by dropping such heated balsam upon them until drops 3 or 4 mm. deep and as wide as the cover-slip are formed. Let these harden. If small bubbles appear in such drops, place the slides in an incubator or a paraffin oven for some hours; the bubbles rise to the surface and may be skimmed off or burst with a flame.

5. To mount a tadpole in such a drop, heat the drop by holding it inverted over a flame for a moment, then with a scalpel previously dipped in xylol make a groove in it of suitable size and shape to fit the tadpole. The sides of the groove should press against the object in such a way as to hold it in the desired position. Some tadpoles should be mounted to present a dorsal, others a lateral, aspect.

6. When the object is in place, fill in the space about it with a drop

of soft balsam and gently but firmly press a heated cover-slip down upon the surface of the drop. Guard against including air bubbles.

7. Examine the preparations after a week or more and, by pushing the cover-slip toward one side or the other, adjust any of the objects which may have shifted from the desired position.

MEMORANDA ON AMPHIBIAN MATERIAL

1. **To Study Amphibian Eggs Entire,** use a hand lens or dissecting microscope. Place the eggs on a bit of absorbent cotton under 70 per cent alcohol in salt cellars. The eggs are fragile; consequently, to manipulate them, use a soft-hair pencil or a current from a pipette. Use the same egg for surface view and for sectioning when possible.

2. **Special Egg Pipettes** for handling delicate objects should be prepared by breaking off the tip of an ordinary pipette to enlarge the orifice. After rounding up the broken edges in a flame, cover the broken end with a small piece of soft-rubber tubing.

3. **Amphibian Eggs in General** may be fixed (in masses of 15 or 20) in various modifications of Tellyesnicky's fluid (5, p. 232) or in Worcester's aceto-formol-sublimate mixture (25, *b*, p. 240). Chromic acid (13, p. 235) brings out surface views well, but the material becomes very brittle and does not take stains readily. If surface views alone are desired, formalin-preserved material will answer.

4. **To Remove the Gelatinous Coats of Eggs,** roll them over and over on a bit of blotting paper. Either fresh or preserved eggs may be handled in this way. To prevent very soft eggs from drawing down and adhering tightly to the blotting paper, roll them off on to a paper of harder texture just before the last trace of gelatinous film has been removed.

Whitman (*American Naturalist*, XXII, 857) recommends putting the fixed eggs into a 10 per cent solution of sodium hypochlorite diluted with 5 or 6 volumes of water and leaving them until they can be shaken free. This requires only a few minutes. Rinse the eggs in 35 per cent alcohol. It is advisable to remove the albuminous coats before hardening in alcohol.

Child (*Zeitschrift für wissenschaftliche Mikroskopie*, XVII [1900], 205) states that the albumen which surrounds many ova becomes transparent and dissolves if after fixation (in any way except with chromic acid) the ova are passed up through the grades of alcohol to 80 per cent, hardened, and then passed down again through the alcohols into water which has been slightly modified with any acid except chromic.

5. **Amphibian Eggs Are So Friable** that they are ordinarily sectioned in celloidin. If they are cleared from 95 per cent alcohol (avoiding absolute) into cedar oil or oil of wintergreen they are less brittle. However, good sections are obtainable with the paraffin-rubber method (memorandum 10, p. 45) or by the more tedious paraffin-celloidin method (memorandum 8, p. 73). Old-

er embryos are readily sectioned in paraffin according to the method already given (p. 137).

6. **Whole Eggs May Readily Be Cut into Halves** with a safety-razor blade. They are often more serviceable for study along with observations on the external changes of cleavage, blastula formation, gastrulation, etc., than the most elaborate serial sections.

7. **The Germinal Layers** are much more distinctly seen in sections of young embryos of Ambystoma than in those of the frog.

8. **Very Thick Sections** are sometimes useful. A 7 to 9 mm. tadpole, for example, cut into three sagittal sections is excellent for studying the general topography of organs. Also older tadpoles with the skin removed from one side before mounting are serviceable.

9. **Eggs Mounted Entire or in Halves,** according to B. M. Allen's gelatin method (*Kansas University Bulletin*, IX, No. 8 [December, 1914]), are very successfully studied as opaque objects. The method, also useful for preserving free-hand sections of various kinds, fine dissections, etc., is as follows:

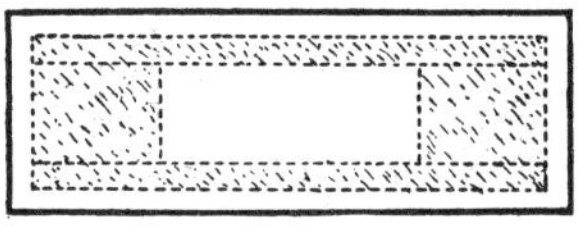

a) Dissolve thymol in distilled water with the aid of heat until a saturated solution is obtained. Filter until clear.

b) Soak gelatin in the thymol water until it has absorbed all it can hold, then drain off the excess of water. The gelatin, now ready for melting, may be kept in corked test-tubes.

FIG. 41.—Method of Mounting Embryos (after B. M. Allen).

c) Prepare cells for mounting large objects by placing strips of glass upon a slide as indicated by the shading in the diagram (Fig. 41).

d) Liquefy some of the gelatin by placing a test-tube full into a warm water-bath, then pour it into the newly made cell.

e) When the gelatin has set, melt small areas in it with a hot needle and insert the objects (e.g., a series of frog eggs in different stages of segmentation) to be mounted. Each of these must be held in proper position with the needle until the gelatin solidifies.

f) Flood the cell with gelatin heated just enough to be a liquid and place a slightly warmed slide on top as a cover. Avoid air bubbles, and see that there is a complete film of gelatin between the cover-slide and the glass strips. The cover should be held firmly in place till the gelatin solidifies.

g) With a toothpick or similar object thoroughly clear out every trace of gelatin from the grooves formed by the projecting edges of the slides and the strips between them. Dry well.

h) Run some cement such as gold size into the groove and set the preparation aside to harden. Add more cement from time to time until the groove is completely filled up. Keep the preparation out of direct sunlight or away from radiators, as it must not be subjected to heat.

If the objects have previously been hardened in formalin, so much the better, as the formalin will gradually diffuse out into the gelatin and harden it.

Hollow-ground slides designed for use with a hanging drop may be used if preferred. It is sometimes desirable to solidify the gelatin more rapidly after the object is mounted, by placing it on ice.

10. For Artificial Fecundation of Amphibia Eggs see 16, *b*, p. 153.

THE CHICK

A setting hen or an artificial incubator is necessary. In many ways the latter is more convenient as it may be kept in the laboratory and is ready at all seasons of the year. There are many kinds of good incubators on the market at present which may be had for a small sum.

Whatever method of incubation is employed, the eggs must be fresh and must not have been subjected to rough handling. The date and hour at which incubation is to begin should be written on the shell of each egg in ink. If late stages of development are desired, the egg must be turned every few days. All products of combustion from the lamp or burner should be kept from the eggs and the supply of fresh air and moisture carefully maintained. The temperature should be maintained at 39° C. (102° F.). Should it rise above 40° C., embryos will be destroyed.

Prepare at least 5 embryos as directed in the practical exercise, 2 for *in toto* preparations and 3 for sections.

1. Place an egg which has been incubated for between 46 and 54 hours, while it is yet warm, in a vessel which contains sufficient normal saline warmed to 39° to cover the egg. In the fowl the embryo always makes its appearance as a germinal disk or *cicatricula*, as it is termed, situated on one side of the yolk, which is the real egg of the hen, the white being simply a nutritive mass added in the oviduct. This disk or *blastoderm* in the early stages of incubation always turns uppermost no matter in what position the egg may be placed. Moreover, it has been found that the embryo in nearly every instance lies in such a position that when the blunt end of the egg is toward the left, the head of the chick is directed away from the operator. This fact affords a very reliable means of orienting the embryo, especially in the very young stages when the anterior and posterior ends are not easily recognized by the observer.

2. Break through the shell at the broad end over the air chamber by tapping it sharply, and let out the air, or the broad end will tilt up.

3. Begin at the hole made in the end and with blunt forceps remove

the shell and shell membrane bit by bit from the upper surface of the egg until the embryo comes plainly into view. Remove with a pipette the thin layer of albumen which lies above the blastoderm.

4. With as little agitation of the liquid in the vessel as possible by means of fine scissors cut rapidly around the blastoderm well outside of the vascular area. Leave sufficient extra-embryonic (4 to 5 mm.)membrane around it.

5. Carefully float the blastoderm into a thin watch-glass, keeping it as flat as possible. Shake it gently to remove the piece of vitelline membrane covering it, or any yolk which may adhere. The aid of a needle may be necessary to remove the vitelline covering. Run a pipette gently around the margin and suck away the fluid so that the edges will flatten out.

6. With a pipette remove all excess of fluid from the watch-glass but do not let the embryo become dry. In order to keep the edges from curling up and obscuring the embryo, in the center of a piece of filter paper the size of the complete preparation cut a circular hole the size of the embryonic area and place it over the embryo.

7. Flood with the fixing fluid (picro-sulphuric fixer; reagent 32, p. 242) until the embryo is completely immersed. The fluid should be allowed to act for from 2 to 3 hours.

The paper may be left in place through subsequent treatment to 95 per cent alcohol for embryos which are to be sectioned, or to xylol for those which are to be mounted whole. The whole preparation can be lifted with forceps.

NOTE.—Some prefer to fix the embryo before removing it from the egg. After some of the albumen is drawn off, the fixing agent is squirted on to the blastoderm. As soon as it is opaque the latter is then removed to a vessel containing the fixing agent. Andrews (*Zeitschrift für wissenschaftliche Mikroskopie*, XXI [1904], 177) injects picro-sulphuric acid (1) between the vitelline membrane and the blastoderm and (2) between the blastoderm and the yolk by means of a pipette which has a fine upcurved point. The blastoderm may then be readily freed from the yolk. This operation should be performed before the egg has been subjected to the action of any reagents.

8. Wash in repeated changes of 70 per cent alcohol. Pass down through 50 and 35 per cent alcohol to water. Stain in alum-cochineal for 24 hours (Conklin's hematoxylin may be used if preferred).

9. Wash the object in water and transfer it through 35 and 50 per cent alcohol, leaving it 30 minutes in each. Decolorize the embryo

Some Embryological Methods 143

slightly in weak acid alcohol, then wash in 70 per cent alcohol, and leave it there until ready to proceed.

10. Transfer the object through 95 per cent (1 hour), absolute alcohol (2 hours) to cedarwood oil, where it should remain about 2 hours or until it ceases to appear opaque. (McClung and Allen recommend synthetic oil of cassia, or cinnamic aldehyde, above everything else for clearing embryos because it causes the least shrinkage).

Mount two embryos entire, one with the ventral, the other with the dorsal, side uppermost. Put bits of broken cover-glass or threads of glass under the edges of the cover to avoid crushing them.

The three remaining embryos are to be so sectioned (steps 11 ff.) that the student will have a complete series of sections in each of the three different planes of the body with reference to the axis of the spinal cord: viz., transverse, sagittal, and frontal. Read carefully memorandum 1 on orienting serial sections.

CAUTION.—*Before sectioning any embryo always make an outline drawing of the entire embryo; then rule lines across the drawing parallel to the plane of section. Unless this is done great difficulty will be experienced frequently in understanding the sections.*

11. Infiltrate the embryo with paraffin in the usual manner by leaving it in melted paraffin for 2 or 3 hours. A paraffin melting at about 48° C. should be employed and sections should be cut 20 to 30 microns thick.

12. Imbed and cut in the usual way (chap. v). Mount the entire series.

MEMORANDA

1. **Directions for Orienting Serial Sections.**—*a)* In mounting *transverse sections* (sections across the main axis of the object), the sections beginning at the anterior end of the object are laid on the slide in the same sequence as the reading on the page of a book. In order to have right and left sides and dorsal and ventral surfaces in proper relation to the observer, mount the object in such a way that, in cutting, the knife will enter it on the left side and at the anterior end. Leave room at one end of the slide (see p. 53) for a label and also a small margin at the opposite side.

b) To get proper orientation of *frontal sections* (sections lengthwise of the object in a plane including right and left sides), arrange the object so that the knife will enter it on the right side and slice off the dorsal surface first. Mount sections, with their posterior ends toward the upper edge of the slide, placing the first section of the series to the left end of the upper row. This throws left and right, dorsal and ventral, into their proper position as viewed through the compound microscope, and the observer looks from the dorsal toward the ventral aspect of the object.

e) To mount *sagittal sections* (sections lengthwise of the object in a plane including ventral and dorsal sides), arrange the object in such a position that the knife enters the ventral surface and slices off the right side first. Mount with the posterior end toward the upper edge of the slide, placing the first section of the series at the left end of the upper row. Through the compound microscope the observer views the object from the right toward the left. The head will appear to be toward the upper end of the slide, the dorsal surface toward the left.

It is frequently advantageous to have the imbedding-mass trimmed unsymmetrically by leaving the edge which first comes in contact with the knife longer than the opposite edge. One may thus readily discover if a section or part of a series has been accidentally turned over.

2. Orientation of Objects in the Imbedding-Mass so that sections can be cut accurately in definite planes is frequently difficult to accomplish. The following methods are useful in many instances:

I. *For paraffin sections.*—With a soft pencil rule the strip of paper which is to be used for making the imbedding-box into small squares or rectangles. After imbedding, upon removal of the paper a copy of the pencil marks will be found upon the block of paraffin. If the object has been arranged in the melted paraffin with reference to these lines, it is easy so to arrange the block in the microtome as to cut the object along any desired plane. It is frequently an aid to orientation by this method to have one of the central ruled lines broader than the others, or double.

Small objects which cannot conveniently be oriented in melted paraffin may be properly oriented and fixed to a small strip of paper ruled as above, before they are placed in the paraffin bath, by a mixture of clove oil and collodion of about the consistency of thick molasses, as in Patton's method (*Zeitschrift für wissenschaftliche Mikroskopie*, XI [1894], 13). One or a number of small objects which have previously been cleared in oil of bergamot or cloves are mounted in small separate droplets of the reagent and oriented under a dissecting lens with reference to the ruled lines. The paper is then placed in turpentine which washes out the clove oil and fixes the object in place. The paper with objects attached is then passed through melted paraffin and imbedded in the ordinary way. Upon removal of the paper from the hardened block a sufficient number of pencil marks remain to be used as a guide in sectioning. Instead of pencil marks Patton employed ribbed paper.

II. *For celloidin sections.*—

Eycleshymer's Methods.—*a*) For imbedding, metal boxes made of two L's (Fig. 29) are used. The L's are held together by overlapping strips. The ends and sides of the box are perforated at regular intervals by small holes which have been drilled opposite one another in such a way that threads drawn through them are parallel. Threads of silk are run through the holes from side to side, drawn taut, and cemented to the outside of the box with a drop of celloidin. Each piece of thread should have an end two or three inches long hanging outside the box. A piece of heavy blotting paper is used as a bottom for the box. The object is oriented on the parallel threads and the imbedding-mass poured in and hardened. The loose ends of the threads are then soaked in a solution of thin celloidin which contains lamp-black, the celloidin drops holding the threads taut are dissolved by a drop of ether-alcohol, and the blackened ends are drawn through the block of celloidin. The lamp-black leaves distinct black lines through the mass which will serve for properly orienting the celloidin block on the microtome.

This method is valuable also in reconstructions from sections (see chap. xix). In such work it is very desirable to establish "reconstruction points" to guide in fitting the wax plates together properly. The black rings of lamp-black left in the sections answer admirably for this purpose.

b) For small objects in which reconstruction points are not required Eycleshymer uses fine insect pins from which the heads have been clipped and the headless ends loosely inserted in handles. The objects are mounted on the points of the pins and oriented in the desired position. Each pin is then removed from its handle, and the free end is inserted from below into a small perforation which has been made by passing a somewhat larger pin lengthwise through a cork. A number of pins may be mounted on the same cork. To prevent the objects from becoming dry, the cork must frequently be inserted into the mouth of a vial full of alcohol in such a way that the objects are immersed. If desired, the objects may be sketched *in situ* under alcohol by weighting the cork with lead and placing it in a beaker of alcohol. To pass the objects through the various grades of alcohol, etc., simply transfer the cork bearing them to successive vials of proper size containing the different fluids. For imbedding in celloidin use the method given on p. 68, steps 2 ff. When the celloidin mass has hardened, the paper is removed and the pins are drawn out through the cork, thus leaving the objects in place ready for sectioning.

3. **In Measuring the Length of Embryos** some embryologists (e.g., Minot) measure the greatest length of the embryo along a straight line (limbs not included) when the embryo is in its normal attitude; consequently in some early stages where the embryo is greatly flexed the neck-bend would be the point to which to measure instead of the tip of the head, because it is the most anterior region; in stages where the embryo is straight, the head would be included. Other embryologists (e.g., His and German authors in general) make use exclusively of the so-called "neck-length"; that is, the distance in a straight line between the neck-bend and the caudal-bend. Still others, in the case of human embryos, use the so-called "sitting height" and "standing height."

4. **For the Embryology of Teleosts** the following are the most useful mounted stages:

I. *Whole mounts.*—The 2-, 4-, 8-, 16-, 32-, and 64-cell stages (only the blastodisk segments); early periblast; late periblast; early germ-ring; embryonic shield; various stages of early embryos, such as embryos of 45, 50, and 60 hours.

II. *Sections* (paraffin).—Of 4, 16, and 32 cells (vertical sections parallel to the first plane of cleavage); late cleavage (vertical sections); early, mid, and late periblast (vertical sections); transverse and sagittal sections of early germ-ring, embryonic shield, early embryo, late germ-ring, and closing of blastopore, respectively.

All stages may be fixed in picro-acetic (reagent 29, p. 241) or Bouin's fluid for 30 to 40 minutes. The eggs are finally preserved in 83 per cent alcohol. Child finds that fixation for about a minute in 10 per cent acetic acid saturated with corrosive sublimate, followed by 10 per cent formalin, gives good results without the yolk becoming hard. The ova of the Salmonidae must be removed (after fixing and hardening) from their envelopes before the embryo can be studied.

Before the preserved material can be mounted *in toto* or sectioned, the essential part (the blastoderm) must ordinarily be dissected off under a dissecting lens by means of sharp needles. If the blastoderms are to be mounted entire they may be passed down through the alcohol (see Walton's device, memorandum 6, p. 31), stained in Conklin's hematoxylin (reagent 58, p. 252) then dehydrated and mounted in the usual way. To avoid crushing the objects, the cover-glass should be supported by means of bits of broken cover or glass threads. Material which is to be sectioned may be stained *in toto* or the sections may be stained on the slide. In the latter event, to facilitate orientation, it is necessary to tinge the blastoderms slightly with Bordeaux red or some other cytoplasmic stain unless the fixing reagent has already done so. For the same reason it is best to imbed the material in a watch-glass, arranging it near the bottom of the paraffin-mass so that one can see with a microscope how to shape the paraffin block in order to cut sections in the proper plane. The immersion in the melted paraffin should not be longer than 5 or 10 minutes. The paraffin is best hardened under 95 per cent alcohol. The sections may be stained by any of the hematoxylin methods; iron-hematoxylin (p. 54) yields excellent results.

5. **To Study Living Eggs of Teleosts,** a thin, flexible piece of sheet celluloid or mica should be used instead of a cover-glass. The egg must be rotated from time to time, and this is easily accomplished with such a flexible cover.

6. **For Artificial Fecundation of Teleost Eggs,** see p. 153.

7. **Tilting the Microscope into a Horizontal Position** and examining the egg in its normal medium by direct light is an excellent method of studying blastodisk formation in such forms as Ctenolabrus, for instance. Inasmuch as the blastodisk forms on the lower side of the egg, it appears to be on top when viewed through the compound microscope.

8. To Preserve Teleost Eggs in Convenient Form for Demonstrating discoidal cleavage, embryonic shield, germ-ring, etc., Smith (*Transactions of the American Microscopical Society*, XXXIII, No. 1 [January, 1914]) seals pieces of ⅛-inch glass tubing at one end by holding in a flame. A series of eggs fixed in corrosive-acetic mixture and preserved in formalin is placed in each tube and the opening plugged with cotton. The tube may be held in the hand and examined with a lens or dropped into a watch-glass filled with water and examined under a lens or binocular microscope.

9. For the Average Course in Embryology of the Chick the following mounted stages are the most useful:

I. *Mounted in toto.—*

Approximately,
48 hours viewed from above and below
36 " " " " " "
30 " " " "
24 " " " " " "
18 " " " "
12 " " " "
64–72 " " " "
96 hours (studied in alcohol under the dissecting microscope)

II. *Sections.—*

48 hours, transverse, sagittal and frontal
36 " " "
30 " " "
24 " " "
18 " "
10 " "
72 " " " " "
96 " " "

The number of embryos needed for the above-mentioned preparations is as follows:

5 embryos of 48 hours (27–29 somites)
4 " " 36 " (15–18 ")
3 " " 30 " (10–14 ")
4 " " 24 " (4–6 ")
2 " " 18 "
1 " " 12 "
1 " " 10 "
4 " " 64–72 hours (cervical flexure formed)
3 " " 96 hours

10. **To Mark Anterior and Posterior Ends of Young Chick Embryos** in blastoderms which still have a homogeneous aspect, Duval's osmic-acid method is very useful. With a strip of paper 5 mm. wide by 50 mm. long a triangular bottomless box with narrow base is constructed. This is placed on the yolk inclosing the blastoderm in such a position that the base of the triangle corresponds to what will be the anterior region of the embryo (for orientation of embryo in the egg, see step 1 of the practical exercise). Press the box down against the yolk and fill it with a 0.3 per cent aqueous solution of osmic acid. In a short time the preparation begins to darken and the osmic acid should be removed. The blastoderm may then be removed in the ordinary manner and fixed as desired (Duval used chromic acid for fixing). However, it is very difficult to separate the blastoderm from the egg during the first 24 hours of incubation, and it is advisable, therefore, to fix and harden both together and to remove the blastoderm later (see note under 7, p. 142). The blackened area affords a convenient means of orienting the preparation for sectioning.

11. **For the Stages of Maturation, Fertilization, and Segmentation in Mammals** white mice will prove most useful because these processes are better known in them than in other mammals; furthermore, an abundance of material may be procured. The ovum, however, is extremely small, measuring only about 59 microns in diameter. It is surrounded by a very thin zona pellucida (1. 2 microns). Long and Mark find a modified Zenker's fluid the most satisfactory for fixation. They make up two solutions: (1) a 4 per cent aqueous solution of potassium bichromate; (2) a 4 per cent aqueous solution of corrosive sublimate and 20 per cent acetic acid. The two solutions are mixed in equal proportions when needed for fixing.

Female mice are in heat soon after parturition. They tend to ovulate every 21 days during the spring months. The maturation process requires from 4 to 15 hours and usually occurs between 13 and 29 hours after parturition. The second polar spindle usually forms immediately before ovulation, but the second polar body may not be extruded unless the egg is fertilized. Ovulation occurs between 14 and 29 hours after parturition. The eggs are easily visible at first in a fold of the oviduct near the ovary. Insemination is most successful when it occurs between 18 and 30 hours after parturition. The spermatozoa reach the eggs in the upper end of the Fallopian tube in from 4 to 7 hours.

For details and bibliography see (1) Kirkham, *Biological Bulletin*, XII, No. 4 (1907); also *Transactions Connecticut Academy of Arts and Sciences*, XIII; (2) Long and Mark, "The Maturation of the Egg of the Mouse," *Publications of the Carnegie Institute of Washington, D. C.*, 1911; also, "The Living Eggs of Rats and Mice with a Description of Apparatus for Obtaining and Observing Them," *University of California Publications in Zoölogy*, IX, No. 3 (Feb. 23, 1912); (3) Daniels, "Mice, Their Breeding and Rearing for Scientific Purposes," *American Naturalist*, October, 1912; (4) see also Danforth, *Anatom-*

ical Record, X, No. 4 (February, 1916), for some very practical suggestions regarding use in classes; and (5) Allen, "Mouse Embryos," in McClung's *Microscopical Technique* (1929), pp. 211-12.

The phenomena of maturation and fertilization in the albino rat are described in a paper by Sobotta and Burckhardt in *Anatomische Hefte*, XLII, 1911 (summarized in Huber's paper cited on p. 151). See also Kirkham and Burr, "The Breeding Habits, Maturation of the Eggs and Ovulation of the Albino Rat," *American Journal of Anatomy*, XV, 1913.

12. **For Early Stages of the Mammalian Embryo** rabbits are commonly employed because they breed readily, especially in the spring of the year, and the observer can note the exact time when the female is covered if she has been kept separate from the buck. The period of gestation is 30 days and impregnation may be secured again immediately after littering. The two uteri of the rabbit diverge as two anterior horns from the single median vagina and each terminates in front in a narrow, coiled tube, the oviduct or Fallopian tube. To obtain the early stages the abdomen is slit open from pubis to sternum, the intestinal tract is cut away or pushed to one side, and each uterus and oviduct carefully removed and stretched out along a glass plate. Segmentation stages are found in the oviduct up to about 70 hours from the time of copulation. After that period of time they must be looked for in the uterus. Eggs from large races of rabbits segment more rapidly than those from small races, although this differential rate is scarcely detectable before the 32-cell stage. Ovulation occurs about 10 hours after coition and fertilization must take place within the next 2 hours. At the time of fertilization the first polar body has been formed; during the next 10-hour interval the second polar body is extruded, and fusion of the male and female pronuclei occurs. While in the oviduct, with the aid of a lens the ova may sometimes be seen through its walls. A segmenting ovum once located, a transverse cut is made to one side of it through the wall of the oviduct, and the ovum, which is very small, is gently squeezed out by compressing the oviduct behind it. With a spear-headed needle or the point of a scalpel the ovum is conveyed to the fixing fluid. In case segmenting ova are not visible from the exterior of the oviduct, the latter must be slit open carefully with a pair of fine-pointed scissors, and the eggs sought for by means of a lens. In case no red corpora lutea are visible on the surface of the ovary, indicating a recent discharge of ova from the Graafian follicles, further search is useless.

At ovulation the rabbit ovum is a spherical cell about 0.1 mm. in diameter. It is filled with a dense, yellowish yolk and is covered by a tough, elastic zona pellucida secreted by the follicular wall. When the follicle ruptures, frequently some of the follicular cells remain attached to the surface of the discharged ovum and may persist for several days before they disintegrate. In the oviduct an albuminous coat secreted by epithelial cells of the wall is gradually laid down in concentric rings around the zona pellucida.

The first cleavage occurs about 22 hours after copulation; the second, about 2½ hours later. Between 32 and 33 hours after coition the 8-celled stage arises. This stage may persist 8 hours or more. By 42 hours after copulation both 8- and 16-blastomere stages are usually found. From this point cleavage becomes increasingly irregular; by 6 hours later some 32-cell stages will have appeared. About the time the segmenting mass passes into the uterus the first trace of the cavity of the blastodermic vesicle is seen as a small cleft between the outer layer of cells which are to form the trophoblast and the inner layer which will form the embryo. By 90 hours after coition the segmentation cavity is established, and a rapid expansion of the blastocyst follows. At about the beginning of the eighth day after copulation the inner cell-mass begins active proliferation, and the area involved becomes organized into somites during the latter part of this day. The blastocyst also begins to attach itself to the uterine wall.

The earlier stages (up to 70 hours) may be fixed for from 5 to 8 minutes in a 0.3 per cent aqueous solution of osmic acid, stained in picro-carmine, and transferred to a mixture of glycerin and water, equal parts. They should remain in this fluid for a week under a bell-jar so that the water gradually evaporates. The object may then be mounted in formic-glycerin (formic acid 1 part, glycerin 99 parts). To avoid pressure of the cover-glass, the object should be mounted in a cell or between two slips of paper or pieces of cover-glass. If the preparation is to be permanent the cover-glass should be sealed (see p. 108).

To render the cell outlines distinct, stages of from 70 to 80 hours are best treated, after rinsing in distilled water, with a 1 per cent aqueous solution of silver nitrate for 3 minutes and then exposed to light in a dish of distilled water until they become brown. They are then treated with water and glycerin and mounted in formic-glycerin as in the case of younger stages.

For sections the embryos should be placed in Bouin's or Zenker's fluid for one or two hours, then washed in the customary way for these methods, stained in alum-cochineal, and sectioned in paraffin.

In opening the uterus, the incision should always be made along the middle of the free side, opposite the insertion of the peritoneal fold, because this line of insertion marks the region of attachment of the embryo within the oviduct. By the seventh or eighth day the developing ova have taken up positions at intervals along the inner walls of the uterus and have become so firmly attached to the mucous membrane that they can no longer be detached unmutilated. For further particulars regarding the embryology of the rabbit, the reader is referred to E. Van Beneden and Charles Julin's "Recherches sur la formation des annexes foetales chez les mammifères," *Archives de Biologie*, V (1884), 378. See also Assheton, "A Reinvestigation into the Early States of the Development of the Rabbit," *Quarterly Journal of Microscopical Science*,

XXXVII (1895); and Hammond and Marshall, *Reproduction in the Rabbit* (Oliver and Boyd, 1925).

With the aid of Huber's paper (*Journal of Morphology*, XXVI, No. 2 [1915]; also *Memoirs of the Wistar Institute of Anatomy and Biology*, No. 5), which covers the development of the albino rat from the pronuclear stage to the end of the ninth day, it is now feasible to use the rat for early embryonic stages. Huber found Carnoy's fixing fluid (reagent 2, *b*, p. 230) the most satisfactory. He fixed tissues for several hours, washed them in several changes of absolute alcohol, and stored them in the latter. He found that sectioning ovary, oviduct, and uterus *en masse* was more satisfactory than isolating and sectioning the separate ova, although he used both methods. For staining he employed mainly hemalum, followed by Congo red. His methods are described at some length in the paper.

13. **For Older States of the Mammalian Embryo** pig embryos are commonly employed. They may often be procured in large numbers and with little trouble at the larger pork-packing establishments. The most valuable stage for study is an embryo of from 10 to 13 mm. in length. In most laboratories it is customary to make a detailed study of an embryo of about this stage and then a more general survey of both smaller and larger sizes.

Early stages are much more difficult to obtain than advanced stages. Embryos of 6 mm. length and over may usually be readily located by the enlargements which they cause in the uterine walls. The uterus should be handled carefully and opened as soon as possible. The embryo is best removed by means of fine forceps and a horn spoon. It is very delicate and should not be handled roughly. The chances are that in removing the embryo the membranes will be ruptured and the amniotic and allantoic fluids will escape. Larger embryos should have the body cavity punctured to admit the fixing fluid.

Submerge the embryo without removing the membranes in a bountiful supply of Kleinenberg's picro-sulphuric acid (reagent 32, p. 242), moving it about gently to rinse off any coagulum that may form on the surface. Lavdowsky's fluid (23, p. 239) is also a good fixing agent for pig embryos and is to be preferred for the older ones.

Leave embryos of 6 to 9 mm. $2\frac{1}{2}$ hours; 12 to 15 mm., 4 hours; 20 to 25 mm., 6 to 8 hours.

For washing and subsequent treatment see reagent 32, p. 242. Embryos may be stained *in toto* in alum-cochineal or borax-carmine.

For studying the uterus, placentation (diffuse in the pig), and embryonic membranes in place, formalin-hardened material may be used after first thoroughly washing it in water.

For gross dissection of embryos the specimen should be studied in alcohol under the dissecting microscope.

Because of the asymmetry of young embryos it is impossible to secure strictly *transverse, sagittal,* and *frontal* sections. Minot recommends, therefore, that for practical purposes the plane of section be taken with regard to the head alone irrespective of how it may cut the other parts of the body, and suggests the floor of the fourth ventricle of the brain as the guide for orientation. In his *Laboratory Text-Book of Embryology* he especially recommends that each student prepare sections of the following stages of pig embryos: 9 mm., transverse and sagittal, frontal of the head; 6 mm., transverse, frontal of the head; 17 mm., transverse and sagittal, frontal of the head; 20 mm., transverse and sagittal, frontal of the head; 24 mm., frontal of the head.

14. Human Embryos of all ages are very valuable material for scientific purposes. Physicians and surgeons are urged to preserve such material properly and turn it over to some competent embryologist. Very young human embryos are exceedingly desirable. Fill the containing vessel completely with fluid in order to avoid shaking.

An excellent fixing reagent, the ingredients of which a physician can usually readily procure, is Lavdowsky's mixture (reagent 23, p. 239). The embryo should remain in this fluid from 12 to 48 hours, according to size, and then be preserved in 80 per cent alcohol (or commercial alcohol to which has been added about one-fifth its volume of distilled water). Use a wide-mouthed bottle with tightly fitting stopper.

Zenker's fluid (reagent 6, p. 232) is better for larger-sized embryos. Material should be left in it from 18 hours to several days. For washing and preserving follow the directions given under the description of the fluid. For fetuses use a fruit-jar of such a size that the embryo can be kept in about 10 times its volume of fluid.

In case the above-mentioned fluids are not available, the material may be placed in 10 per cent formalin (1 part of commercial formalin to 9 parts of distilled water) and left indefinitely. As a last resort, if no other fixing reagent is available, the embryo may be placed in the strongest alcohol which can be secured and later transferred to 80 per cent alcohol for preservation.

The specimen should not be handled nor allowed to lie in water. When the proper reagents are not at hand, carefully wrap the object in cloth and keep it on ice if possible until they can be secured. Very small embryos may be fixed and preserved with membranes intact; older ones (6 weeks to 3 months) should have the membranes ruptured. To secure the best fixation of fetuses (2 months and beyond), the specimen should be divided, or at least the body cavity should be opened.

15. For Micro-Dissection of Small Embryos, after fixation Heuser stains for 24 hours in alum-cochineal diluted with 5 times its volume of water. He then fixes the embryo to a small square of thin ground glass with celloidin cement (about 0.75 per cent solution of celloidin) and dissects in alcohol under the binocular microscope. The dissected specimen may be preserved at any stage by placing it, still attached to its glass support, in alcohol in a shallow vial of suitable size. Necessary data may be written with a pencil on the ground surface.

Streeter (*American Journal of Anatomy*, IV, No. 1, p. 87), after dehydrating in absolute alcohol attaches the embryo with a drop of thick celloidin to a smoked isinglass strip coated with thin celloidin and places it in 80 per cent alcohol. The black serves as a good background for the embryo and may be written upon. During dissection the isinglass strip is clamped to a glass slide which has been cemented with balsam to one facet of a cut-glass polyhedral paper-weight. The object can thus be placed in any desired plane and dissected in alcohol under a binocular microscope.

16. **Artificial Fecundation** when it can be practiced is the most convenient means of securing early stages of development. This is possible with many worms, coelenterates, echinoderms, cyclostomes, teleosts, and anuran amphibia.

a) In echinoderms (e.g., sea urchin) the female is cut open and a number of the living eggs transferred to a watch-glass which contains fresh sea water. The testes of a male are teased out in sea water and a drop of the mixture is conveyed by means of a pipette into the dish containing eggs. Immediately upon fertilization a membrane forms around each fertilized egg. In about 40 to 50 minutes after fertilization the signs of the first cleavage should appear. The blastula forms in about 6 hours, and the gastrula in about 12 hours. For the study of fertilization, etc., the following stages should be fixed in Bouin's fluid (p. 30) for 30 minutes and stained in iron-hematoxylin (p. 54); 5 minutes after fertilization, nucleus giving off polar bodies; 30 minutes after fertilization, approaching pronuclei; 50 to 55 minutes after fertilization, division of nucleus (mitotic figure) in the first cleavage.

b) In amphibia (e.g., frog) both male and female are cut open, the vasa deferentia or testes are teased out in a watch-glass full of water, and the ova are then removed from the lower ends of the oviducts and placed in this water. After fertilization the eggs should be placed in glass dishes in not over 4 inches of water. Many eggs should not be placed in one dish. See also memoranda, pp. 143–45.

c) In teleosts the eggs are obtained by stripping the female when she is in spawning condition. At such times the eggs are loose in the body cavity and may be pressed out by gently manipulating the belly of the fish. The head of the fish should be held in one hand, the tail in the other, and the thumb or the thumb and forefinger used to press out the ova into a clean, dry finger-bowl. The milt of the male is obtained in the same manner in a dish containing a little fresh or sea water, depending upon the habitat of the fish. When the water becomes milky with sperm pour the mixture over the eggs. Eggs and sperm are then gently stirred about by means of a feather to insure thorough mixing. However, in some teleosts (e.g., stickleback, Fundulus) it is necessary to kill the male and tease out the testes. In the cunner (Ctenolabrus) 10 minutes after fertilization the formation of blastodisk and polar bodies may be observed; 30 to 33 minutes after fertilization the two pronuclei may be found in close approximation.

If other than the very early stages are required, the fertilized eggs must be transferred to a hatching-box or jar, depending upon the kind of egg. This is best done by means of a horn spoon and a feather. Dead eggs, recognizable by their opacity, should be removed at least once a day. The conditions under which the eggs of different species thrive are so varied that the reader must be referred for details to such special publications as those of the United States Bureau of Fisheries or the fish commissions of the various states.

17. For the Study of Early Cleavage in Living Material the eggs of some of the water snails afford an abundance of excellent material. By watching aquaria which contain snails the fresh material can easily be obtained during the spring and summer. Twigs and bits of board to which the egg-masses may be attached should be placed in the aquaria.

If one is at the seashore, the sea urchin, starfish, squid, various marine annelids, mollusks, and coelenterates afford an abundance of material. Of the marine fishes Fundulus and Ctenolabrus are excellent. Of fresh-water fishes the whitefish (spawning in November or December) and the pickerel (in April) show cleavage well, although in the whitefish it is very slow.

18. For Quick Preparation of Cleavage Stages for study *in toto*, in forms where there is considerable yolk, Spaeth finds useful often a mixture of equal parts of glacial acetic acid, glycerin, and water, to which enough Delafield's hematoxylin is added to make it a light tan color.

19. Chinese Black added to the water on the slide in which eggs with very transparent jelly (e.g., Nereis) are being examined outlines the egg distinctly and shows the path of the spermatozoön through the jelly.

20. For the Study of the Formation of Polar Bodies, Fertilization, and Early Cleavage in Sections nothing surpasses the eggs of Ascaris. The Ascaris (*A. megalocephala*) from the horse is preferable, although *A. lumbricoides* from the pig will answer.

The ovisacs, two in number, are very long convoluted tubes. Different regions contain eggs in different stages of development. The thicker tubes toward the anterior end of the animal contain cleavage stages; back of these are cells showing extrusion of the polar bodies and fertilization stages. The material must be fresh; either bring the live Ascaris to the laboratory or take the fixing fluid to the place for obtaining the material. Slit open the abdominal wall of the worm and remove the ovisacs and after separating the numerous convolutions somewhat, fix them entire for 24 hours in a mixture of absolute alcohol 4 parts, glacial acetic acid, 1 part, or for 15 to 25 minutes in acetic-alcohol-chloroform (reagent 2, b, p. 230) saturated with corrosive sublimate. Preserve in 80 per cent alcohol. To locate eggs of the desired stage tease out eggs at intervals along the ovisacs, stain with acid carmine (reagent 44, p. 248), and examine. The proper region once located, cut out small lengths of the tube, imbed it in paraffin, and make thin transverse sections. In order to keep the eggs from shriveling, the bath in hot paraffin must be curtailed. Use the method for delicate objects (p. 57). Stain by the iron-hematoxylin method (p. 54). Ascaris eggs when smeared on a slide in thick albumen fixative, which is then coagulated with formalin, will go on developing if put into an incubator.

21. **The Cultivation of Removed Embryonic Tissues** in clotted lymph, plasma, nutrient agar, bouillon, and in various salt solutions is an important phase of embryological technique which has been developed largely in recent years, but the subject is too extensive to treat in detail in limited space. A general method is given and the reader is left to look up modifications and other methods in the papers listed at the end of this memorandum.

All dissections must be carried on under aseptic conditions. Sterilize all instruments, pipettes, slides, covers, and vaseline in a Bunsen flame. For cultivation of tissues use a Locke's solution to which dextrin has been added. It is made as follows:

To 100 c.c. of distilled water add:
- $NaCl$ .... 0.900 per cent
- $CaCl_2$ .... 0.025 per cent
- KCl .... 0.042 per cent
- $NaHCO_3$ .... 0.020 per cent
- Dextrin .... 0.250 per cent

Remove an 8- or 10-day chick embryo, under aseptic conditions, to about 10 or 20 c.c. of the sterilized solution heated to 39° C. Cut out bits of intestine, kidney, liver, heart, or spleen a few millimeters in diameter and place into another dish which contains 10 to 20 c.c. of the solution at 39° C. Cut each small piece up into smaller pieces a fraction of a millimeter thick. Draw these up into a sterilized fine pipette, one at a time, with some of the solution and make hanging-drop preparations (Fig. 40) of them on sterile cover-slips which are thoroughly clean and free from every trace of grease. Invert each cover-slip on to a vaseline-ringed, hollow-ground slide which has been sterilized. For rings use a vaseline melting at about 46° C. A bit of paraffin may be added to ordinary vaseline to stiffen it.

Incubate the cultures at about 39° to 40° C. Growth begins within 10 to 20 hours and, as indicated by the number of mitotic figures, reaches its maximum on the second or third day. When it is desired to examine the living tissue do so on a warm stage. The margins of the growing regions are the best points to examine because there the cells are only one or two layers thick. (*Method of Lewis and Lewis.*)

When permanent preparations are desired the cover-slip is removed from the vaseline ring and the film of tissue is fixed, on the cover-slip, by means of osmic-acid vapor. After fixation the denser central piece of tissue is torn away, leaving only the thin film of new growth which is treated on the cover-slip as one treats sections on a slide. Stain in iron-hematoxylin or in Ehrlich's hematoxylin and eosin.

Further details of tissue culture *in vitro* and bibliographies will be found in the following papers: Harrison, *Anatomical Record*, I (1907); *Journal of Experimental Zoölogy*, IX (1910); Carrel and Burrows, *Journal of Experimental Medicine*, XIII (1911); Lewis and Lewis, *Johns Hopkins Hospital Bulletin*, XXII (April, 1911), 241; *Anatomical Record*, VI, Nos. 1 and 5 (1912); *American Journal of Anatomy*, XVII, No. 3 (March, 1915; *Anatomical Record*, X, No. 4 (February, 1916); and in Lee's, *The Microtomist's Vade-Mecum* (9th ed., 1928), pp. 376–89.

22. **The Living Embryo of the Chick** may be kept under observation for some hours while still in the egg by employing one of the so-called "window" methods. The simplest method is to cut out a disk of the shell on one side under as nearly aseptic conditions as possible, so that the embryo is exposed. A bit of the white is removed and a film of celloidin is laid over the opening to form the window. It must adhere firmly at every point around the margin. The embryo will continue to develop for some time if the egg is put back into the incubator.

McClung and Allen, *Microscopical Technique* (1929), p. 212, recommend cementing a ¾-inch rubber ring on to the shell. The outline of the ring is first made on the side of the egg with a pencil and within it the shell is thinned by means of a file. The ring is cemented in place with shellac or collodion solution. When this is hard the inclosed shell and membrane are removed and the space is filled up with albumen from another egg. A cover-glass is placed on top. It soon becomes attached by the drying of the excess albumen and this seals the opening. The egg is then incubated. The embryo may be observed at any time by turning the window to the upper side.

23. **Injection of Embryonic Vessels** is often resorted to (see memorandum 5, p. 101, and memorandum 14, p. 104).

24. **To Prepare Mounts of Chick Sections for the Hardships of Class Use** Hann (*Stain Technology*, III [January, 1928], 1) removes the slide from absolute alcohol and with a pipette applies a few drops of thin celloidin (4 grams dry celloidin in 200 c.c. absolute alcohol and ether, equal parts) evenly over the sections. The celloidin is hardened by leaving the slide on supports in a covered Petri dish exposed to chloroform fumes for 2 minutes. The slide is then immersed in chloroform for 2 minutes, cleared in carbol-xylol for 30 minutes, and mounted in balsam.

25. **A Paraffin Method for Serially Sectioning a Minute Object in a Known Plane** is described by Henry J. Fry in the *Anatomical Record*, XXIV (January, 1927), 4.

26. **Shrinking and Hardening of Embryos in the Paraffin Technique** is usually caused by the higher (above 80 per cent) alcohols. To avoid this, try the dioxan method (chap. vii), or use anilin oil as a dehydrating and clearing agent. For example, embryos fixed by any of the usual fluids such as Bouin's or picro-sulphuric and preserved in 70 per cent alcohol are transferred to a mixture of 70 per cent alcohol 2 parts, anilin oil 1 part, then to one of 95 per cent alcohol 1 part, anilin oil 2 parts, and, next, to pure anilin oil until cleared. They are then passed through anilin oil and xylol, half-and-half (if material tends to shrink, closer intergrades are used), to pure xylol for an hour or more; thence into a xylol-paraffin mixture; and finally into paraffin.

27. **Bouin's is One of the Best Fixing Fluids for Vertebrate Eggs and Embryos.**—The length of the fixation varies for different animals and materials, and ordinarily has to be determined by actual trial. Where fixation tends to make eggs too hard or friable for satisfactory sectioning, the fluid should be diluted with an equal amount of water.

28. **For Staining the Skeletons of Cleared Embryos** the Alizarin Red method, as developed by Dawson (*Stain Technology*, I, 4) and later by Lipman (*Stain Technology*, X, 2), has been found excellent in our own laboratory. Dr. Esther L. Boyer proceeds as follows with a 1-day-old rat:

a) Fix in 95 per cent alcohol for 72 hours or longer. To prevent maceration, longer fixation is desirable with larger embryos.

b) Clear in a 2 per cent solution of KOH for 7–10 days, or until the bones are clearly visible.

c) Stain in a solution of Alizarin Red S, 1 part to 10,000 parts of 2 per cent KOH, for 24 hours, or until the desired intensity of color is secured. If tissue other than bone is stained, it may be decolorized by placing in an acid alcohol (1 per cent sulphuric in 95 per cent alcohol), but it should be watched carefully to prevent decolorization or decalcification of the bone.

d) Clear in increasing concentrations of glycerin (20 per cent, 2 days; 40 per cent, 2 days; 80 per cent, 2 days).

e) Store in pure glycerin to which a small crystal of thymol has been added as a preservative.

CHAPTER XVIII
SOME CYTOLOGICAL METHODS

In the very many cytological methods which have been in vogue during the past few years two classes of fixing fluids, Flemming's strong (or some other chrom-osmic-acetic mixture) and Bouin's (or some of its modifications), and two stains, iron-hematoxylin and safranin, stand pre-eminent as of general utility. Iron-hematoxylin with or without a counterstain may be used successfully after either of the fixing fluids mentioned. The safranin is more likely to prove successful after Flemming's although in some materials good preparations can be made with it after Bouin's fluid. Two other fixing fluids, Gilson's mercuro-nitric and the acetic-sublimate mixture of Carnoy and Lebrun, are also of wide application, especially when followed by iron-hematoxylin as a stain. In recent years more attention has been devoted to proper cytoplasmic fixation, and in this connection various fixing fluids free from acid or with the acid much reduced have been devised.

In preparing tissues for cytological work it is imperative that pieces should be small, not more than 3 or 4 mm. thick where practicable, to insure thorough and even fixation. They should be fixed immediately after killing the animal. To secure penetration from all sides, it is well to place a few layers of filter-paper in the bottom of the vessels in which tissues are fixed and to shake the tissues about a little from time to time. Mechanical injury may be avoided by binding a bit of clean linen on the ends of the forceps with which tissues are handled.

Any student who contemplates undertaking a problem in cytology should read carefully the chapter on "Cytological Methods" (pp. 181–209) in McClung's *Microscopical Technique*. Also, Kornhauser's review of cytological staining in *Stain Technology*, V (1930), 117, is profitable reading.

I. MITOSIS

For general study of cell structures, and particularly cell division, I have found nothing which can readily be obtained in quantities sufficient for class use that surpasses the crayfish testis, the blastodisk of the whitefish, the epidermis and testis of Ambystoma and Necturus, and the maturation and cleavage stages of Ascaris. Ascaris material has already been discussed (p. 154).

Robertson (*Journal of Morphology*, XVII [June, 1916]) finds that in grasshoppers of the family Tettigidae, taken before the last moult, cells dividing mitotically may be found in large numbers in the mesenteron, proctodaeum, fat-body, hypodermis, and the follicles of the gonads. The columnar epithelium of the mesenteron seems to be the most favorable region for finding such divisions. Inasmuch as the members of this family have only thirteen or fourteen chromosomes, the material should prove to be exceptionally valuable for purposes of class demonstration.

Testis of Crayfish

The testicular cells of the crayfish (*Cambarus virilis*) will be found in active proliferation from the middle of June to the middle of July. The chromosomes are too small and too numerous for satisfactory individual study, but the spindles and centrosomes are distinct and the general pictures of representative stages are clear-cut and easily found.

Section Method.—1. Fix small bits of the testes, 3 to 4 mm. thick, in Flemming's fluid (14, p. 235) for 24 hours. Wash in running water 6 to 12 hours, dehydrate, and imbed in paraffin according to the methods for delicate objects (p. 57) or the "drop" method (6, p. 174).

2. Cut sections 5 to 7 microns thick and mount several slides by the water-albumen method.

3. After removing the paraffin with xylol and running the slides down through the alcohols, stain some of the sections by the iron-hematoxylin method (p. 54) and counterstain with acid fuchsin or orange G. Run the slides up through the alcohols, clear in xylol, and mount in a thin balsam. Use a No. 1 cover-slip if the preparation is to be studied with an oil-immersion lens.

4. Place others of the slides in safranin (79, p. 260) for 24 hours, then rinse in water and run up through the alcohols to 95 per cent. Counterstain for 30 seconds in a 0.5 per cent solution of light green (*Lichtgrün* S.F.) in 95 per cent alcohol. If the sections are left too long in the green stain the safranin will be washed out. Pass the slides through absolute alcohol into clove oil for a few minutes, rinse in xylol and mount in thin balsam under a No. 1 cover-slip.

Smear Preparations.—Remove the fresh testis to a slide and tease somewhat with needles in order to rupture the cysts which inclose the germ cells. Spread the mass evenly over the slide with the end of another slide and then flatten it between the two slides in a very thin film. Avoid any considerable pressure. Separate the slides by slipping them apart. Each should bear a very thin coating of the material. Plunge them into Flemming's fluid and leave for 24 hours. Wash in running water for 6 to 12 hours, then with forceps pick or scrape off all lumps of tissue which might later keep the cover-slip from fitting closely. Stain and mount as if the films were sections.

If preferred, some of the slides can be fixed in Bouin's fluid for an hour or two, washed in 50 per cent alcohol, then stained in iron-hematoxylin and counterstained in acid fuchsin or orange G.

Blastodisk of Whitefish (Coregonus)

1. Spawn the females and fertilize the eggs (in early December) as directed in memorandum 16, p. 153.

2. Select eggs in the 32- to 64-cell stage of cleavage (40 to 60 hours after fertilization) and fix for 6 or 8 hours in Bouin's fluid.

3. Wash in repeated changes of 50 per cent alcohol, then in several changes of 70 per cent alcohol..

4. With needles carefully dissect off the blastodisks under a binocular or other dissecting microscope.

5. Dehydrate, section in paraffin (method, p. 38), stain in iron-hematoxylin with or without a counterstain, and mount as usual. Sections should be about 7 microns thick.

The eggs of the pickerel, obtainable in April, may be handled with equal ease. They cleave much more rapidly than do those of the whitefish.

Testis of Necturus

The cells of the testes will ordinarily be found undergoing rapid proliferations in late July and early August. Those toward the posterior end of the testis show the most advanced stages of spermatogenesis, those toward the anterior end the least advanced stages. Both cells and chromosomes are very large. The spindle usually shows up well and the chromosomes exhibit considerable variety in shape and size.

1. From different regions of the testis fix some bits of testis in Bouin's fluid (6 to 8 hours) and others in Flemming's (24 to 36 hours). Wash out the Bouin as in 3, page 160, and the Flemming according to directions in step 1, page 159.

2. Dehydrate, imbed, and section according to the usual paraffin method.

Iron-Hematoxylin Preparations.—3. Stain sections of each kind of material according to the ordinary iron-hematoxylin method with or without a counterstain (pp. 54–56).

Safranin-Gentian-Violet Preparations.—4. Also stain some of the Flemming material according to the safranin and gentian-violet method (80, p. 261).

Safranin-Gentian-Orange Preparations.—Use saturated aqueous solutions of safranin, gentian violet, and orange G respectively. Rinse and stain in gentian violet 2 to 5 minutes (time determined by trial). Pipette absolute alcohol over the sections until the violet is out of the cytoplasm, then follow with orange G, pipetting it on and removing it again almost instantly. Wash off with absolute alcohol, dip in oil of cloves, clear in xylol, and mount in thin balsam under a No. 1 cover.

Somatic Cells of Ambystoma

Epidermal Cells.—Cut off the tails of several one-month-old Ambystoma larvae into Flemming's fluid. At the end of 2 to 4 hours strip off bits of the epidermis from the tails and fix these strips for some 20 hours longer in the fluid. Wash in running water 6 to 8 hours, stain some according to the iron-hematoxylin method (p. 54) and others with safranin and light green (step 4, p. 159). Dehydrate and mount as usual.

If Bouin's is used instead of Flemming's fluid, the peeling off of the epidermis need not be done until the end of fixation (6 to 8 hours).

Peritoneal Cells.—Parmenter recommends larger larvae than those used for epidermis. He cuts away the side walls of the body cavity, pulls out the intestine, and fixes the remaining tissue in the region of the spinal column in Flemming's or in Bouin's fluid as above. Bits of the peritoneum on either side of the dorsal mid-line are stripped off and prepared as was the epidermis.

Either of these kinds of preparations show splendid polar views of cell-division stages, though little or nothing of lateral views. They are especially favorable for showing longitudinal splitting of chromosomes.

Living Cells.—Curarize Ambystoma or other young amphibian larvae by adding (according to size of larva) 5 to 10 drops of a 0.5 per cent solution of curare in equal parts of glycerin and water to a watchglassful of water. After 40 minutes remove for half an hour to a 1 per cent solution of sodium chloride in water. Wrap in blotting paper and examine the tail fin on a slide under the microscope. Cell divisions may be seen in progress.

If replaced in fresh water such larvae recover after some hours.

After curarization some workers prefer to cut the larva in two parts in front of the hind limbs, studying only the tail. The gills also show interesting cell activities. If curare is not at hand a 3 per cent alcohol or ether may be used, although not so successfully.

For an *intra-vitam* technique for the study of the living cells of insects, see Baumgartner and Payne, *Science*, LXXII (1930), 199.

II. MITOCHONDRIA

Recent studies tend to show that mitochondria occur more or less extensively in nearly all kinds of tissue. They were largely overlooked in the past because many of the fixing fluids in use contain strong organic acids, such as acetic, and these dissolve mitochondria. They are sometimes stained with great sharpness by iron-hematoxylin, following fixation in Flemming's strong mixture (14, p. 235) in which, instead of 1 c.c., only 3 to 6 drops of glacial acetic acid are used for every 15 c.c. of chromic acid. For their careful study, however, cytologists are using special methods. These are too numerous and complex to be reviewed in an elementary guide to technique. Bibliographies and discussion of the technical details will be found in the publications of Bensley and particularly of Cowdry. See Bensley, *American Journal of Anatomy*, XII (1911), 297–388; Cowdry, *Internationale Monatsschrift für Anatomie und Physiologie*, XXIX (1912); *American Journal of Anatomy*, XVII, No. 1 (November, 1914); *ibid.*, XIX, No. 3 (May, 1916); *Contributions to Embryology*, No. 11 (Carnegie Institution of Washington), pp. 198–205, McClung's *Microscopical Technique* (1929).

The beginner will find pancreas probably the best tissue on which to practice. The mitochondria of the acinous cells are filamentous and of large size. Four methods of wide application are as follows:

Regaud's Method.—1. Fix for 4 days, changing every day, in a formol-bichromate mixture made as follows:

Bichromate of potassium, 3 per cent..........	80 parts
Formalin (commercial)......................	20 parts

2. Mordant in 3 per cent potassium bichromate solution for seven days, changing every second day.

3. Wash in running water 24 hours, dehydrate, clear, imbed in paraffin, and cut thin sections.

4. Mount and pass sections down to water in the usual way.

5. Mordant in 5 per cent iron alum at 35° C. for 24 hours; rinse in *distilled* (not tap) water and stain for 24 hours in a "ripened" (a few weeks) hematoxylin stain made by dissolving 1 gram of pure hematoxylin crystals in 10 c.c. of absolute alcohol, and adding 10 c.c. of glycerin and 80 c.c. of distilled water. Do not rinse too long before staining or the mordant will be extracted.

6. Differentiate in 5 per cent iron-alum, watching under the microscope.

In our laboratory we find this the most reliable stain for mitochondria in other than embryonic tissues. It gives good results even with the usual iron-hematoxylin method. H. W. Beams reports that treatment overnight in 1 to 2 per cent osmic acid solution, after removal from the formal-bichromate mixture, tends to fix fat in the sections so that it is not easily dissolved during dehydration and clearing.

Benda's Method.—1. Fix for 8 days in a modified Flemming fluid made as follows:

Chromic acid, 1 per cent	15 c.c.
Osmic acid, 2 per cent	4 c.c.
Glacial acetic acid	3 drops

2. Wash in water for 1 hour, then for 24 hours in a mixture of equal parts of pyroligneous acid and 1 per cent chromic acid.

3. Transfer to a 2 per cent potassium bichromate solution for 24 hours, run up through the grades of alcohol to xylol, and finally infiltrate with paraffin and section. Sections should be about 5 microns thick.

4. After removal of paraffin from sections run them down to distilled water and place them in a 4 per cent iron-alum solution for 24 hours.

5. Wash thoroughly in water and transfer to a solution of Kahlbaum's sulphalizarinate of soda (made by taking 1 part of a saturated aqueous solution of the stain to from 80 to 100 parts of distilled water) for 24 hours.

6. Rinse the slide in distilled water and flood it with a crystal violet anilin-water solution (equal parts of anilin water and a 3 per cent solution of the dye in 95 per cent alcohol). Warm until the solution steams, keeping it heated for about 3 minutes.

7. Wash in distilled water, transfer to 30 per cent acetic acid for 1 or 2 minutes, then wash in running water for 5 or 10 minutes.

8. Dry the slide with filter paper, dip it for a minute into absolute alcohol, place in oil of bergamot until cleared, then transfer it through xylol and mount in balsam in the usual way.

A successful preparation should show chromatic elements a deep purple and the cytoplasm a light red with mitochondria violet.

Wildman (*Journal of Morphology*, XXIV, No. 3 [1913]) modifies the method by transferring slides from the alizarin solution, after rinsing, into a 3 per cent solution of crystal violet (3 c.c. of anilin stain in 100 c.c. of distilled water) for 10 minutes; rinsing and passing into 80 per cent alcohol for 5 seconds; passing through 95 per cent and absolute alcohol; and, when properly differentiated clearing and mounting in the usual way.

Bensley's Acid-Fuchsin, Methyl-Green Methods.—Fix tissues for 24 hours in the following:

Osmic acid, 2 per cent	2 c.c.
Potassium bichromate, 2.5 per cent	8 c.c.
Glacial acetic acid	1 drop

Sections should be 4 microns or less in thickness. Mount by the water method (p. 24), remove paraffin with toluol, then pass through absolute alcohol to water. Treat for from 30 seconds to 1 minute (determined by trial) with a 1 per cent solution of potassium permanganate, then for the same length of time with a 5 per cent solution of oxalic acid. The permanganate extracts the mordanting elements of fixation and the oxalic acid removes the permanganate. Thoroughly wash in water.

Stain for 5 minutes in Altmann's acid fuchsin (acid fuchsin 20 grams, anilin water 100 c.c.) which has previously been warmed to 60° C. Wash thoroughly in distilled water, dip for an instant into a 1 per cent solution of methyl green, then wash, rapidly dehydrate in absolute alcohol (avoiding alcohols of intermediate strength), clear in toluol, and mount in balsam. Toluidin blue can be substituted for methyl green.

If the material does not stain well with the acid fuchsin, or if the methyl green or toluidin blue obliterates it, treat the sections with a 2.5 per cent aqueous solution of potassium bichromate for about half a minute and rinse in water just before staining in acid fuchsin.

In spinal ganglion cells, for example, mitochondria should appear bright red; Nissl substance, green (or blue); neurofibrils in the axon hillock, light brown; and the canalicular systems should be revealed.

One of the very best fixing fluids for mitochondria is Bensley's formol-bichromate-sublimate mixture (Appendix B, reagent 24).

Bensley's Copper-Chrome-Hematoxylin Method.—Fix materials in acetic-osmic-bichromate mixture and prepare for staining as in the preceding method. Wash for 1 hour in distilled water, then thoroughly dehydrate. Leave in absolute alcohol for 24 hours, then pass through equal parts of bergamot oil and absolute alcohol (1 hour) into pure bergamot oil for 3 hours, followed by equal parts of bergamot oil and paraffin (1 hour), then by paraffin melting at 60° C. (2 to 3 hours). Cut sections 4 microns thick and fix to the slide by the water method. Remove paraffin with toluol and pass down through the alcohols to distilled water.

Place sections for 5 minutes in a saturated aqueous solution of copper acetate, wash in several changes of water, and transfer for a minute to a 0.5 per cent aqueous solution of a well-ripened hematoxylin. Wash in water and transfer for 1 minute to a 5 per cent aqueous solution of neutral potassium chromate. The sections should turn a blue-black color. If they are of only a light blue shade, place them again in the copper acetate and repeat the operations from there on.

Wash several minutes in water, then differentiate under the microscope in Weigert's borax-ferricyanide mixture (borax, 2 parts; ferricyanide of potassium, 2.5 parts; water, 200 parts) diluted with 2 volumes of water. Wash 6 to 8 hours in tap water, then dehydrate, clear in toluol, and mount in balsam. Mitochondria should appear a bluish black against a clear background.

Mitochondria in Living Cells stain specifically with Janus green. There are several Janus greens; but of these, Cowdry reports that only Janus green B (diethylsafraninazodimethylanilin chloride) of the Farbwerke Hoechst Co. (obtainable from L. A. Metz & Co., New York) will give the desired reaction. He finds that mitochondria will stain in human lymphocytes in a dilution of Janus green in normal saline solution of 1:500,000. Ordinarily, for living tissue, a dilution of 1:15,000 or 1:20,000 in normal saline, in Locke's solution, or in Ringer's solution, is employed. Janus blue, G and R, may also be used as a vital stain for mitochondria.

While fresh tissues may be stained by immersion in the dye, much better results are obtained by injection through the blood vessels, in normal saline, after the vessels have been thoroughly flushed out with normal saline solution.

Mitochondria in Tissue Grown "in Vitro" (see p. 155) may be studied readily, according to M. R. and W. H. Lewis (*American Journal of Anatomy*, XVII, No. 3 [1915]), who observed their changes, growth, and division in embryonic tissues of the chick.

III. GOLGI APPARATUS

For demonstration of the Golgi apparatus, workers in my own laboratory (notably Beams and Wu) have found Ludford's modification of the Mann-Kopsch technique the most reliable. Beginners will probably find pancreas or the spinal ganglion cells of young mammals the best material with which to experiment. For a full review of various methods see Gatenby's account in Lee's *Microtomist's Vade-Mecum* (9th ed., 1928), chap. xxvii.

Ludford's Modification of the Mann-Kopsch Method for Golgi Apparatus.—1. Fix the tissue for 18 hours in Mann's mercuro-osmic fluid (equal parts of a 1 per cent osmic acid solution and a saturated solution of corrosive sublimate in normal salt solution).

2. Wash for 30 minutes in distilled water.

3. Transfer to 2 per cent osmic acid (just enough to cover tissues) and keep in an incubator at 30° to 35° C. for 3 to 5 days, depending upon the nature of the material.

After the third day, remove bits of tissue from time to time, crush thoroughly and examine under the microscope (as suggested by Nassonov) to determine the degree of impregnation. Do not let the process go too far. If the smell of osmic acid disappears and the solution turns black, pour off the old solution, rinse in distilled water, and add new 2 per cent osmic acid solution.

4. Transfer to water and keep in the incubator for another day.

5. Dehydrate, clear, imbed in paraffin, section (3 to 6 microns) and mount as usual.

6. After the slides have dried overnight, place one in xylol to remove the paraffin, then mount in balsam. In a successful preparation Golgi apparatus,

fat and yolk should appear black, mitochondria, nuclear structures and cytoplasm, yellowish or brownish.

7. If mitochondria as well as Golgi apparatus have been blackened, using another of the slides, remove the paraffin with xylol, then place the slide in old turpentine. Watch the effect of the turpentine under the microscope from time to time until the mitochondria disappear and the black fades from fat globules and yolk while yet remaining in the Golgi apparatus (about 15 minutes). Pass back through xylol and mount in balsam. Some technicians prefer to do this destaining in weak peroxide of hydrogen.

8. If a counterstain is desired, after removing paraffin with xylol pass unbleached sections down as usual to water. Stain for half a minute or less in a solution of dilute neutral red made as follows:

Distilled water	1,000 c.c.
Neutral red	1 gram
Glacial acetic acid, 1 per cent	2 c.c.

9. Rinse with distilled water, shake off the excess water, and pour absolute alcohol on the sections. When sufficiently differentiated, clear in xylol and mount in balsam.

Beams's Technique for Demonstrating Golgi Bodies in the Lumen of Glands.—1. Fix pieces of lactating mammary gland of the white rat for 2 days in Brouha's modified Flemming fixing fluid, made as follows:

Sol. A. Saturated solution of corrosive sublimate	600 grams
Glacial acetic acid	40 grams
Sol. B. Osmic acid	1 gram
Chromic acid	1 gram
Distilled water	100 grams

To 4 parts of solution A add 1 part of solution B.

2. After fixation wash the tissue in water for from ½ to 1 hour.
3. Dehydrate, clear in xylol and imbed in paraffin.
4. Cut sections from 3 to 5 microns and fix on slides by the albumen method.
5. Transfer sections through xylol and decreasing strengths of alcohols into distilled water.
6. Then place them in 0.2 per cent solution of gold chloride plus 1 drop of acetic acid to every 10 c.c. of solution until sufficiently bleached.
7. Wash slides in distilled water and place in 5 per cent solution of sodium hyposulphite for 2 minutes and wash over again in distilled water.
8. Dehydrate and clear in xylol.
9. Mount in balsam.

Says Beams:

"Slides prepared by the above described technique show the Golgi apparatus in the lumen in the form of small round bodies. These bodies are apparently located in the presumably cytoplasmic layer surrounding the fat droplet which is derived from the cytoplasm of the secreting epithelium. When the fat is dissolved the Golgi apparatus remains as granular-like bodies marking the outer limiting membrane of the droplet. Large numbers of these figures may appear in one lumen which, before the fat is dissolved, presents a deeply stained mass varying from approximately 2 to 15 microns in diameter. In some cells which were fixed just before extrusion of the fat droplet the Golgi bodies may be seen in the bulging limiting membrane of the cell describing a convex arc into the lumen of the gland."

For the Sevringhaus technique (a very valuable method) see memorandum 15.

IV. STAINING OF LIVING OR FRESH TISSUES

Intra-vitam staining, so called, has come more and more into prominence during the past few years. It is questionable if staining of really "vital" elements ever occurs, although undoubtedly various granules in cells may be stained while the cells are yet living. Certain stains also may be used with fair success with fresh or lightly fixed cells.

For *intra-vitam* staining, neutral red, Bismarck brown, Janus green, and methylen blue are the dyes most commonly employed. They are used in the proportion of about 1 part of the dye to 10,000 or 20,000 parts of some normal fluid, such as normal saline, Ringer's solution, or Locke's solution.

For lightly fixing and staining fresh cells methyl green acidulated to about 0.75 per cent with acetic acid is in common use. Also for the study of fresh cells tissues are teased in a solution of Ripart and Petit (12, p. 235) to which 0.1 per cent osmic acid has been added, and then stained in methyl green. Acid carmine is frequently used for the study of chromosomes in fresh cells (see memorandum 2, p. 173). For the use of Sudan III with living animals see p. 169. Trypan blue will make certain parts of the living body take on an intense blue color, but the color seems to be due wholly to the engulfment of the colored particles by certain phagocytic cells, particularly in the connective tissues of the body (Evans and Schulemann, *Science*, XXXIX [1914], 443–54). Thus, 1 c.c. of a 0.5 per cent solution injected into the peritoneal cavity of a mouse will rapidly blue it from ears to tail without noticeably interfering with its normal activities.

Equal parts of glycerin, 95 per cent alcohol, and distilled water is a useful examining-medium in which fresh tissues may be kept for a long time without marked deterioration.

V. TESTS FOR CERTAIN CELLULAR STRUCTURES

The following tests while not always specific are serviceable in helping to identify some of the more usual cellular contents:

Acidophil Granules are red after hematoxylin and eosin stain; yellow after iron-hematoxylin and Van Giesen's; yellow to orange after Mallory's.

Amyloid Substances are red after Hanstein's rosanilin-violet.

Archoplasm stains intensely with acid fuchsin or light green.

Basophil Granules are blue after hematoxylin and eosin; black after iron hematoxylin and Van Giesen's; red after Mallory's.

Calcification may usually be detected by means of 3 to 5 per cent hydrochloric acid. When treated with this solution carbonate of lime emits bubbles of carbon dioxide, while phosphate of lime simply dissolves.

Cell Walls are usually well defined by acid fuchsin when used as a counterstain with some of the hematoxylins.

Centrosomes are best shown by the iron-hematoxylin long method (p. 54). Heidenhain finds that they are more sharply defined if the sections, previous to mordanting in iron-alum, are stained for 24 hours in a weak solution of Bordeaux red. For demonstration in the living cell Huettner and Rabinowitz recommend the pole cells of developing *Drosophila* eggs (*Science*, LXXVIII [1933], 367).

Chromatin.—In fresh cells, *methyl green* stains only chromatin when it colors any part of the cell. Absence of coloration with this dye does not necessarily mean absence of chromatin. Congo red followed by picro-anilin blue (p. 259) is also sometimes serviceable with fresh tissue. Moderate *digestion* with gastric juice (at about 40° C.) will remove albumins and leave chromatin. Prolonged treatment with 1 per cent caustic potash or with fuming hydrochloric acid will remove all chromatin from a nucleus. A 10 per cent solution of sodium chloride swells chromatin and may dissolve it.

After sublimate fixations, thin (3-micron) sections stained by the Ehrlich-Biondi method (47, p. 248) or Auerbach's fuchsin-methyl-green method (52, p. 250) should show "active" chromatin or chromosomes green, linin, and plasmosomes red.

The following "Newton-Gram method" is widely used for chromatin:

(1) Mordant for from 30 to 45 minutes in a mixture of iodine, 1 gram; potassium iodide, 1 gram; 80 per cent alcohol, 100 c.c. (2) Rinse rapidly in water and stain for from 15 to 20 minutes in a 1 per cent solution of crystal violet. (3) Rinse again in water and differentiate in the iodine-potassic iodide solution (step 1), moving the slides gently in the liquid. Rinse in water. (4) Counter-stain lightly with orange G if desired. Dehydrate by passing through the alcohols including absolute. Clear in xylol and mount in balsam.

Chromosomes.—The best single stain in fixed material is either iron-hematoxylin or safranin, although chromatoid bodies and mitochondria, when present, may also stain by these reagents. In fresh tissue, acid carmine (44, p. 248) colors chromosomes, as does also methyl green (67, p. 257). For tests by Feulgen's reaction see Margolena in *Stain Technology*, VII (1932), 9–16.

Fat.—As the fat of tissues is dissolved by xylol, alcohol, and other reagents used in the paraffin and celloidin methods, only teased, free-hand, or frozen-sectioned, fresh material, or material fixed in some non-fat solvent fixer such as formalin or Müller's fluid, can be used.

Osmic Acid (26, p. 240) is the commonest test for fat. It stains most but not all fatty bodies brown or black. Pure palmitic and stearic acids and their glucosides do not reduce osmic acid; it is reliable only for detection of oleic acid and olein. Osmicated fats are rendered sufficiently insoluble to permit of dehydration and mounting in balsam, if absolute alcohol is avoided and cedar oil is used instead of xylol for clearing. However, euparal (p. 58) may be used.

Sudan III is a specific stain for fat (see 83, p. 262). Large fat drops stain from a brilliant red to an orange; small ones may be yellowish red. The fat of animals fed with the dye will become intensely colored by it. The fat in the layers of yolk laid down in hens' eggs while the fowls are fed on the dye (Riddle, *Science*, XXVII [1908], 945) is stained red, and eggs so colored (Gage) hatch into chicks with the body fat colored pink.

Scharlach R is superseding Sudan III as a stain for fat. It stains fat orange to red. If a permanent mount is desired, frozen sections may be fixed 10 minutes in formalin vapor, stained for 12 hours in a saturated filtered solution of the dye in 70 per cent alcohol, washed in water, counterstained in alum-hematoxylin, washed and mounted in glycerin or glycerin-jelly. However, see 81, p. 262.

Fat may be removed from tissues ordinarily by treatment with alcohol, ether, or chloroform.

Free Acid in tissues may be detected by Congo red, the solutions of which become blue in presence of free acid. Neutral red is turned bright red by acid, yellow by alkalies.

Glycogen.—Readily soluble in aqueous media, hence tissues should be fixed and hardened in 95 per cent alcohol. Gage (*The Microscope*) states that a Lugol's solution made of 1.5 grams of iodine crystals, 3 grams of iodide of potassium, 1.5 grams of sodium chloride, and 300 c.c. of water gives a differential stain (a mahogany red) for glycogen in sections. For very soluble glycogen he recommends that 50 per cent alcohol be substituted for the water in the stain. He deparaffins with xylol, mounts in yellow vaseline, and seals with shellac or balsam.

Hemoglobin, after proper fixation, stains a characteristic, clear, deep red color with eosin. For crystals see p. 121.

Intra-cellular Reduction Processes may be detected by Janus green used as an *intra-vitam* stain (p. 167). With reduction the color changes from blue or green to red.

Lecithin may be distinguished from fat by its less solubility in ether and its greater capacity for stains. Formalin-fixed material if brought into acetone has the fat dissolved, but not its lecithin. The latter may be stained by osmic acid, hematoxylin, orange G, acid fuchsin, methyl green, or toluidin blue, although the tissue should be dehydrated in acetone and left as little as possible in alcohol.

Mineral Ash in isolated cells, particularly in protozoa, is shown by means of micro-incineration. For a good simplified method see MacLennan in *Science*, LXXVIII (1933), 367.

Mitochondria (see pp. 162–65).—The most nearly specific single stain for mitochondria in fresh tissue is probably Janus green (p. 165).

Mucin in cells, after sublimate fixation, stains with *basic* but not with *acid* anilin dyes. Either thionin or toluidin blue stains mucin reddish, surrounding elements blue. Methylen blue and safranin are also good stains for mucin. See also muci-carmine and muci-hematin (69 and 70, p. 258).

Nissl's Granules (tigroid substance).—For the methylen-blue method see p. 256.

Oxidase Reaction (Schultze's).—The presence of an oxidizing ferment in cells may be disclosed by the following method:

Solution 1. Boil 1 gram of α-naphthol in 100 c.c. of distilled water until it melts. Add pure potassium hydrate (about 1 c.c.) until the naphthol is completely dissolved. The solution should pass from yellow to yellowish brown.

Solution 2.—Make a 1 per cent aqueous solution of dimethyl-p-phenylendiamin (Merck) at room temperature. Filter.

Use frozen sections of formalin-fixed material or cover-glass preparations fixed in vapor of formaldehyde. Move the preparations gently back and forth in solution 1 for about 3 minutes, then do the same in solution 2. Wash in distilled water and mount and examine in water or in glycerin-jelly. Oxidase granules are stained deep blue.

Plasmosomes remain unstained in fresh material treated with acid methyl green (67, p. 257) which stains chromatin. With the Ehrlich-Biondi stain they stain red, sometimes orange; with safranin, gentian violet, and other chromatin dyes, in regressive staining (p. 25), the plasmosomes frequently retain the stain more tenaciously than resting chromatin does. Because of greater refractivity they are frequently demonstrable in unstained preparations and sometimes in living cells.

Secretion Antecedents.—See Bensley, *American Journal of Anatomy*, XII, No. 3 (1911); XIX, No. 1 (1916).

Spindle Fibers are frequently well stained by acid fuchsin when used as a counterstain after such fixers as Flemming's, Gilson's, or Bouin's fluids. See also "Euparal," p. 58.

VI. SPECIAL METHODS

1. Allen's Method for Mammalian Tissues.—[1]

Fix tissues in Allen's B-15 fluid, which is made up as follows:

Picric acid, saturated aqueous solution	75 c.c.
Formalin (c.p.)	25 c.c.
Glacial acetic acid	5 c.c.

Just before using heat to 37° C. and add 1.5 grams of chromic acid crystals, agitating the mixture vigorously until the crystals are dissolved. Then add 2 grams of urea crystals. During fixation keep the liquid heated to 37° or 38° C.

---

[1] *Anatomical Record*, X, No. 9 (July, 1916).

Pieces of brain 0.5 c.c. in volume fix in 1 hour. Bits of young mature testes require a little longer and pieces of older testes 2 to 3 hours.

Let the fluid and the contained tissue cool to room temperature and dehydrate gradually by the drop method (memorandum 6, p. 174) with alcohol up to 75 per cent alcohol, then finish dehydration with anilin oil. Regulate to about one drop per second, or less if the quantity of fixing fluid is small. Bring pieces of soft tissue, some 0.5 c.c. in volume, up to 75 per cent alcohol in about 1 hour. Harder tissues require more time.

To wash out all picric acid thoroughly, replace the ordinary 75 per cent alcohol on the object with 75 per cent alcohol containing a few drops of a saturated aqueous solution of lithium carbonate. Keep up agitation with a very slow current of air and continue the washing until the yellow color ceases to appear in the fluid. As soon as possible after washing, to avoid shrinkage in alcohol, start replacement by anilin oil, letting it drop in slowly. To insure rapid mixing a stronger current of air may be required; when nearly pure anilin is reached leave the tissue in it until it is clear like amber.

Replace the anilin with bergamot oil or synthetic oil of wintergreen, following the same method. Change the oil once after the tissue has arrived in pure oil.

Warm the oil and tissue slightly and add to it every 10 minutes a few drops of melted paraffin, which must be thoroughly mixed with the oil by means of a pipette. When the mixture is about 85 or 90 per cent paraffin, transfer the object to pure paraffin with a melting-point of 52° to 55° C. If bergamot oil has been used, make at least four changes of paraffin. Leave the tissue in each about 30 minutes and in a fifth paraffin about an hour. Testes material requires longer time. Imbed and section in the usual way.

Painter (*Journal of Experimental Zoölogy*, XXV, No 1 [Jan. 1922]) offers the following alternative to the drop method of dehydration: replace fixer with 35 per cent alcohol; this with 50 per cent alcohol and anilin oil, equal parts; followed by 70 per cent alcohol and anilin oil, equal parts; then by pure anilin oil; and finally by oil of wintergreen. This technique eliminates the hardening effects of the higher alcohols. Several workers in my own laboratory find it more satisfactory than the drop method.

Allen has pronounced another fixer, which he designates B-20, an improvement on B-15 for mammalian chromosomes. Make up as for B-15 but use only half the amount of chromic acid called for; to each 50 c.c. add 1 c.c. of a solution of 1 gram of osmic acid in 50 c.c. of 1 per cent chromic acid. Mix at the time of using. After-treatment is the same as for B-15.

2. Hance's Method for Mammalian and Avian Material.—

1. Fix very small bits of tissue for from 4 to 12 hours in chilled, freshly prepared Flemming's strong solution to which has been added 0.5 per cent of urea. Keep vials or bottles containing the material in cracked ice and preferably in a refrigerator. Have 10 to 15 c.c. of fluid for each piece of tissue.

2. Wash in running water for 12 hours and then dehydrate very gradually (e.g., "drop method"). If tissue is to be left for several days dehydrate to 70 per cent alcohol, and leave it there. Hance prefers to dehydrate to 95 per cent alcohol, however, and imbed the material at once.

3. The 95 per cent alcohol is changed twice, the tissue remaining in the last change 1 or 2 hours.

4. Clear in cedarwood oil or oil of bergamot, preferably by the drop method, although satisfactory results may be obtained by adding fractions of the clearer, from time to time, to the 95 per cent alcohol which contains the tissue. When the transition has been completed change the tissues to fresh oil two or three times. They may be left overnight in either oil.

5. When ready to infiltrate with paraffin, wash out the oil with chloroform (30 to 60 minutes). Transfer to fresh warm chloroform. Add melted paraffin at intervals during a period of 2 to 4 hours and finally transfer to pure paraffin and allow to remain 2 to 4 hours.

6. Cut avian tissues about 5 microns and mammalian tissues about 10 microns thick.

7. Affix and prepare the sections for staining as usual.

8. Bleach the sections of their osmic acid stain by placing them for 12 hours or longer in a mixture of 4 parts of 70 per cent alcohol and 1 part of commercial hydrogen peroxide.

9. Stain in iron-alum hematoxylin and mount in balsam according to the usual procedure.

VII. PHOTOGRAPHING CELLULAR STRUCTURES

1. Select a slide in which the part to be photographed is well stained in iron-hematoxylin. The background should be unstained, as a sharp contrast in the slide gives better results in the picture. Eastman Wrattan "M" plates, or panchromatic films, have proved satisfactory. The series of Eastman Wrattan filters is a valuable adjunct to microphotography. For ordinary hematoxylin-eosin slides, the "G" filter is best.

2. Using an apochromatic lens and a compensating or projecting ocular, bring the part selected for the picture into sharp focus under the microscope.

3. Place the microscope containing the object in focus under the camera. Adjust the bellows to any desired length. A lower ocular with a longer bellows, at the same magnification, produces a better picture than a higher ocular with a shorter bellows. In using a camera mounted in a vertical plane there is less likelihood of jarring the object out of focus when the microscope is placed under the camera than when it is adjusted in a horizontal plane.

4. Focus the object under the microscope upon the ground-glass screen of the camera. To insure sharp focus, a focusing glass should be used.

5. Foot and Strobell (*Zeitschrift für wissenschaftliche Mikroskopie*, XVIII [1901], 421–26) developed the following method, which obviates focusing up-

on the ground glass of the camera every time a picture is taken. Procure from an oculist several concave spectacle lenses (—1 to —10 diopters). When steps 1 to 4 have been completed, remove the microscope from under the camera without the least shift of focus. *Do not touch the micrometer or fine adjustment screws of the microscope.* Place over the ocular one after another of the spectacle lenses until one is found through which the object appears in as sharp focus as it did upon the ground-glass screen. Note the number of the particular lens used. Thereafter, when an object is to be photographed with the same ocular and objective, the same tube and bellows length, one need only place the dioptric lens over the ocular, focus until the points desired in the finished photograph stand out sharply, remove the dioptric lens, place the microscope under the camera without shift of focus, and take the picture. Since the dioptric lens corrects the focus for that bellows length, the necessity of refocusing upon the ground glass is removed. Whenever a different combination of ocular objective and bellows length is used, the proper dioptric lens must be found.

6. It is good practice to allow the microscope to stand for a time after the object has been focused through the dioptric lens in order to be sure that the focus does not shift. After the picture has been taken, by replacing the lens over the ocular one can see if the focus has been held throughout the time. Even with the utmost care the focus will sometimes change. A good chapter on "Photography with the Microscope and with Projection Apparatus" will be found (pp. 206–45) in Gage, *The Microscope* (14th ed.).

MEMORANDA

1. **Accessory Chromosomes** (sex chromosomes, X-elements) are perhaps best demonstrated in the testes of some species of the short-horned grasshoppers taken about the time of the last moult. In these forms mitotic figures are large and chromosomes usually distinct. Among the Hemiptera, the squash bug (*Anasa tristis*) for single X-element, and the stink bug *Euschistus*, for X- and Y-elements, are recommended. Flemming's or Bouin's fluid may be used for fixing, and iron-hematoxylin or safranin for staining.

2. **For Quick Determination of the Chromosomal Condition** of cells aceto-carmine preparations (44, p. 248) are useful. The entire testis of an insect is put on the slide in a drop of the stain and the cells separated by slight pressure on the cover, or smears from the testes of larger animals are made and flooded with the stain, after which a cover-slip is added. The edges of the cover should be sealed with vaseline, or preferably paraffin, to prevent evaporation. Chromosomes are stained in a few minutes. Such preparation should be carefully checked by observations on well-fixed and stained materials, however, since the great amount of acetic acid in the acid carmine swells chromosomes and is likely to lead to erroneous conclusions regarding details.

To render aceto-carmine preparations permanent, Buck turns over the slide bearing the stained material and supports it face down on two thin glass rods in a Petri-dish containing equal parts of xylol, absolute alcohol, and glacial acetic acid, and leaves it until the cover comes off (5 to 30 minutes). It is advisable first to crack off as much of the paraffin seal as possible. The slide is rinsed in the solvent for 5 minutes, drained, and freed of as much of it as possible, passed through two changes of a mixture of equal parts of absolute alcohol and xylol (5 to 10 minutes in each), then through xylol (10 to 15 minutes), and mounted in balsam. Metz and Gay, for paraffin-sealed smears of *Sciara* salivary gland, soak off the cover of the stained preparation in equal parts of clove oil, 95 per cent alcohol, and glacial acetic acid, pass through two changes of 95 per cent alcohol (up to 30 minutes), into absolute alcohol for 5 minutes, then clove oil for 10 minutes, followed by xylol for 5 minutes, and then mount in balsam.

3. **Protoplasmic Currents** in cells may be seen to good advantage in Rhizopods, in the plasmodia of Myxomycetes, and in the stamens of Tradescantia.

4. **Celloidin Instead of Paraffin** is used by various cytologists to avoid the bad effects of hot paraffin. For example, Danchakoff (*Zeitschrift für wissenschaftliche Mikroskopie*, XXV [1908]) starts with very thin celloidin, changes to somewhat thicker, then to still thicker (about 3 per cent), and finally, having arranged the tissues, lets the solvent evaporate very slowly throughout 4 or 5 days to a week. The mass should become opaline, homogeneous, and about as hard as vulcanized rubber. It is stored in 80 per cent alcohol. Thin sections can be cut if the mass is sufficiently hard.

5. **Urea** and other organic substances such as maltose are used very successfully in such fixing fluids as Flemming's and Bouin's by Professor C. E. McClung and his associates (see Allen's method, VI, p. 170). From 1 to 3 or more grams per 100 c.c. of fixer is the quantity used. The exact proportions for any particular tissue can be determined only by trial.

6. **Very Gradual Changing of Fluids** is recommended in the treatment of tissues to be used for cytological studies. This may be accomplished very successfully according to the drop method described by Allen in the *Anatomical Record* for July, 1916. The apparatus he designed for the purpose is shown in Fig. 42.

A 2- or 3-liter aspirator bottle ($W.B.$) is filled with water and its stop-cock opened slightly until the water begins to drop into the tightly corked pressure bottle ($P.B.$). As the air is compressed in this bottle bubbles will begin to issue from the air tube through the liquid in the small container (C) in which the tissue lies. To secure a steady stream of air through C, the end of the air tube is drawn out into almost a capillary and the rubber tube connecting it with $P.B.$ is clamped nearly shut. The purpose of the current of air is to insure quick and thorough mixing of liquids in C.

The alcohol or other replacing fluid in the supply bottle ($S.B.$) is started to dropping at the desired rate, the flow being regulated by a stop-cock or clamp. A siphon on the side of *C* removes the excess of fluid into a waste jar, so that the concentration of the liquid which is being added steadily rises.

For higher alcohols or oils the air should be dried by passing it through a tube containing calcium chloride or through sulphuric acid (see figure). Two liters of water in *W.B.* should last all night at the rate of a drop a second.

Dehydration by Dialysis may be accomplished simply by placing a piece of thick parchment paper as if for filtering in a stemless funnel. The object to be dehydrated is placed in the apex of the parchment cone in just enough water to cover it. The tip of the funnel is then placed in a jar containing 95 per cent

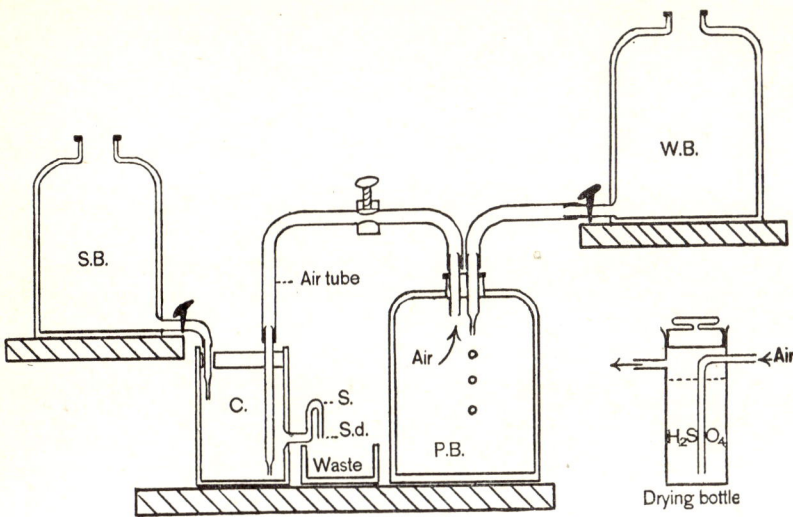

FIG. 42.—Apparatus for Gradual Change of Liquids (after Ezra Allen)

alcohol in such a way that the level of the alcohol outside the funnel is considerably higher than that of the water within. The alcohol dialyzes through the parchment until, in about 12 hours, the level of the liquid inside and outside is the same. By that time the object is in about 85 per cent alcohol and may then be transferred to 95 per cent alcohol in the usual way and finally to absolute alcohol. See also note on p. 58.

7. **The Use of Anilin Oil in Place of the Higher Alcohols** for completing dehydration is strongly advocated by Allen (*op. cit.*, p. 170) because delicate tissues are less likely to shrink in it. Since anilin oil does not mix with paraffin it must be followed by some clearing oil. Allen prefers oil of bergamot or synthetic oil of wintergreen. See also Painter's suggestion, p. 171.

8. **The Mature Testes of Young Mammals or Birds** are better for studies in spermatogenesis than those of older animals.

9. **For Dissection of Living Cells** an apparatus (Fig. 72) designed by Rob-

ert W. Chambers (manufactured by E. Leitz, Inc., American office, 60 E. Tenth St., New York City) is widely used in American laboratories. The technique of manipulation, an elaborate and exacting one, will be found discussed in detail by Chambers in McClung's *Microscopical Technique* (1929), pp. 39–73, together with its applications to the fields of bacteriology, protozoölogy, experimental embryology, and cellular anatomy and physiology. A micro-manipulator for similar purposes, designed by Janse and Péterfi (manufactured by Carl Zeiss, Inc.; New York office, 485 Fifth Ave.), is widely used in European laboratories. See further comments on p. 219.

10. **For the Estimation of Very Minute Quantities of Carbon Dioxide,** Tashiro has designed several pieces of accurate and delicate apparatus which are described in full in the *American Journal of Physiology*, XXXII, No. 2 (1913), 107–45, and the *Journal of Biological Chemistry*, XVI, No. 4 (1914), 485–94. These various types of apparatus may be obtained from the Eimer & Amend Co., of New York City.

11. **Euparal** (VIII, p. 58) is of great value in the study of certain achromatic cellular elements.

12. **Hyrax,** a new synthetic resin derived from naphthalene, is highly recommended (*Science*, July 5, 1929) by G. Dallas Hanna, of the California Academy of Sciences, as a mounting medium for such objects as diatoms where high refractive index (1.70 to 1.80) is desirable. It is soluble in xylol and is practically colorless in mounts.

13. **Gold Chloride as a Bleacher** following fixation or impregnation with osmic acid has been found valuable by H. W. Beams. He uses 0.5 to 1 per cent solution for from 12 to 24 hours.

14. **For Extremely Refractory Materials,** such as grasshopper eggs, which become hard and gritty when subjected to ordinary fixation and the usual after-treatment in the paraffin technique, Slifer and King (*Science*, LXXVIII, No. 2025 [October 20, 1933]) fix in Carnoy-Lebrun fluid, following the procedure of McNabb (*Journal of Morphology and Physiology*, XLV, No. 1 [1928]). Allowing 1 c.c. of fixing fluid for each grasshopper egg, the eggs are placed in this fluid for about 5 minutes at room temperature; the chorion of each egg is then punctured on one side with a fine steel needle; and the eggs are left for another 5 minutes. They are then transferred to weakly iodized 70 per cent alcohol for 24 hours, or until the solution (several changes) is no longer decolorized. Such eggs may then be left indefinitely in 70 per cent alcohol. After washing in the iodized alcohol, Slifer and King cut grasshopper eggs of the desired stage in halves and store the micropyle halves in 70 to 80 per cent alcohol until needed. When ready to proceed, they soak the material in 4 per cent phenol in 80 per cent alcohol for 24 hours, and dehydrate in 95 per cent alcohol. The eggs are then cleared in carbol-xylol, infiltrated with paraffin, and each one mounted with the cut end out. The paraffin is trimmed

away from the face of the block until the yolk is exposed. The whole is soaked in water from 24 to 48 hours. The eggs then cut as readily as ordinary material. The method is recommended for other similarly difficult materials.

15. **Sevringhaus' Technique for Demonstration of Golgi Apparatus,** together with separation of mitochondria from acidophilic granules (*Anatomical Record*, LIII [1932], 1), is excellent. The following (with certain shortenings of times and omissions of various intergradations of reagents which experience in our own laboratory makes us believe are unnecessary) is his method:

Fix small pieces of the tissue (e.g., anterior pituitary) for 18 to 24 hours in the following modification of Champy's mixture:

2 per cent osmic............................	1 part
3 per cent potassium dichromate..............	2 parts
1 per cent chromic acid......................	1 part

(Dr. Pearl E. Claus of our laboratory gets better results with 2 parts of chromic acid.)

Divide the material (see II), and with one part continue as in I.

I

1. Wash 30 minutes in several changes of distilled water.
2. Place in a mixture of 1 part pyroligneous acid and 2 parts 1 per cent chromic acid for 18 to 24 hours.
3. Wash 30 minutes in several changes of distilled water.
4. Mordant for 3 to 6 days in 3 per cent potassium dichromate.
5. Wash 24 hours in running water.
6. Run up through very gradual stages of alcohol to 95 per cent during a period of 6 hours.
7. Absolute alcohol, 25 minutes; absolute and cedar oil, half and half, 30 minutes; cedar oil, 12 to 24 hours; cedar oil and xylol, half and half, 1 hour; pure xylol, 1 hour; xylol-paraffin, 1 hour; 45° paraffin, 30 minutes.
8. Transfer to 58° paraffin for 30 minutes and imbed.
9. Cut 3-micron serial sections and mount. Paraffin must be kept cold for successful cutting.
10. Run sections down to water and stain for 5 minutes in 20 per cent acid fuchsin in anilin water. The slide should be gently heated to steaming two or three times, allowing the fuchsin to cool each time.
11. Wash the slide with distilled water and flood with 1 part saturated solution of picric acid in absolute alcohol and 7 parts 20 per cent alcohol. Differentiate under microscope (15 seconds to 1 minute usually sufficient).
12. Rinse in distilled water and flood the slide for 1 minute with 1 per cent solution of phosphomolybdic acid.
13. Wash thoroughly in water and add the following counter-stain (30

seconds to 1 minute usually sufficient): 7 parts 1 per cent methyl green, 5 parts 1 per cent acid violet, 5 parts 50 per cent alcohol. Rinse in distilled water and blot.

14. Pass into 95 per cent alcohol for 5 seconds and then differentiate finally with 1 part absolute alcohol and 3 parts oil of cloves. Blot and pass rapidly through absolute alcohol to xylol and mount in balsam.

If the procedure has been successful, basophilic granules will be blue; acidophilic granules, orange red; mitochondria, brilliant fuchsin; nucleoli, red; red blood cells, usually yellow orange.

II

Treat the material set aside as follows:

Wash 24 hours in running water, rinse in distilled water, and place in 2 per cent osmic acid at 40° C. for 8 to 10 hours, followed by 4 to 6 days at 35° C. Wash 24 hours in running water and proceed as in step 6. If the tissue is so greatly blackened that destaining is necessary, treat sections for 1 minute each, first with 0.125 per cent solution of potassium permanganate, and then, after rinsing, with 1 per cent oxalic acid. Wash thoroughly and proceed as in step 10.

These preparations reveal the Golgi apparatus and often give more contrasting differentiation of the mitochondria from the acidophilic granules than do the unosmicated sections.

NOTE.—In our own laboratory we find that this technique can be very materially shortened by using the dioxan method (chap. vii).

16. **The Freezing-Drying Method** of studying cell structure takes advantage of the fact that material frozen and dried by evaporation at a temperature of −20° C. retains its form. Such a technique is particularly valuable in microchemical investigations. See Gersh, *Anatomical Record*, LIII (1932), 309–37; Bensley and Gersh, *ibid.*, LVII (1933), 205–38, 369–85; LVIII (1933), 1–15.

17. **For Demonstration of Two Classes of Acidophiles in the Anterior Pituitary Body** of the female rabbit or the female cat, Dawson's azocarmine modification of Mallory's connective tissue stain (*Stain Technology*, Vol. XIII, No. 1, 1938) is outstanding when successfully applied. Ordinary acidophiles stain with the Orange G; the new type of acidophile is a deep red; chromophobes are light pink to colorless; basophiles, deep blue to light blue; and the chief cells of the tuberalis, light pink to colorless, with the colloid of the vesicles blue.

CHAPTER XIX

RECONSTRUCTION OF OBJECTS FROM SECTIONS

In investigating objects which possess complex internal cavities or complicated structures it is frequently very difficult to gain an adequate idea from the direct study of serial sections, or by means of macerated or teased preparations; consequently various methods of plastic or geometrical reconstruction from the sections are resorted to. For such reconstruction, sections must be of uniform thickness, serial, and they must possess similar orientation.

RECONSTRUCTION IN WAX

Born's method of constructing wax models of objects from serial sections is widely used for both embryological and anatomical subjects. The thickness of the sections, the magnification of the microscope, and the plane of section must be known.

Wax plates are prepared as many times thicker than the actual sections as the latter will be magnified in diameters. For example, if the serial sections are $\frac{1}{30}$ of a millimeter thick ($33\frac{1}{3}$ microns), and they are to be magnified 60 diameters, then the wax plates must be made 60 times as thick as the sections, or 2 mm. thick. This is a thickness commonly used. Count the number of sections to be reconstructed, and prepare an equal number of plates.

Preparation of the Wax Plates

a) The hot-water method.—1. Prepare the wax according to the following formula:

Beeswax	6 parts
Paraffin (melting-point 56° C.)	4 parts
White lump (not powdered) rosin	2 parts

Melt together and thoroughly mix.

2. To prepare plates of the proper thickness (2 mm.), use shallow straight-walled rectangular tin pans which will afford a water surface of 3×4 feet; 2,000 grams of the prepared wax poured on very hot water in such a pan will give a plate 2 mm. thick. Air bubbles which form in the wax may be driven off before it cools by playing a Bunsen flame over the surface. The wax should spread evenly over the surface of the water if both wax and water are sufficiently hot. If gaps remain, close them by drawing a glass slide over the surface of

the wax. To prevent the plate from splitting while cooling, after it has stiffened somewhat cut the edges free from the walls of the pan. When the water has become tepid, remove the wax plate to a flat support and leave it to harden.

b) The wax-plate machine method.—Several instruments have been devised for making the plates more rapidly and more accurately than by the hot-water method. Huber's apparatus, for instance, consists of a heavy-cast-iron plate with movable side pieces which can be adjusted to a height corresponding to the desired thickness of the wax plates. The whole instrument is supported upon three adjustable legs, by means of which it can be made exactly level. Melted wax slightly in excess of the quantity necessary for a wax plate is poured on to the iron plate in an even layer, and rolled out with a hot roller until the roller comes to run directly on the side pieces of the instrument. When the wax plate is cool enough to handle, it may be placed in a pan of cold water to harden.

Practical Exercise.—When possible an outline drawing of the part to be reconstructed should be made before it is sectioned.

1. Reconstruct the heart of a chick at the end of the third day of incubation, under a magnification of 60 diameters. For this magnification, if it is desired to use a wax plate 2 mm. thick, the original sections should have been 33.3 microns thick.

2. Place a sheet of blue tracing-paper on the wax plate with the colored side toward it. Over the tracing-paper place a sheet of ordinary drawing-paper. With the aid of a camera lucida or other projection apparatus, outline on the drawing-paper the part to be reconstructed. In doing this the outline is also traced in blue on the wax. Number each drawing, and also indicate the number of the section on the slide to which it corresponds; also number the wax plates with reference to the drawings.

3. Lay the wax plate on a suitable flat surface, and cut out the outlined parts with a sharp, narrow-bladed knife. Leave bridges of wax to hold in place the parts that would otherwise be separate pieces. Pile up the successive sections in proper sequence as they are cut out.

4. In finally putting the model together, accurately adjust the parts (for reconstruction points see memorandum 13, p. 151, and II, *a*, p. 145), and build up the model in blocks of five sections each (Bardeen's suggestion). If necessary, unite the essential parts by means of pins or fine nails. Remove all temporary wax bridges (see 3) by means of a hot knife.

When all blocks are properly adjusted and united, smooth over the surface by means of a hot spatula.

MEMORANDA

1. **Geometrical Reconstructions,** first described by Professor His, are often all that is necessary to give one the desired information about internal organs. Before sectioning, an outline drawing of the object is made in a plane at right angles to the intended plane of section, and under the same magnification that will be used for the reconstructed drawing. For example, if the sections are to be transverse, the outline drawing of the object would be a profile view from the side. After sectioning the object, each section is drawn under the same magnification as was used for the outline drawing.

To reconstruct any special part of the object, draw a median line on the outline drawing corresponding to the long axis of the object. At right angles to this line draw a series of equidistant parallel lines corresponding in positions to the sections that have been made. For example, if the magnification is 100 diameters and the sections 10 microns thick, then the parallel lines must be 1 mm. apart. Then, beginning with the first section, indicate by dots in the proper plane in the profile drawing the relative distances of the part in the sections above or below the median line along the proper one of the parallel lines. All of the sections having thus been plotted, connect the dots of corresponding parts in the successive zones. It is frequently sufficient to reconstruct only every fifth or even every tenth section. When the plane of section is not quite at right angles to the axis of the object, an equal alteration of angle must be made between the median line of the outline drawing and the parallel lines.

Such a reconstruction as that above would give lateral views of the various internal parts. To get their aspects as seen from above or below, the original outline drawing of the specimen as a whole should have been made from this point of view instead of from the side. In actual work one should make reconstructions in both planes.

For a modification of Weber's method of graphic reconstruction, see Scammon, *Anatomical Record*, IX, No. 3 (March, 1915). For suggestions on profile reconstructions see Streeter, *American Journal of Anatomy*, IV, No. 1 (1904), 86.

2. **A Special Drawing-Table** for rapid and convenient drawing of sections for reconstruction has been devised by Bardeen. For details, see *Johns Hopkins Bulletin*, XII, 148.

3. **Sheets of Blotting Paper** instead of wax are recommended by Mrs. Gage (*Anatomical Record*, I, No. 7 [November, 1907]). Models are finished by coating with paraffin. The advantages claimed for this method are lightness, durability, and ease and cleanliness of production. The method is also given in Gage, *The Microscope* (14th ed.), pp. 413–16.

4. **Plates with the Paper of the Drawing Rolled into the Wax,** following directions in Karl Peter, *Methoden der Rekonstruktion* (Gustav Fischer, Jena), have been found very satisfactory by Rice. A thin drawing-paper, smooth on the drawing side, porous on the other, is used and drawings are duplicated by

means of a carbon sheet. One copy is kept for reference, the other is pressed into the plate. To accomplish the latter, the slab on which the wax is rolled out is smeared thoroughly with turpentine and the drawing laid face down on it. The melted wax is then poured upon the back of the drawing and, when beginning to harden, is covered with tissue paper, which is then rolled into it. The two papers thus present good surfaces for building up the plates. The thicker paper with the drawing on it gives a fixed contour which is a helpful guide when it comes to smoothing down the model.

Twisted wires in short lengths are used for supports. The ends are spread to afford anchorage where they lie between the plates. Two longer wires should be twisted together, then cut into proper lengths.

5. **Photography of Sections upon Large Plates** has been resorted to by Warren H. Lewis (*Anatomical Record*, IX, No. 9 [September, 1915], 719–29) as a substitute for the laborious and time-consuming method of drawing each section on paper. He uses line bromide or azo G hard (matte) prints. He considers the photographs far superior to drawings and maintains that, when time is taken into account, the method is less expensive than the old method of tracing. Lewis' paper is full of valuable suggestions and should be read by everyone who contemplates doing much work in reconstruction. The chief points emphasized are: the use of photographs; the use of series of guide-lines which coincide with planes that are at right angles to each other and perpendicular to the plane of the sections; and the use of plaster-of-Paris casts.

6. **For Rapidly Cutting Out the Wax Plates,** Chester H. Heuser, of the Wistar Institute of Anatomy and Biology, has devised a series of metal styli of varied design which when electrically heated are handled much as one would handle a pen in writing. The cutting instrument proper (copper, iron, or brass wire, pointed or flattened at one end according to need) is coated, except for about 15 mm. at the working end, with a thin layer of an insulating asbestos paste. A piece of No. 32 German silver wire about 20 cm. long is then wrapped around the coated surface, with the coils kept well insulated from each other, and finally covered with the paste. The ends of the German silver wire are attached to small copper wires which run to a lamp-board bearing several incandescent bulbs with sockets connected in parallel. The lamp-board serves as a rheostat, so that any desired temperature can be obtained in the stylus by altering the number of lights and thus regulating the current which passes through the German silver wire. Styli of different kinds of metal attached in common to the same lamp-board afford different temperatures for cutting points and smoothers. Heuser contemplates inserting small rheostats in the system to regulate the temperatures of the individual instruments more accurately.

Professor Mark (*Proceedings of the American Academy of Arts and Sciences*, XLII, No. 23 [1907]) uses an electrically heated wire moved rapidly by a modified sewing machine for cutting out the wax plate.

CHAPTER XX
DRAWING

I should make it absolutely necessary for everybody, for a longer or shorter period, to learn to draw. It gives you the means of training the young in attention and accuracy, which are the two things in which all mankind are more deficient than in any other mental quality whatever.—*Huxley*.

Drawing is an important part of the work in most biological sciences. The essential phases of a subject can be condensed into a few pages if the drawings accurately represent the dissections or microscopical preparations studied. The following simple directions are written mainly to aid the student in preparing his notebook, but it is hoped that they may also be useful to individuals preparing manuscripts for publication.

Materials for Class Work.—All the materials needed for ordinary class work can be selected from the list here given with the approximate price attached.

Pencils—one 4H, one HB, one 2B, 10 cents each.
Pens—Gillott's lithographic penpoint, No. 290, 5 cents each.
Ink—water-proof India ink made by Charles Higgins & Co., retails at 25 cents per bottle.
Ruler—celluloid, 10 cents.
Ruby eraser—10 cents.
Crayon pencils—red, blue, and yellow, 5 cents each.
Loose-leaf notebook with bond paper, 40 cents.

The pencils may be of any standard make; the 4H is a hard pencil for line work; the HB, a medium pencil, is useful in placing outlines and shading. The 2B, for black shading, should not be used unless the drawings are afterward fixed to prevent rubbing. The ruler graded in centimeters on one edge and inches on the other is indispensable. The eraser is for erasing pencil lines; for erasing ink a sharp knife is best. A larger assortment of colored pencils will often prove useful, but the three primary colors will answer most purposes. Two-ply bristol board at 2 cents per sheet may be used in the notebook instead of the bond paper. Where classes are large, bookstores will make up bound notebooks for 35 cents each, containing a good grade of paper upon which drawings can be made. Where pencil drawings only are required, pens and ink may be omitted from this list. On account of the small amount of locker or drawer space usually available for one student, an elaborate drawing outfit should be avoided. The excellence of student drawings is judged by the exactness with which the preparations are depicted.

I. METHODS OF REPRESENTATION

Outline.—In beginning a drawing, the field which the picture will occupy should be marked off with dots. The placing of two faint lines which cross at right angles in the middle of the drawing-field is a great help to beginners, especially in drawing bilaterally symmetrical objects. Next determine the relation of the length of the object to the breadth; then calculate the size of the drawing. If the object is large, a reduction will be necessary; if small, it can be represented better 5 or 10 times its original size. The important points can be indicated in the drawing-field by dots which, when connected by light lines, roughly block in the object in correct proportion and size. This crude picture may then be worked over until angles are removed and a neat outline results. For the preliminary mapping of the object an HB pencil is best, as the lines are easily erased. The outline when finished must consist of a continuous line of uniform thickness with no overlapping edges where the pencil has been removed from the paper and put down again. Outline is the most important part of the drawing, for "a good outline may redeem bad finish, but no amount of excellence in finish can save a picture that has been incorrectly outlined." What details to include depends upon the purposes of the drawing. The principal points of an anatomical drawing stand out more clearly when they are not obscured by unnecessary details. In histological drawings, details are essential, but they should still be kept subordinate to the general effect of the picture.

Depth.—Usually the third dimension, depth, is not considered in drawings made from sections. When drawings of reconstructions or whole mounts are made, however, all three dimensions must be indicated in the drawing. Likewise, drawings of such things as digestive canal, nerve cord, heart, lungs, and kidneys stand out better when depth is represented. This can usually be done by indicating degrees of light and shade.

Ink Drawings.—In drawings which are to be inked, the outline should be carefully drawn in pencil and as many corrections as possible made before ink is applied. Place the ink upon the pen by means of the quill attached to the cork of the ink-bottle. If the original outline is even, the inking can be done readily; a line uniform in thickness results from the applications of firm, steady pressure. The pen will give a ragged line if held so that one nib bears more heavily upon the paper than does the other, or if it becomes sticky with dried ink. A smoother line will be obtained if the penholder is held at a wide angle to the paper and only the very point of the pen is allowed to touch.

Shading.—Where differentiation of parts is desired, shading may be used. This can be done either by stippling or by lines. In making the dots in a stippled drawing, the pen must be grasped firmly and only the point placed upon the paper. If the pen strikes the paper at an acute angle, three-sided instead of round dots result. The dots must all be of the same size. To indicate degrees of shade, vary the number of dots, not their size. A heavy shading can be accomplished by placing the dots close together, whereas dots farther apart give the impression of light shading. Lines can be used with good effect upon large drawings. Let the lines, placed an even distance apart, follow the shape of the shadows. To indicate heavy shadows, lines in an opposite direction can be placed across the first set. Be careful not to cross-hatch until the first lines have dried, otherwise blots will occur.

Pencil Drawings.—When correctly executed, pencil drawings are more artistic and permit of more subtle differentiation in detail. Minute points can be shown by the use of stippling. In stippling with a pencil, follow the same procedure as in the use of a pen. The pencil-point should be sharp and rounded on all sides. In large drawings, lines can be evenly placed to outline shadows, but they are not as effective as shadows put in by blending graphite, obtained by rubbing the pencil over the paper. For the latter method a stub is necessary. This can be made from a strip of paper 1 inch wide and 5 inches long in the following manner: Begin to roll the paper at one end and let each turn overlap the preceding turn slightly, until an elongated coil results. The pointed end of this can be used in spreading graphite evenly over a surface. The graphite is placed on the part of the drawing where the darkest shadows occur with an HB or 2B pencil, and is worked over with the end of the stub until the sharp edges of the shadow gradually grade out into the lighter parts. The 2B pencil ordinarily should not be used as the source of the graphite, as shadows can be darkened by repeating the application of the HB pencil.

Shadows.—In most pictures of biological subjects, one cannot stand off and observe where the light falls upon the object and what part is in shadow. For that reason a knowledge of where the shadows occur if an object is illuminated from any one direction is necessary. To gain such knowledge from a description is impossible; it is therefore advisable for students wishing to shade their drawings to consult an artist who can give usable information in the form of demonstrations. A careful study of textbook illustrations will aid materially. Shading requires practice, and even then it may not be successful. In most cases where not imperative it had better be left out entirely.

Fixing pencil drawings.—Where pencil drawings are made with soft pencils which are liable to rub, they must be fixed. This is done with a fixing solution and a special atomizer which can be bought at any art store. To prepare the fixative, make a saturated solution of white shellac in alcohol. Allow this to stand for a day or so; dilute one-half; then filter off the liquid. To prevent evaporation when not in use, this must be kept in a tightly stoppered bottle.

The drawing should be placed in an upright position, about 2 feet from the spray. In order to avoid a glossy surface spray lightly.

Wash-Drawings.—After a faint outline of the section or object has been made with a hard pencil, fasten the paper to a board with thumb tacks, and with a large brush dampen the entire surface, removing the surplus water with the brush or a blotter. Mix the wash as follows: With a wet brush remove some pigment from a cake of Winsor & Newton's Charcoal Grey or Ivory Black, and put in the water in the mixing-pan. Repeat this process until the wash is slightly darker than the desired background tint. Next put a wash of this over the entire background; it will dry lighter. Allow the paper to dry partly before darker tones are put in. Where a very dark portion is confined to a small area, the paper should be quite dry, otherwise the wash will run into the surrounding part of the drawing. If details are to be put in by stippling or linework, make a wash (from the same cake) the color and consistency of ink. This can be applied with a pen or brush after the paper dries.

Some artists use a dry paper which, however, requires more skill in applying the wash. The darkest tones can be applied first, gradually working up to the lightest. After experimenting, use that which gives the best results for the purpose in hand.

Where several wash-drawings are to be made with the same tones it will be found simpler to put in all the backgrounds first. Mix up plenty of wash for this purpose, as it is not easy to duplicate the exact shade at another mixing. Wash allowed to stand becomes darker upon evaporation of the water; hence, if after the backgrounds are put in the work must be deferred until later, the same wash will do for darker tints. Do not redampen the whole surface, as that will lighten the background. The darker tones can be blended into the background with a clean wet brush.

MEMORANDA

1. **Cleanliness.**—Even where the drawings are correctly made as to size and proportion, the notebook will not present a good appearance if the pages contain finger-marks and blots to mar their whiteness. With sufficient diligence finger-marks can be erased, but the best way is not to make them in the

A B C D E F G H I J K L M N O P Q R S T U V W X Y Z

a b c d e f g h i j k l m n o p q r s t u v w x y z

0 1 2 3 4 5 6 7 8 9

FIG. 43.—Simplified Gothic or "Shop Skeleton" Letters and Figures Used in Labeling Drawings

beginning. Where laboratory work requires dissection, rough sketches can be made upon scrap paper and later copied into the notebook. Blots can be avoided if care is used in placing the ink upon the pen; a quill is attached to the stopper of the ink-bottle for this purpose.

2. **Size and Arrangement of Drawings.**—Uniformity in size and arrangement of drawings should be preserved wherever possible. For example, in the development of the frog's egg, no increase in the size of the egg takes place from the single-cell stage to the end of gastrulation, hence all drawings representing this series of development must be the same size. The neural tube, largest in circumference in the head, decreases gradually toward the posterior end of the body, and must be drawn correctly in cross-sections, otherwise one will have an erroneous idea of its structure. The drawings must not be crowded upon the page. Exact margins and spaces equidistant between drawings, though not artistic, give the impression of neatness desirable in scientific work.

Labeling.—The results will not equal expectation if the labeling is poorly done. Too few students consider this item in making a notebook. The pencil-point must be rubbed upon fine sandpaper until it is smooth and conical. The beginner should draw 3 parallel lines, 2 mm. apart, upon which to place the letters. The letters may be placed close together or farther apart, but in either case the space between the letters must be kept the same. Larger spaces are left between words. Letters may be straight or slanted to suit the individual taste. If one has never done any labeling, time should be taken to practice upon a piece of paper before finishing a drawing. Many letters like *b*, *d*, and *g* contain an *o* combined with a straight line, hence it is necessary to learn to make an *o* properly. In making *b*, *d*, *p*, and *g*, be careful that the up or down stroke is straight and joins the curve smoothly. The *b*, *d*, *f*, *h*, *k*, and *l*, extend above the other letters to the height of capitals, while *g*, *j*, *p*, *q*, and *y* extend just as far below the line. The *t* falls between the stem letters and the short letters. The Gothic style given in Fig. 43 is easy to learn and will answer all purposes.

II. PREFERABLE MODES OF REPRESENTATION FOR SPECIAL COURSES

General.—Methods of representation vary as the subject-matter in each instance requires different handling. The modes here described have been found practical in different biological courses; however, each may be altered to suit individual requirements. In elementary courses in general zoölogy unshaded ink-drawings are best. They are more accurately executed by elementary students because defects are so obvious that they do not pass unnoticed; moreover, the student exercises more care in making the drawing because of the greater difficulty of changing it after it is once drawn. Such drawings may be made more or less diagrammatic, depending upon the ability of the student and his previous training. *Students with no former experience in representing upon paper what they see need not be discouraged, for clear-cut outline drawings can be made by anyone, if due consideration is given to the points enumerated under the foregoing paragraphs.*

Embryology.—The first step in embryological drawing is a careful outline with a 4H pencil. As an aid in drawing complex sections—a cross-section of an old tadpole, for example—a cover-glass ruled into squares can be fastened into the ocular. The drawing-paper is ruled into the same number of squares as the cover-glass; each square as many times the size of one square of the cover-glass as the intended magnification. Parts of the object under the squares in the ocular can be located in corresponding squares upon the paper (Isaacs, *Anatomical Record*, IX, 711–13). In such drawings cells are not usually indicated. It is a decided advantage to have the parts colored, especially if organs from the same germ layer are colored alike. If three colors are chosen, one for each of the germ layers, ectoderm (blue), endoderm (yellow), and mesoderm (red), and if these are consistently used throughout the sections, one can see at a glance the development of the organs. Crayon pencils work up rapidly and give good results if the color is put on lightly, so that in the finished drawing the colors blend. To spread crayon evenly, use a blunt, rounded point and make long strokes with the side of the crayon. For example, in a cross-section of a frog embryo, the neural tube with its optic cups is ectodermal in origin and the blue color immediately indicates its relationship to the ectoderm which is similarly colored; the mesoblastic somites and mesenchyme are red; the lining of the archenteron yellow. In older embryos where many parts develop from the mesoderm, other colors may be used for special parts—as brown for kidneys. The shade of red may likewise be varied, a dark red being used for blood vessels, obtained by putting on a heavy layer of crayon; while a light red may be used for mesenchyme. By this method a student has constantly before him the layers from which various organs develop, while the instructor can immediately see that the student has or has not a clear conception of the manner in which the organism is built up. Water colors can be used instead of crayons, but in the hands of most students they are less satisfactory. Where cellular structure is put in, colored inks may be employed. However, drawing the individual cells in an embryo is too time-consuming a process for ordinary class work in embryology.

Histology.—Histological drawings are best executed in pencil. Details must be shown. In general, these, especially nuclear differences can be put in by stippling. In intercellular matrices, such as connective tissue, the texture of the tissue should be represented by irregular lines. Light lines give the effect of fibers and fibrils very well. Where different tissues of an entire organ are to be distinguished, the whole drawing may be covered with a light ground-substance of blended graphite and the details worked up with pen and ink. This combination is effective and has the added advantage of quick execution.

Cytology.—Cytology requires even more detail than histology. Stippling or wash is used chiefly. A different arrangement of the dots gives the granular, alveolar, or reticular appearance of cytoplasm. Solid lines should be avoided as much as possible; fibers can be represented by dots placed close together in a linear row. Chromosomes can be stippled, made solid, or blended with graphite over their entire area. Where the cell cytoplasm is homogeneous, a light ground-coat of graphite can be placed over the entire cell and the cell parts stippled upon this with pen or pencil. A wash can be substituted for the graphite; but as it requires more careful application it is not recommended for class work.

Crayons used judiciously in both histological and cytological drawings are often effective. Secretory granules, where present, can be put in with color if the crayons are sharpened to a fine point. Ground-substance of tissues can be depicted with crayon and the characteristic features of the tissue added with pen or pencil. Granules of white blood corpuscles have definite color reactions upon the basis of which they are classified; such granules should therefore be colored in a drawing.

III. DRAWINGS FOR PUBLICATION

To make illustrations for publication in a book or scientific journal, one must not only understand form, color, perspective, and composition, but also know something about the technique of reproducing drawings in printed form. Manuscripts which contain drawings that can be inexpensively reproduced are more readily accepted for publication than those which require expensive plates. Before undertaking a series of drawings for publication make a careful study of similar work in standard journals, particularly in the journal in which you expect to publish, and if possible consult a capable printer.

Materials for Manuscript Drawings.—In general, drawings for publication should be made in black and white, because this style can be reproduced most inexpensively. For working with ink, a waterproof India ink, such as Higgins', is best. This can be applied with pen or brush. Gillott's penpoints are satisfactory. Inasmuch as each person not only handles a pen differently, but uses different degrees of pressure in working, he must determine by experience the number of the pen best suited to his needs. For fine linework and stippling, the writer has found that Gillott's lithographic pen No. 290 gives the best results. Fine red-sable brushes may also be used for this work, although for line-process reproduction (p. 191) the pen drawing is likely to prove more successful.

Drawings should be made upon a good quality of paper. Bristol board, either 2- or 4-ply, or Whatman's hot-pressed (smooth) water-color paper, either of which can be obtained at any store carrying a complete line of stationery materials, can be used. Whatman's paper is of the same texture throughout; moreover, it can be used for ink, wash, or pencil-work. Anvil drawing-paper, No. 105, is excellent for large illustrations or charts. This is a cloth-backed paper which can be rolled without injury. A disadvantage is the large size and expense of a roll, which contains 10 yards of paper 36 inches wide. It is put out by the Keuffel & Esser Co., of New York. Cloth-backed papers can be obtained from other firms. A stipple-board (Ross board), manufactured by the Charles J. Ross Co., consists of a chalk surface upon a paper back. The advantage of this paper is the rapidity with which drawings can be made. A stipple effect is obtained simply by rubbing the flat side of a pencil-point back and forth over the chalk surface. Different stipple effects are obtained by using different grades of the paper. For all general uses No. 8 is best.

Water colors may be procured in a variety of forms and makes for wash-drawings. On account of the cost of reproduction, however, in papers which are to be published colored drawings should be avoided wherever possible. Winsor & Newton's Ivory Black or Charcoal Grey, which come in cakes, are best for black-and-white wash-drawings. Never use blue in a drawing to be photographed, as it does not take well. Remember also that yellow and brown appear as black. Likewise keep in mind that the results are in no appreciable degree dependent upon the number or kind of tools used, but upon the skill shown in execution.

Camera Lucida.—As an aid in making drawings of microscopic objects, an instrument known as the camera lucida is often employed. With such an instrument the image of an object under the microscope can be projected upon the drawing-paper (see p. 212)

Reduction of Drawings.—In making drawings for publication, it is advisable to make them larger than they will appear in the finished cut, as in the reduction many irregularities are lessened. But under no circumstances make a crude drawing with the idea that in the print it will appear perfect, for while reduction minimizes, it does not obliterate defects. The original drawing should not ordinarily be more than twice the size of the intended cut, while a reduction of one-fourth or one-third will probably give a better result. If one is not careful about the spacing of lines and dots used in the drawing, reduction tends to make them run together.

Line-Process.—Line-process is not only the least expensive form of reproduction, but likewise the most accurate. Prints reproduced in this manner contain contrast not gained by any other method. The drawing is photographed directly upon a zinc plate, which is treated so as to make the lines and dots of the design impervious to acid. The plate is now placed in a bath of acid which eats away the unprotected portions. The metal plate, containing the design in relief, mounted type high, is set up and printed with the type upon text-paper if the details are not too fine and if no colors are required. Otherwise it must be printed on special paper and inserted as a "plate."

Lines.—For this type of reproduction use ink as the working medium and apply it with pen upon smooth white bristol board or water-color paper. Lines should not be extremely fine, and care should be taken that they are far enough apart to reproduce well. Likewise, cross-hatching, if employed, must be coarse enough to stand the reduction, otherwise it will appear as a solid black mass. While ink thinned to a gray gives a difference in tone in the original drawing, it must be borne in mind that this behaves erratically in reproduction. Gray lines often appear in the print as broken black lines, or they may be entirely lost. To avoid graying black lines the pen must be wiped frequently and filled with a fresh supply of ink. To get effects in tone vary either the thickness of the lines or the distance between the lines. Lines in the foreground should be farther apart and heavier than those in the background.

Dots.—If drawings are stippled, the dots should be made with the amount of reduction in mind. Too fine stippling necessitates etching upon copper, a tedious process, which doubles the cost of production. To secure degrees of light and shade, vary the distance between the dots, not the size of the dots. Lithographic crayon used upon stipple-board (see p. 191) will reproduce by line-process if the drawing is coarse enough to stand a reduction to one-half its size.

Graphs.—Graphs for line reproduction should be made upon co-ordinate paper in which the lines are in light blue. As light blue does not photograph, the co-ordinates, both perpendicular and horizontal, which are to appear in the cut must be inked in black.

Color.—A separate block and printing are required for each color used in line engraving. In a drawing in which several colors are to appear, the original drawing in black and white must be accompanied by a colored sketch. From a process-plate of the black-and-white drawing a print is taken. For each color used in the sketch, the printer then makes a separate block, which necessitates a separate run on the press for each color. Avoid fine gradations in color, as thereby the expense of reproduction is increased.

Half-Tone.—Shaded and wash-drawings and photographs are

ordinarily reproduced by the half-tone process. A screen containing 133 to 175 meshes per square inch is placed between the camera and the drawing when the latter is photographed. This is done to break up the flat tones into dots, otherwise they would print as black masses. The photograph thus made is transferred to a zinc or copper plate and etched in the same way as a line-process plate. Where the details of the picture are not too delicate, the print can be made upon text-paper. For proper reproduction, drawings which require a fine screen must be carefully printed upon special coated paper.

It must be remembered that the screen on account of its fine dots introduces shadow into the lighter parts of the picture; while the white spaces between the dots render the darker areas lighter. These changes demand stronger contrasts in the original drawings and photographs which are to be reproduced by half-tone than those desired in the print.

Wash-drawings.—Wash-drawings (p. 186) for reproduction should not contain solid black lines, as the screen breaks up the lines into dots. Be careful that no pencil lines remain in the drawing and avoid smudges and finger-marks. If clear white spaces are desired in the print, the shadows produced by the screen in these places must be tooled out of the plate to give the required effect—a process which is slow and expensive.

Combination with wash.—Drawings can be made in which certain parts are put in with pen or brush lines upon a wash-background. This makes a satisfactory combination for half-tone prints. On the other hand, fine pencil striae upon a wash-background are usually lost in reproduction, since they offer insufficient contrast with the background.

Pencil.—Pencil-work does not reproduce satisfactorily by half-tone. Where blended graphite is used for a background instead of wash, a reproduction can be made if the print is made upon good coated paper. However, the results are so inferior to the originals that it pays to take the time to learn how to make good wash-drawings.

Color.—Color can be applied upon half-tone prints, although a separate block and printing are necessary for each color. Plates are sometimes made as described under "Line-Process," and these printed in color upon a half-tone background (see p. 191).

Photographs.—Photographs are usually reproduced by half-tones. A hard-finish, glossy paper which brings out strong contrast between the blacks and whites of the picture is best for prints. Azo hard X, glossy white Velox paper made by the Eastman Kodak Co., and solio paper of a brownish tinge reproduced well. Photographs will stand some reduction, but on the whole it is better to have prints the same size as the intended cut. Prints should be squeegeed. This is done with a roller which, passed over the wet print, removes the moisture and gives a hard finish to the gelatin film when it dries.

Lithography.—Lithography is the most faithful but also the most expensive form of reproduction. Most journals will not accept drawings which have to be lithographed unless the author or artist pays the extra cost of reproduction. Intricate drawings of many colors which cannot be reproduced by line-process or half-tone come out well by this method, if cost is not an item to be considered. Likewise stippled pencil drawings make excellent illustrations with this sort of reproduction.

Arrangement of Drawings for Reductions.—Line-drawings where not used as text-figures should be arranged in the form of plates. Half-tones are usually so arranged, since they generally require a special coated paper. To arrange drawings in the form of a plate, one must know the exact amount they are to be reduced when printed. If one-half, for instance, they must be arranged as a plate twice as high and twice as wide as the journal page. The drawings can either be made directly upon bristol board in the order in which they are to be printed, or they may be pasted upon the board in such order. The latter method is usually practiced, but it should be noted that in the case of half-tones the edges of the pasted parts are reproduced in the cut. Due allowance must be made for lettering and the margins of the page.

Lettering.—All the original drawings should be so lettered that the letters will be of the proper size when reduced. Letters can either be pasted on or printed by hand (p. 187). A drawing presents a neater appearance if the lettering is parallel to the base line. A cut is more legible if, instead of abbreviations, the names printed in full are connected to the proper part of the drawing by leaders. Gummed sheets containing letters, numerals, and such words as "Plate" and "Fig." in several sizes can be bought for use in this work. Likewise publishers of some journals will print letters and words which can be pasted on.

For a good chapter on all kinds of drawing connected with microscopy see Gage, *The Microscope* (14th ed., 1927), pp. 160-205.

A Manual of Drawing for Science Students by Justus F. Mueller (New York: Farrar & Rinehart, 1935) is a valuable small guide for biological students.

APPENDIXES

APPENDIX A

THE MICROSCOPE AND ITS OPTICAL PRINCIPLES

For an understanding of the optical principles involved in microscopy, four things must be borne in mind with regard to a ray of ordinary daylight:

1. It has an appreciable breadth.
2. It travels in a straight line in a homogeneous medium.
3. It is bent (*refracted*) in passing obliquely from one medium into another of different density.
4. It is in reality a composite of a number of different colored rays, ranging from violet to red, and each of these has a different refrangibility.

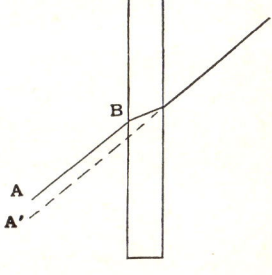

Fig. 44

The amount of refraction undergone by light in a given case depends upon the difference in density of the two media which the light traverses. Thus, glass is denser than air; hence, in passing from air obliquely through a glass plate (Fig. 44), a ray of light AB would be bent out of its original course. On reaching the air again, however, it would resume its original direction, although it would be displaced to an amount equal to the distance between A and A'. It is on account of such displacement that an object in water, for example, appears to be at a different point from where it really is.

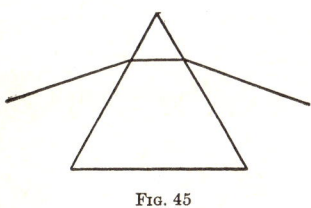

Fig. 45

On the other hand, after traversing a prism, a ray does not resume its former direction, but takes a new course upon leaving as well as upon entering the prism (Fig. 45). This new direction is always toward the base of the prism, and the amount of deviation depends upon the shape and density of the prism. If the base is down, then the ray is bent downward; if the apex is down, the ray still deviates toward the base, that is, it is bent upward.

Lenses.—Each of the two principal forms of lenses is in effect practically two prisms, (1) with the bases placed together (Fig. 46, *a*, *convex lens*), or (2) with the apexes together (Fig. 46, *b*, *concave lens*).

In the convex lens, since rays of light are refracted toward the bases of the respective prisms, they will converge; in the concave lens, for the same reason, they will diverge. The terms *converging lens* and *diverging lens*, therefore, are used frequently as synonymous with the terms "convex lens" and "concave lens." All lenses are modifications or combinations of these two types.

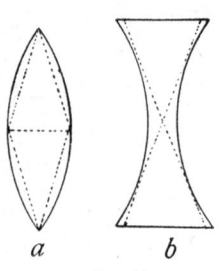

Fig. 46

If parallel rays of light pass through a convex lens (Fig. 47) they are so refracted as to meet in one point F, which is termed, in consequence, the focal point or principal focus. If, on the other hand, the source of light be placed at the focal point, then, after traversing the lens, the rays of light will emerge parallel. If parallel rays of light came from the opposite side of the lens, manifestly there would be a second focal point at F'. The two principal foci are termed *conjugate foci*, and will be equidistant from the center of the lens when both sides of the lens have equal curvature.

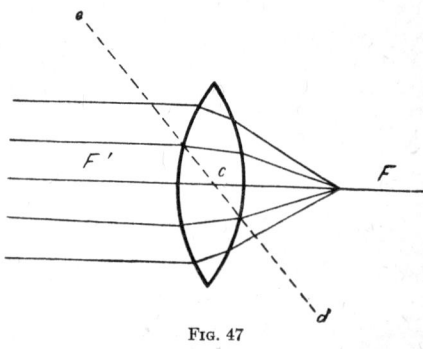

Fig. 47

The ray which passes through the center of the lens (Fig. 47, *c*) and the focal point traverses what is termed the *principal axis* of the lens. The *optical center* of the lens is a point on the principal axis through which rays pass without angular deviation. It may be within or outside the lens, depending upon the form of the latter. Any line (*ed*), other than the principal axis, which passes through the optical center of the lens is termed a *secondary axis*.

In the case of a concave lens, parallel rays will be caused to diverge (Fig. 48) and the principal focus, F, of the lens is determined by the extension of the divergent rays till they meet at a point which lies on the same side of the lens as the source of light. Such a point has no actual existence, and is known, consequently, as a *virtual focus*. The focus of a convex lens, on the other hand, is *real*, and may be determined readily by allowing the sun's rays, which are practically parallel, to pass through it on to a screen. By moving the lens backward and forward, the spot of projected light varies in size and brightness. When smallest and brightest the spot is at the focal point of the lens.

Images.—In Figure 49 the object, represented by an arrow, lies beyond the principal focus of a convex lens as in a photographic camera, for example, or the objective of a compound microscope. Light rays pass out in all directions from any luminous point. Hence, one

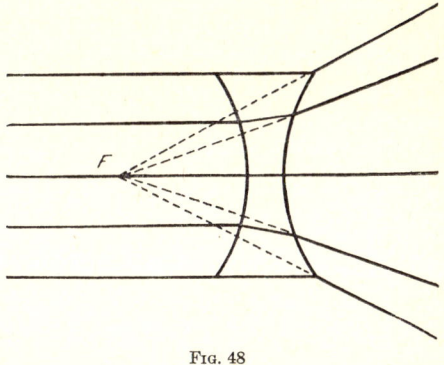

Fig. 48

ray from any point on the arrow, the tip, for instance, will pass through the focal point, *F*, and one will pass through the optical center of the lens. From what was determined above, manifestly the ray through *F* will emerge as one of the parallel rays upon leaving the lens, and the one through the optical center of the lens, since it traverses a secondary

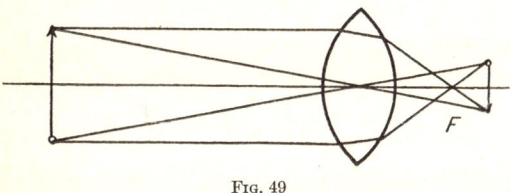

Fig. 49

axis, will not be refracted, hence the two rays must cross. Their point of intersection is the point at which the image of the arrow-tip will be formed. The same fact may be determined, likewise, for any other point of the arrow, for example, the opposite end. Thus the distance from the lens at which the image is formed may readily be determined. In focusing a photographic camera, for example, the image comes sharply into view on the ground-glass plate at the back of the camera when the plate is brought into the plane in which these rays through the focus and the optical center intersect beyond the lens. It will be observed from the figure that the image is reversed. The size of the image diminishes as the object lies farther beyond *F*.

In case the object lies between the lens and the principal focus, as in Figure 50, parallel rays from the object would converge to meet at the conjugate focus F', and an eye at this point would see the image projected and enlarged without being reversed. The plane in which the image is formed is determined by finding the points of intersection of the secondary axes through points of the object with the imaginary elongation of the refracted rays as shown in the figure. The image is magnified because the observer judges of the size of an object by the visual angle which it subtends. The greater the convexity of the lens the shorter the focus, and also, since the rays are bent more, the greater the magnification.

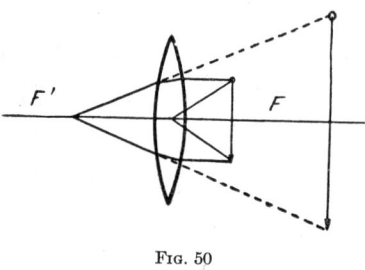

FIG. 50

The Simple Microscope.—The simple microscope (the ordinary so-called magnifier, etc.) operates upon this principle; the image of an object is projected and enlarged but not inverted (Fig. 50).

The question arises as to why there is a *best distance* to hold the simple microscope from an object. Why will not any point answer so long as it is within the focal point? As a matter of fact, the object may be placed at any point within the focus, and it will be found that the nearer it is brought to the lens the less it is magnified. There is one most favorable point for observation, however, which is neither at the point of highest nor of lowest magnification, but an intermediate point, where the lens is freest from chromatic and spherical aberrations.

In reality the eye forms an integral part of the optical arrangement when the microscope is being used, but in our elementary exposition of the subject it is disregarded.

The Compound Microscope.—The general principle of the compound microscope is represented in Figures 51 and 58. The object ab (Fig. 51) lies beyond the principal focus of the first lens or *objective* (really a system of lenses), hence the image AB is reversed. This image, in turn, is viewed through a lens, the *eyepiece* or *ocular* situated nearer the eye of the observer. The ocular acts as a simple magnifier, projecting and enlarging the image but not reversing it again. As a matter of fact, the ordinary ocular of a compound microscope cannot be taken from the instrument and used as a simple magnifier because it is made of two planoconvex lenses which are so adjusted that the image from the objective of the compound microscope is not brought to focus until it has traversed the larger or field-lens of the eyepiece (Figs. 55, 58). The image is really examined, therefore, at a point between the two lenses of the eyepiece. Such an eyepiece is termed a *negative* eyepiece or ocular and is widely used today for microscopical

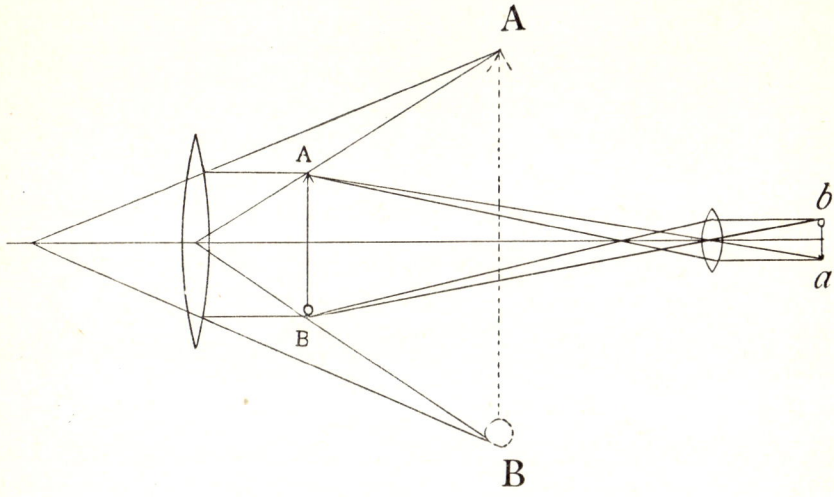

Fig. 51

work. The commonest form, the *Huygenian*, is an adaptation of an ocular designed originally by Huygens for the telescope. By contracting the area of the real image, the field-lens of a negative ocular not only brightens the image but also increases the size of the field that can be examined. It is usually also designed, in conjunction with the eye-lens, to help render the image achromatic.

Positive eyepieces are also made. An inverted image of the object is formed below the system of ocular lenses. Such an ocular operates as a simple microscope.

A good objective is made up of from two to five systems of lenses, as shown in Figure 52. A single system in turn may be a *doublet* (Fig. 57) or a *triplet*, each made of different kinds and shapes of glass. A good objective is a very delicate piece of apparatus and must be handled with great care. Each component is very accurately ground and the systems distanced with extreme precision in order to get a clear image. If not already familiar with the parts of the compound microscope, the student should study Figures 54, 55, and 58, with a microscope before him.

 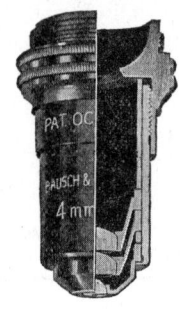 

Fig. 52.—Lens Systems of Various Objectives
Bausch & Lomb 16 mm., 4 mm., and 1.9 mm. oil-immersion objectives, respectively

DEFECTS IN THE IMAGE

Spherical Aberration.—A simple convex lens, unless corrected, will not give a sharply defined image because it does not refract to the same degree all rays passing through it. Those which traverse its edges are brought to a focus nearer the lens (Fig. 53). This results not only in an indistinct image but in a distortion of shape as well. Straight lines, for example, appear curved, and when the parts of the object are in focus in the center of the field, those nearer the margin are hazy and indistinct. This defect is greatest in strongly curved lenses, that is, since magnification increases with increased curvature, in high powers. Spherical aberration is corrected by one or more of the following processes:

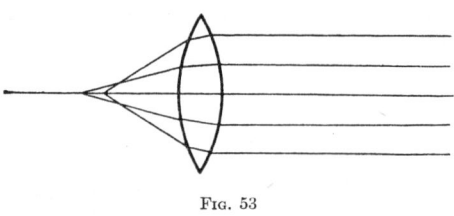

Fig. 53

1. Cutting off the marginal rays.
2. Changing the shape of the surface of the lens.
3. Combining several lenses equivalent to a single lens.

Chromatic Aberration.—As with a prism, ordinary light in passing through a lens is broken up into its component colors. This process is technically termed *dispersion*. Since the colors are not all bent to the

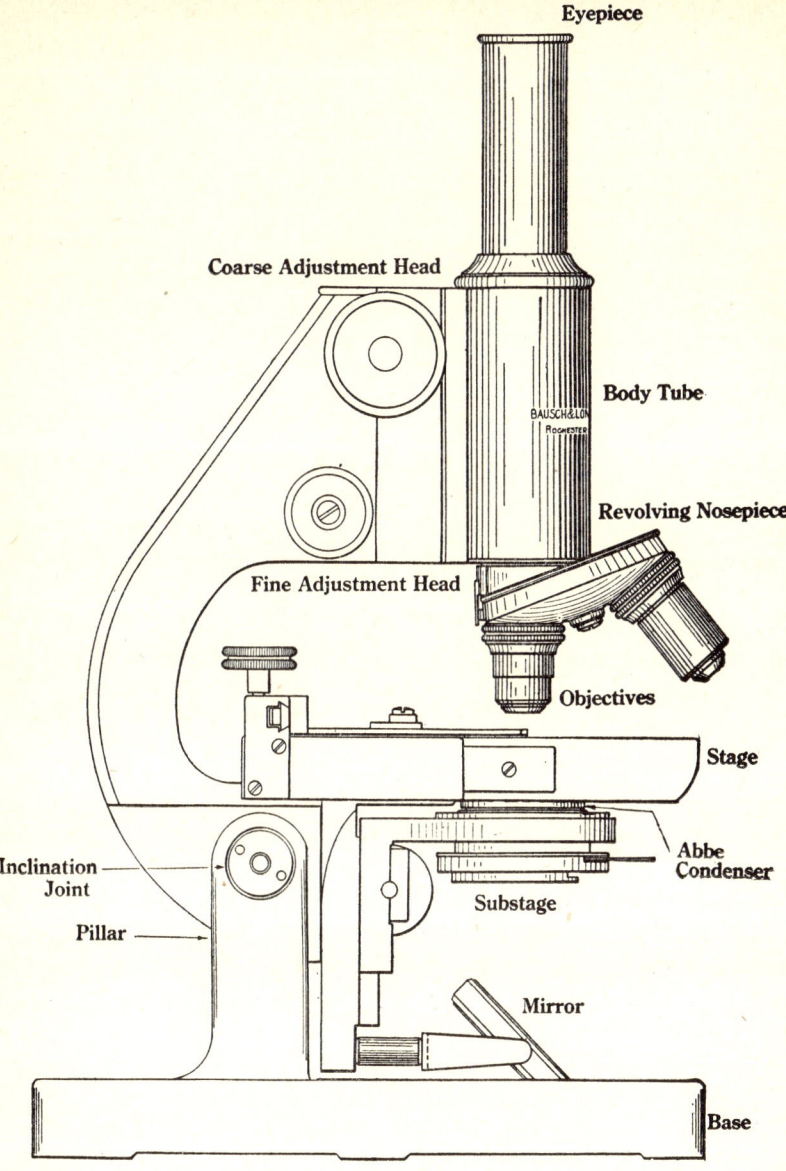

Fig. 54.—A Compound Microscope with Parts Named

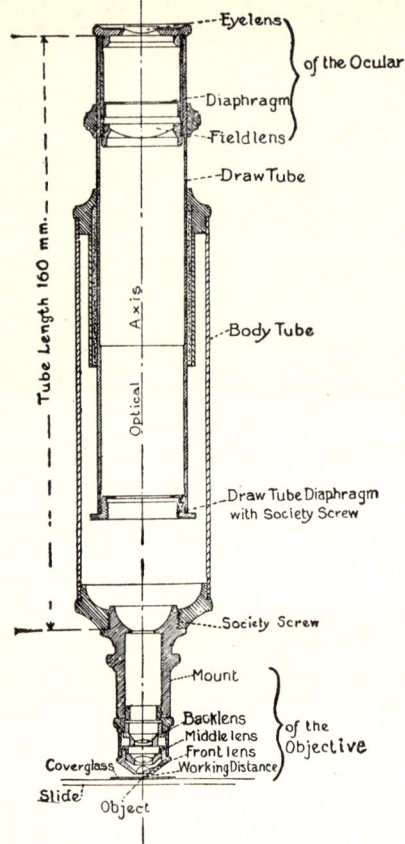

Fig. 55.—Sectional View of Microscope Tube Including Ocular and Objective.

same extent, the result is that each color has a different focus; the ones which are bent most (violet rays) come to a focus nearest the lens, and those which are least affected (red rays) meet at a point farther away (Fig. 56). This failure of the color rays to meet in one focal point is termed *chromatic aberration*, and if uncorrected causes the image of an object viewed through such a lens to be bordered by a colored halo.

The defect is corrected by properly combining glasses of different dispersive powers but of kindred refractive powers. Flint glass (silicate of potassium and lead), for example, has a dispersive power equal to about twice that of crown glass (silicate of potassium and lime), although their refractive powers are nearly the same. By combining a biconvex lens of crown glass with a concave lens of flint glass so constructed that its dispersive power will just equal that of the crown glass (Fig. 57), the error may in large measure be corrected. Such an arrangement does not interfere seriously with the refractive powers of the lens so constructed. Unfortunately no two kinds of glass have been found which have proportional dispersive powers for all colors, so that in the ordinary achromatic objective only *two* of the

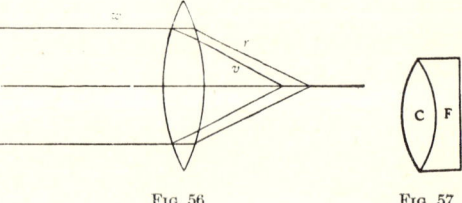

Fig. 56   Fig. 57

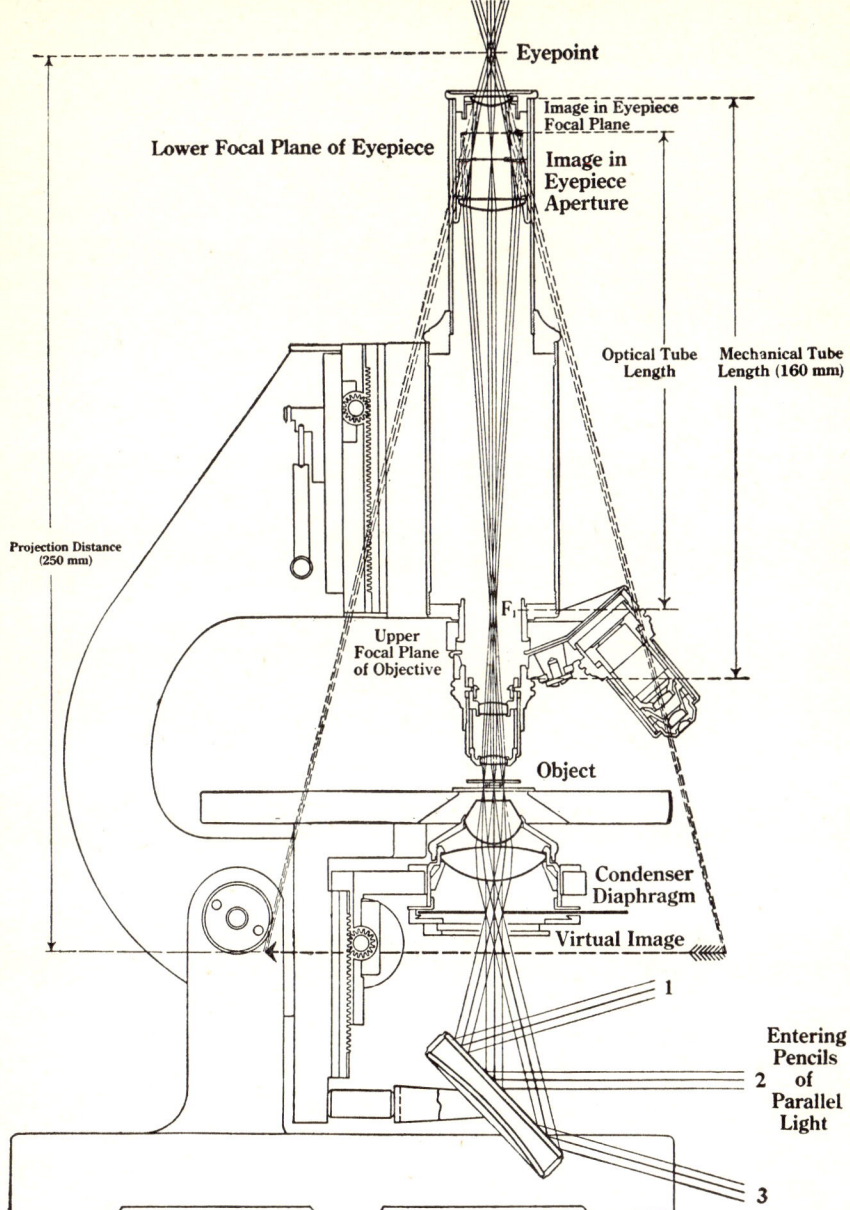

Fig. 58.—Diagram Showing Path of Light Rays through the Compound Microscope, Together with Images (Courtesy the Bausch & Lomb Optical Company)

different colors of the spectrum have been accurately corrected and brought to one focus. The colors left outstanding form the defect known as a *secondary spectrum*. In the apochromatic objectives (p. 209) *three* rays are brought to one focus, leaving only a slight *tertiary spectrum*.

NOMENCLATURE OR RATING OF OBJECTIVES AND OCULARS

Oculars.—Different makers, unfortunately, use different systems in marking their lenses to indicate relative powers of magnification. In the case of lettering the system is wholly arbitrary; the only rule is that the nearer to A the letter is the lower the magnification. When the objective bears a figure it is usually indicative of the magnifying power of the part marked. Thus a $\frac{1}{12}$-inch objective magnifies approximately 120 diameters; a $\frac{1}{8}$-inch, 80 diameters; a $\frac{1}{2}$-inch, 20 diameters; a 1-inch, 10 diameters; a 2-inch, 5 diameters; and so on. This means that an objective which forms an image 10 times the real diameter of the object itself, on a screen placed 10 inches (the conventional distance of vision) from its back lens, is rated as a 1-inch objective. If it formed an image only 5 times the real diameter of the object it would be a 2-inch objective; if 30 times, a $\frac{1}{3}$-inch objective, and so on. Such magnification is termed the *initial magnifying power* of the objective.

The objectives of most manufacturers are now rated in millimeters and the conventional distance of vision taken as 250 mm. An objective of 3 mm. focus, therefore, yields an initial magnification of 83.3 diameters ($\frac{1}{3} \times 250 = 83.3$). Compensating oculars (see p. 213) bear numbers which indicate the number of times the eyepiece, when used at a given tube-length, increases the initial magnification. Ocular 12, for example, with a 3-mm. objective would yield a magnification of $83.3 \times 12 = 1,000$ diameters, with a standard length of tube. Unfortunately this simple system does not apply to most ordinary oculars, which are more or less arbitrarily lettered or numbered.

SOME COMMON MICROSCOPICAL TERMS AND APPLIANCES
(Alphabetically Arranged)

Achromatic Objective.—An objective corrected for chromatic aberration (p. 202). The correction is not absolute.

Achromatism.—Freedom from chromatic aberration.

Angular Aperture.—The angle (measured in degrees) formed at the point of focus (F, Fig. 61) by the outermost rays (aF, bF) which traverse the objective to form an image. This angle is an important consideration because on it depends in large measure the defining or resolving power of the objective. It is evident that the larger the angle is the greater the number of rays of light that will be admitted from an object. Thus the object will be better defined to the

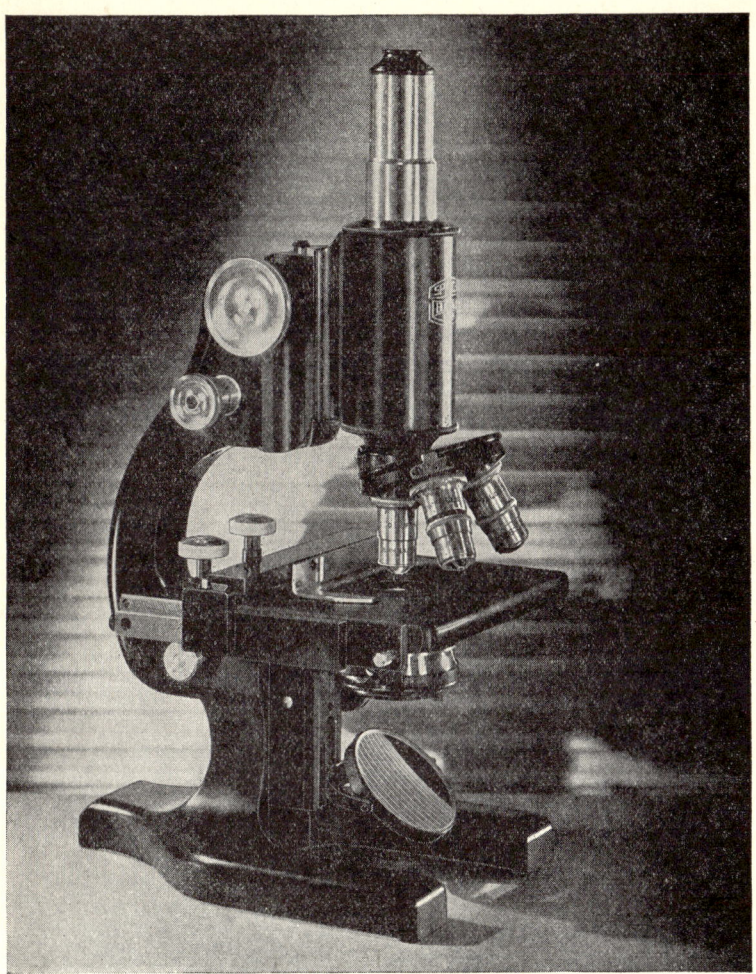

Fig. 59.—The Spencer No. 33 Compound Microscope

eye. In low powers the angle may be very wide, in high powers it must necessarily be small. Two objectives, even though they may possess different powers of magnification, will have the same brilliancy if they are of the same angular aperture; on the other hand, if they have the same magnifying power, but

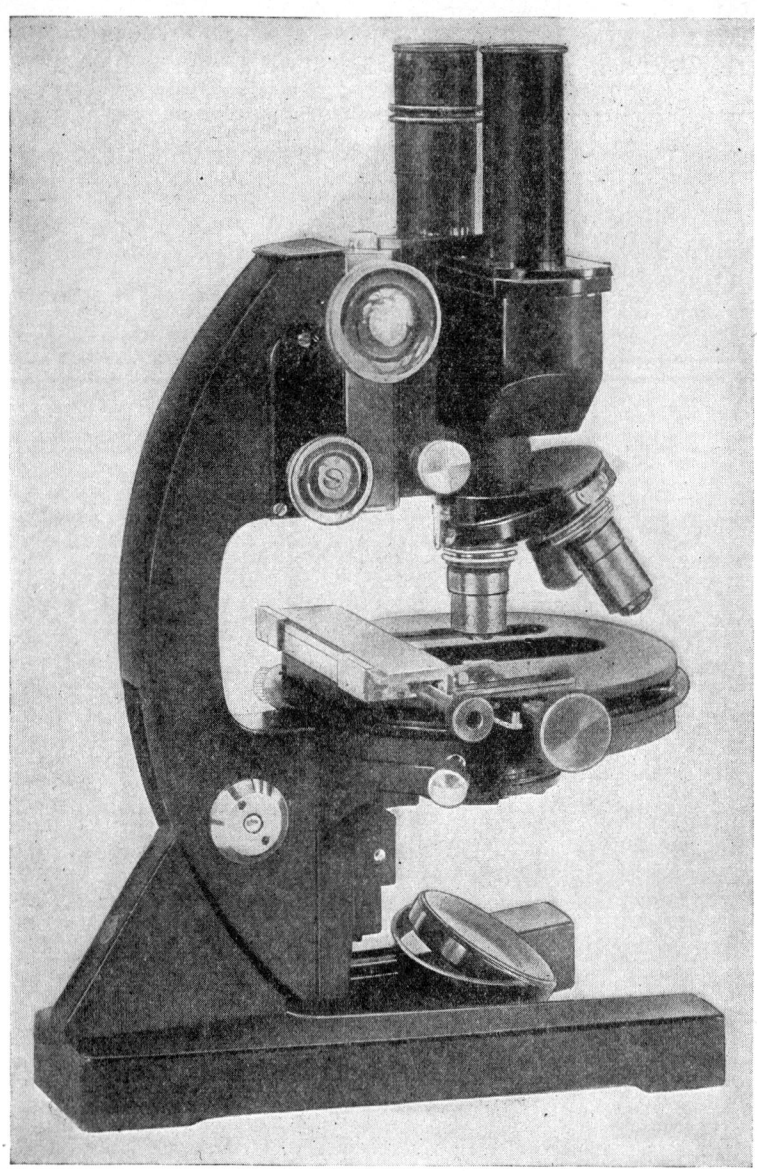

Fig. 60.—The Bausch & Lomb GGDE-8 Compound Binocular Microscope

differ in angular aperture, the brilliancy is reduced in the one of smaller angle. In immersion lenses the liquid used between the lens and the object, by reducing refraction, has the effect of increasing the angle of aperture. See "Immersion Objective," also "Numerical Aperture" (pp. 218, 222).

Apertometer.—An instrument for measuring both the angular and the numerical aperture of objectives. It is fitted to the stage of the microscope.

Aplanatism.—Freedom from spherical aberration (p. 202). The result is a flat field as viewed through the microscope. Aplanatic lenses are usually also achromatic.

Apochromatic Objective.—An improved form of objective which is more exactly achromatic than the ordinary objective because it is corrected for rays of three colors instead of two, and this correction is equally good in all parts of the field. In the ordinary achromatic objective after correction there is a residue of color which is known as the secondary spectrum. In the apochromatic lenses correction is made for a third color, and usually only a slight tertiary spectrum is left uncorrected. Spherical aberration is also more fully corrected. Furthermore, in these objectives the foci of the optical and the chemical rays are identical, hence the lenses are well adapted to photography. In the glasses of the apochromatics, silicon is replaced by boron in the flint series and by phosphorus in the crown series. Fluorite was used in conjunction with the glasses in the earlier forms of apochromatic lenses with the result that the lenses frequently deteriorated in warm, moist climates. Several makers are now able to construct apochromatic objectives without the use of fluorite. Both dry and immersion apochromatics are made.

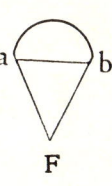

Fig. 61

Binocular Dissecting Microscope.—A microscope adapted to vision with both eyes at once. One of the most important advances in microscopy during the past twenty years has been the development of binocular microscopes with erecting prisms which enable one to carry on dissections, to study thick injected preparations, and to perform other manipulations under higher power and otherwise more advantageously than formerly. In general, they consist (Fig. 62) of two optically distinct tubes so combined that the objectives focus on the same point from different angles. A magnified, stereoscopic vision is thereby provided, so that objects which have depth stand out in pronounced relief. The upper parts of the tubes may be rotated so as to adjust the eye-points of the oculars to the width between the pupils of the observer's eyes. If, upon closing one eye and then the other, an image is not seen by each eye without moving the head, the eye-points are too close together or too far apart. The oculars should be separated or approximated accordingly. When they are correctly adjusted one should get a distinct, stereoscopic appearance. Other adjustments are provided to compensate for differences of focus in the right and the left eye.

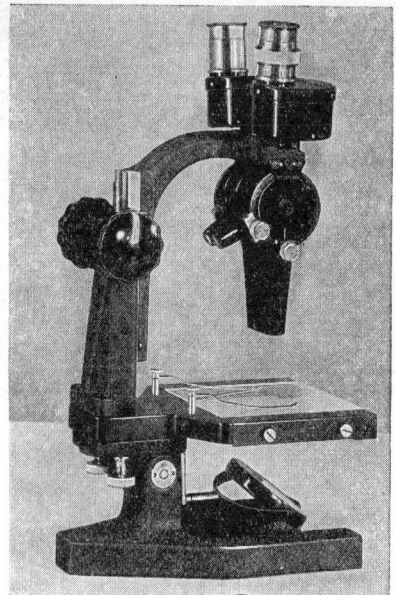
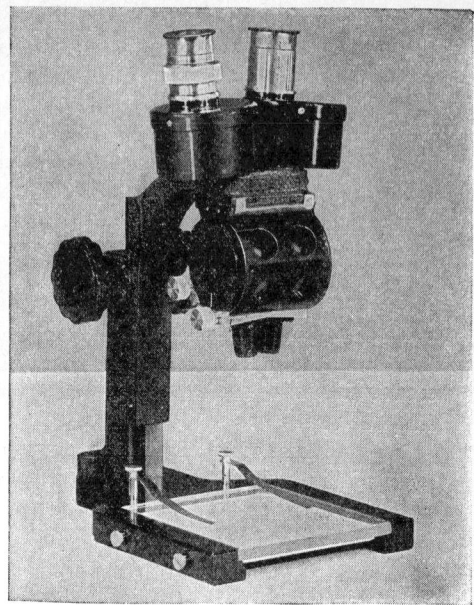
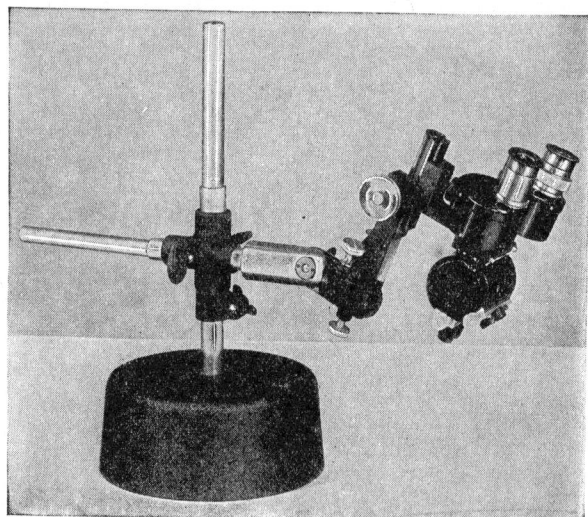

Fig. 62.—Three Popular Arrangements of the Greenough Type Binocular Microscope
(Courtesy the Bausch & Lomb Optical Company)

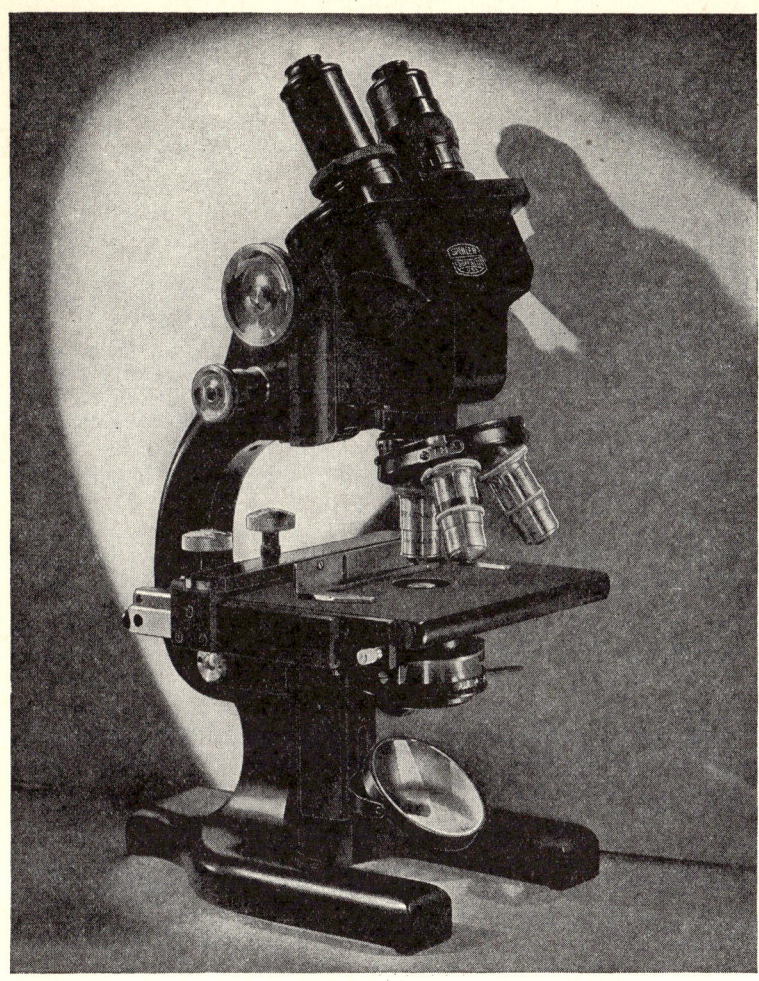

Fig. 63.—Binocular Compound Microscope

A stereoscopic effect is obtained by proper adjustment of the distance between the eyepiece tubes. Interchange between binocular and monocular body-tubes is easily made. A camera lucida can be used on the instrument.

A very simple form of binocular magnifier, known as the Hardy Binocular Loop, may be obtained from F. A. Hardy & Co., of Chicago, Illinois. It is worn like a pair of spectacles. A more elaborate binocular magnifier, to be worn with an elastic headband, may be secured from the Bausch & Lomb Optical Co., of Rochester, New York.

Binocular Single Objective Compound Microscopes (Fig. 63) have been so perfected in late years as to make them available for all powers of the microscope. They constitute an outstanding advancement in microscopy.

Brownian Movement or Pedesis.—An oscillating or dancing motion observable in small particles in a liquid when seen under the microscope.

Calibration of Microscope.—See "Micrometer" (p. 219).

FIG. 64.—Simple Camera Lucida.

Camera Lucida.—An apparatus containing a glass prism or thin glass plate so arranged that when it is placed over the eyepiece of the microscope, the observer may see the image of the object under the microscope projected on to his drawing-paper on the table. The point of the pencil is also visible; consequently the outline of the object may be readily traced on the paper. In the simpler camera lucidas a thin neutral-tint glass slip is so arranged that it is in alignment with the eye-lens of the ocular, except that it sets at an angle of 45 degrees to it. When the microscope is tilted into a horizontal position the observer sees the image of the object reflected from the upper side of the glass slip, but, since the latter is somewhat transparent, he also sees the white paper spread below on the table (Fig. 64).

Another form of simple camera lucida is the Wollaston. To use it the microscope must be inclined. The essential part of the camera consists of a quadrangular prism. The eye of the observer is so placed over the edge of the prism as to receive rays of light from the object with one portion of the pupil, and from the drawing-paper with the remainder.

Some form of the Abbe camera lucida, however, is used by most workers. It consists of a cap which is fitted immediately above the eyepiece and which contains two right-angle prisms cemented together to form a cube (Fig. 65). The lower one of the prisms is silvered along its cemented surface, although a small central opening is left through which the object under the microscope may be viewed; connected with the cap is an arm which bears a mirror, and this mirror may be so adjusted as to reflect the image of the drawing-paper on the table on to the prisms from one side. The prisms are so set that the silvered surface of the lower one reflects this image upward to the eye of the observer which also, coincidentally, is viewing the magnified image of the object through the hole in the silvering. When proper adjustment of the light received from object and paper respectively is made, a pencil-point may be distinctly seen when brought into the field of vision over the paper; consequently the outline of the object may be accurately traced. The secret of success in working with a camera lucida is to have the illumination in the two fields properly balanced. Small screens of tinted glass are provided with the instrument for such regulation. With the Abbe camera lucida the microscope may be used in a vertical or in an inclined position. If the microscope stand is inclined, the drawing-board upon which the paper rests must have the same inclination, or the outline when drawn will be distorted. Likewise, if the mirror of the camera is at any other angle than 45 degrees, an adjustment of the drawing-surface must be made; in short, the axial ray of the image and the drawing-surface *must always be at right angles* to prevent distortion. This means that if the mirror is depressed below 45 degrees the drawing-surface must be tilted toward the microscope *twice as much* as the mirror is depressed. For example, if the mirror is

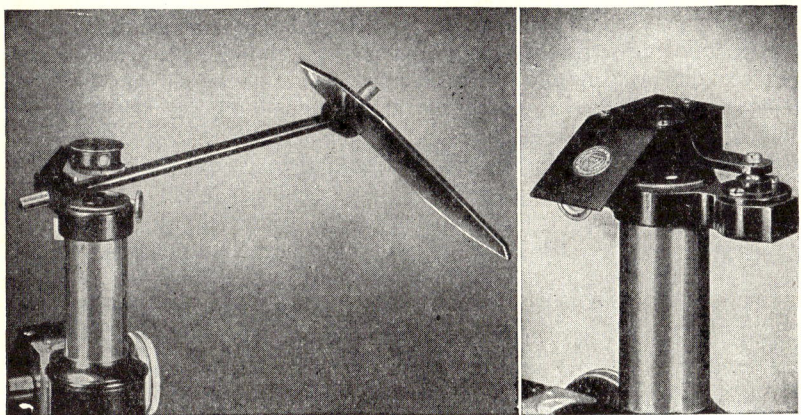

FIG. 65.—Abbe Camera Lucida; Simplified Model at Right

depressed to 37 degrees (8 below 45 degrees), the drawing-board must be tilted (raised) 16 degrees. See also remarks under "Micrometer" (p. 219). When the camera is in proper position the field of the microscope should appear at about the same size as without the camera. If the field is reduced or unevenly lighted, the camera is too near or too far from the ocular, or it is tilted, or the prism is not properly centered.

Compensating Ocular.—A specially designed eyepiece for use with apochromatic lenses. It was found advantageous to undercorrect the objective and then to rectify the aberration by overcorrecting the ocular. The so-called *searching ocular* is a low-power compensating ocular used for the first finding of objects. The object once located in the field, the higher *working oculars* are used in observation.

Condenser.—A lens or a series of lenses mounted in a substage attachment for the purpose of concentrating light upon the object to be examined. They are made in various grades of excellence, non-achromatic, achromatic, and apochromatic. Some wide-angle condensers are used as immersion condensers; the immersion fluid is placed between the upper surface of the condenser and the lower surface of the object-slide. Condensers are especially valuable with high-power objectives and oil-immersion lenses. For the best results the condenser must be accurately centered and the object must lie at the apex of the cone of light formed by it. Unintelligent use of the condenser is a very common fault. Condensers are constructed to receive parallel rays of light, hence the *plane mirror* only should be used with them if the illumination is from daylight. See "Illumination" (p. 216).

Correction Collar.—A device for adjusting the distance between the lens systems of objectives so that the proper corrections may be made for different thicknesses of cover-glass. Low-power objectives are not so sensitive as those of high power to the influence of the cover-glass. Ordinary objectives, however, are mounted in a rigid setting and corrected for a specific tube-length and a standard cover-glass (about 0.18 mm. thick, i.e., a No. 2). With a cover-glass of different thickness correction should be made by altering the tube-length of the microscope, lengthening it for a thinner cover and shortening it for a thicker one. With a 4 mm. ($\frac{1}{6}$ in.) or a 3 mm. ($\frac{1}{8}$ in.) dry objective a deviation of as little as 0.05 mm. in the thickness of the cover-glass, if uncorrected, is sufficient to obliterate fine details of the object. With homogeneous immersion lenses the defect caused by different thicknesses of cover-glass disappears (see "Immersion Objective," p. 218; see also "Tube-Length," p. 224).

Fig. 66.—Adjustable Microscope Lamp

Cover-Glass Correction, Cover-Glass Thickness.—See "Correction Collar" (p. 214).

Daylight Glass.—A specially constructed glass which, when used as a screen with a nitrogen-filled (mazda) lamp, yields a light almost exactly like daylight. It gives very nearly true color values. The light is soft like that from a white cloud, and is more comfortable to work with in microscopy than any other form of artificial illumination. Commonly a 100-watt mazda lamp is used.

Dark Field Microscopy.—See "Ultramicroscopy" (p. 224).

Demonstration Microscope.—A microscope designed to be passed around a class with specimen in place. Most of the compound types are in the form of adjustable tubes which, when in use, are pointed toward a window or a lamp.

Demonstration or Pointer Ocular.—An ocular provided, at the point where the real image of the object is produced, with a delicate rod of some kind which may be rotated to point out objects in the field. A simple type may be made by cementing a hair across the opening of the ocular diaphragm with balsam. When the balsam is hard the hair is cut at the center of the opening and one end is removed. It is necessary to have both ends of the hair supported until the balsam hardens, otherwise the free ends will sag and not be in focus. To use, rotate the ocular.

A double demonstration eyepiece, by means of which the image formed by the objective can be viewed simultaneously by two observers, has also been devised.

Diaphragm.—Opaque plates with openings of various sizes for regulating the illumination of the object to be examined. The iris diaphragm (Fig. 67)

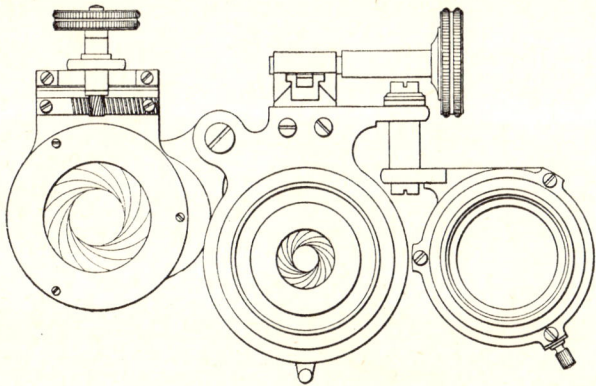

Fig. 67.—Top view of a Substage Attachment with Condenser and Lower Iris Diaphragm Thrown out of Optical Axis.

is the best type. It consists of a series of overlapping plates placed around a central opening the size of which may be varied by means of a lever. Revolving diaphragms are commonly used on the cheap grades of microscopes. They consist of round disks perforated by openings of various sizes which may be rotated between the mirror and the object. The nearer to the object the diaphragm is placed the better the intensity of the illumination can be regulated. Most of the better class of microscopes are provided with two iris diaphragms, one beneath the condenser to be employed when the latter is in use, the other flush with the stage to be used only when the condenser is out. If this second iris diaphragm is lacking, its place is taken by means of a cap-diaphragm which may be fitted into the substage in the place of the condenser. A *central-stop diaphragm* is one with an opaque center and a slit around the edge, so arranged that a hollow cone of light, consisting of rays of great obliquity, will be produced.

Dissecting Microscope.—An instrument so constructed as to enable an operator to carry on minute dissections under magnification. Ordinarily they are simple microscopes mounted on a stand of some kind. The best instruments (Fig. 68) are provided with well-corrected lenses, with glass stage, mirror, black-and-white substage plate, and rests for the hands. The binocular type, Fig. 62, is indispensable for the finer modern technique. See "Binocular Microscope."

Embryograph.—A form of camera lucida for drawing at slight magnification small objects, such as embryos. A camera lucida attached to a simple microscope is frequently used for this purpose.

Eye-Point.—The point above an ocular or lens at which the largest number of rays from the instrument enter the eye. The largest field of the microscope is visible from this point.

FIG. 68.—Dissecting Microscope

Flatness of Field.—See "Aplanatism" (p. 209).

Homogeneous Immersion Objective.—See "Immersion Objective" (p. 218).

Huygenian Ocular.—See p. 201.

Illumination.—Any means employed to direct light upon the object under observation. Light which traverses the object is said to be *transmitted* light. Most microscopical work in biology is done by means of transmitted light, hence the object must be rendered more or less transparent if not naturally so. If the object is symmetrically lighted, the lighting is designated as *axial* or *central* illumination. If one side is lighted more than another, the term *oblique* illumination is employed. In the case of transmitted light, the light which traverses the object is usually light reflected from a mirror because it is generally inconvenient or impossible to hold the instrument directly toward the source of light. Makers of microscopical appliances, however, now supply admirable miniature electric lamps which may be used with the mirror or in place of it.

Light which falls upon the object and is reflected from it to the eye, either directly or through a microscope, is termed *reflected light*. Such illumination is employed but little in ordinary histological work, but it is useful in the examination of opaque objects such as metals, insects, etc. The illumination may be increased by means of a bull's-eye condenser or a mirror. In some microscopes the mirror can be swung above the stage for the purpose of illumining an object which is to be studied by reflected light.

The best light for microscopical work is light reflected from white clouds. Direct sunlight is never used. The light should come from in front of the observer or from one side. Various kinds of artificial light are used for microscopical work, such as an ordinary lamp with flat wick, the Welsbach, or the ordinary electric light. Some of the newer electric lights especially designed for microscopy are excellent. Whatever the source, the rays must be steady

Fig. 69.—Lens Holder with Flexible Arm

and brilliant. If a lamp with flat wick is used, greater brilliancy is secured when the edge of the flame is turned toward the microscope; the object should be lighted directly by the image of the flame. To do this with low powers, the lamp may have to be turned so that the flame is oblique to the microscope.

In artificial light the rays are divergent, not parallel, as in the case of sunlight; hence they will not come to focus at the same point when reflected from the mirror as the latter do. This should be corrected by using a large bull's-eye condenser between the source of light and the mirror, or by sliding the mirror along the mirror-bar farther away from the stage so that the concave mirror will have a longer distance in which to bring the rays to focus. If a substage condenser is used, the same results may be obtained by depressing the condenser somewhat below the level of the stage. Lamps made for the microscope often have a metal chimney with a bull's-eye in one side.

The objectionable yellowness of most artificial light may be eliminated by interposing a piece of green signal-glass between the lamp and the microscope. With most microscopes, round slips of blue glass which fit into the substage mechanism are supplied for this purpose. Some workers still employ as a screen an ammonia sulphate of copper solution in a globular flask. To make the solution, dissolve a small amount of copper sulphate in water and add ammonia. At first a precipitate appears, but if an excess of ammonia is added this is dissolved and a transparent deep-blue liquid results. This should be diluted with water sufficiently to get a blue of just the proper depth to render the transmitted light white as seen through the microscope. The globular flask also acts as a condenser. See comments under "Daylight Glass."

Immersion Objective.—A kind of objective in which a liquid is used between the front lens and the cover-glass. Cedar oil is the most widely used medium. Inasmuch as the optical properties of cedar oil (refraction and dispersion) are almost the same as crown glass, it is often termed a *homogeneous immersion fluid*. A homogeneous immersion lens, therefore, would be one intended for use with such a fluid. The advantage of an immersion over a dry lens lies in the fact that, other things being equal, after leaving the cover-glass rays, which would be so refracted in a rarer medium like air as to miss the front lens of the objective, reach this lens in the case of immersions and traverse the objective. With homogeneous immersions the rays of light are carried without deflection through cover-glass and fluid and into the glass of the front lens. Water has a greater density than air and less than glass; hence, with a *water immersion* more rays of light reach the front lens than with a dry lens, and less than with a homogeneous immersion lens (Fig. 70). The effect of an immersion is practically to widen the angle of the lens (see "Angular Aperture," p. 206).

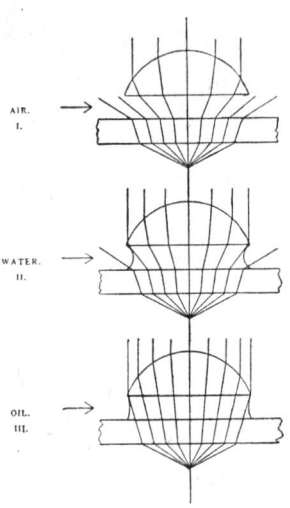

FIG. 70.—Diagram to Illustrate the Relative Amounts of Light Utilized with Dry, Water Immersion, and Homogeneous Oil Immersion Objectives, Respectively (after Bausch).

The value of the immersion objective is enhanced if the immersion fluid is placed between the upper lens of the condenser and the slide as well as between the objective and the cover-slip.

Magnifying Power.—The power of a lens to multiply the apparent dimensions of an object viewed through it. It should be expressed in *diameters*, not in areas. While magnifying power is very important, it is only so in connection with resolving power. If high power were the only essential, a series of single lenses might be used. The impossibility of using such a series for high magnification is due to the fact that proper correction of aberrations cannot be made, and, consequently, a distinct image cannot be obtained. For determination of magnification see "Micrometer" (p. 219).

Mechanical Stage.—A stage attachment (Fig. 71) for the more accurate manipulation of an object or a series of objects which must be moved about under the objective. The best mechanical stages are provided with scales and verniers so that an object once recorded may be easily found again. They are often very serviceable, especially with high powers.

Micro-Manipulator.—An instrument (Fig. 72) for manipulating, injecting, operating upon, or submitting to various physical or chemical influences such objects as individual cells, bacteria, and other microscopical objects (see also p. 176). A detailed description of the Chambers instrument will be found

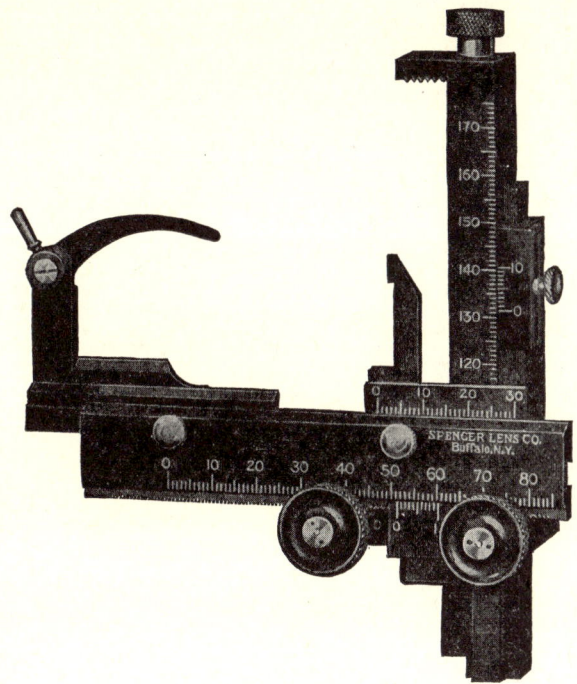

Fig. 71.—Attachable Mechanical Stage for Microscope

in the E. Leitz catalogue. The development of micro-dissection by the aid of such instruments is perhaps the most notable advance in micrology during the past twenty years. The adjustments of the micro-needles and micro-pipettes of the apparatus are so accurate and so delicate that cells only 0.008 of a millimeter in diameter may be impaled and dissected. The microscope may be fitted with one or with two manipulators. The cell or tissue to be dissected lies in a drop of liquid hanging from the roof of a moist chamber which is moved by a mechanical stage.

Micrometer.—A scale for measuring objects under the microscope. The *stage micrometer* consists of a finely divided scale ($\frac{1}{10}$ and $\frac{1}{100}$ mm.) ruled on

Fig. 72

glass or metal. It is commonly mounted on a glass slide of standard size. To determine the actual size of an object with the stage micrometer, it is most convenient to use a camera lucida. The outline of the object to be measured is projected on to a sheet of drawing-paper and marked off. The object is then replaced under the microscope by the micrometer and the micrometer scale is projected on to the paper. Knowing the actual distance between the lines on the micrometer scale, the magnification as well as the real size of the object is readily calculated.

The size of the image projected by a camera lucida on to a piece of drawing-paper at the level of the table, however, does not represent the true magnifying power of the microscope. The latter is really considerably smaller if the microscope is in a vertical position because the magnification of a lens or a system of lenses is calculated in terms of the conventional distance of vision (250 mm., see p. 206) while the distance from the ocular to the table is considerably more than 250 mm. Since the rays of light diverge after leaving the ocular, manifestly the projected image will be larger (possibly by as much as 60 per cent) at the level of the table than at a level just 250 mm. from the point of emergence of the rays from the ocular. To determine the actual magnification of the microscope, therefore, one would have to bring the drawing-surface to within 250 mm. of this point of emergence, sketch the projected scale of the stage micrometer on the paper, and then, by means of an ordinary metric rule, compute the number of times the divisions of the micrometer scale have been magnified. The standard distance of 250 mm., if the Abbe camera lucida is used (with camera mirror at 45 degrees), includes the distance along the mirror-bar from the optical axis of the ocular to the mirror, plus the distance from the mirror to the drawing-surface.

In practical work it is not necessary to make drawings or measurements exactly at this standard distance; one needs only to have a scale made out for the distance from the camera lucida at which the drawings are actually to be made, although it must be carefully borne in mind that any variation in the elevation of the drawing-surface will alter the size of the projected image. A series of carefully prepared scales for various combinations of objectives and oculars should be made and kept for future use. On each should be recorded the tube-length used, the number of the objective and of the ocular, the length of the camera mirror-bar, and the angle of the mirror, for if any one of these is changed the scale is no longer accurate.

When much measuring is to be done an ocular micrometer is used. It consists of a circular glass disk with a scale ruled on it and is inserted in the ocular between the eye-lens and the field-lens. By means of a stage micrometer the value of the divisions of the ocular micrometer is determined for a known tube-length and every combination of lenses it is desired to use in the work of measurement. Suppose that it takes four divisions of the ocular micrometer to correspond to one of the finer divisions of the stage micrometer, then since the divisions of the latter are equal to $\frac{1}{100}$ mm., each space in the ocular micrometer must be equal to $\frac{1}{400}$ mm., that is, 0.0025 mm. A *filar* or *screw micrometer* is a more convenient form of ocular micrometer, which is provided with delicate movable spider lines that can be adjusted to the space to be measured by means of a fine screw with very accurately cut threads (Fig. 73). At the end of the screw is a graduated disk which gives the value of the distance between the spider lines. The pitch of the screw is either $\frac{1}{50}$ inch or 0.5 mm. When once the valuation of this ocular micrometer has been determined by means of a stage micrometer, measurements can be made rapidly and with great precision.

The *step micrometer*, in which the intervals are arranged in groups of ten, each group being conspicuously marked by a black, stairlike notching along one side, is one of the most desirable types of ocular micrometers.

FIG. 73.—Filar Micrometer

Micron.—The one-thousandth part of a millimeter; expressed briefly by the Greek letter μ. It is the unit of measurement in microscopy.

Mirror.—The compound microscope is usually provided with both concave and plane mirrors, which may be rotated or swung in any direction. The plane mirror is used with the condenser; the concave, whenever it is of advantage to have light concentrated upon the object, with the condenser out. The mirror should be capable of being moved up or down the mirror-bar so that it can be accurately focused upon the object. See also "Illumination" (p. 216).

Muscae Volitantes.—Small filaments or specks which float across the field of vision. They are really small opacities in the vitreous humor of the eye.

Numerical Aperture.—A system which expresses the efficiency of an objective by indicating the relative proportion of light rays which traverse it to form an image. With the introduction of immersion objectives, it became evident that angular aperture alone is not sufficient to indicate the real capacity of an objective. For instance, an immersion and a dry lens may be of precisely the same angular aperture, and yet the immersion lens is more efficient because it sends more rays of light through the objective (see "Immersion Lens," p. 218). It was found necessary to take cognizance of the medium which intervenes between the cover-glass and the front lens of the objective.

Professor Abbe, in 1873, proposed the name *numerical aperture* and introduced the formula N.A = n sine u, in which n signifies the refractive index of the medium between cover-glass and objective, and u equals half the angle of aperture. That is, by multiplying the refractive index of the medium by the sine of half the angle of aperture the numerical aperture is obtained. For example, suppose that one had an oil-immersion lens of 90 degrees angular aperture, then half the angle of aperture is 45 degrees, and by turning to a table of natural sines, the sine of 45 degrees is found to be 0.707. The refractive index of cedar oil is 1.52. Then N.A. = 1.52×0.707 = 1.075. Suppose that the lens were a dry instead of an immersion lens; then since the refractive index of air is 1, the formula would read N.A. = 1×0.707 = 0.707. Thus the two products 1.075 and 0.707, respectively, represent the relative capacities of an oil immersion and a dry objective of 90 degrees angular aperture.

Parfocal.—A term ordinarily applied to eyepieces of different powers that may be exchanged in the microscope without very materially affecting the focus of the instrument. The term is also applied to objectives attached to a revolving nosepiece if each is approximately in focus when turned into place.

Pedesis.—Same as *Brownian movement*.

Penetration.—The quality of an objective that permits of "looking into" an object having sensible thickness. It is greatest with low powers and narrow angles and is antagonistic to resolving power. It is the natural consequence of certain conditions in the making of lenses and is reckoned of secondary importance, because practically the same results are obtained by manipulating the fine adjustment.

Photomicrography.—Photography of small or microscopic objects. The subject, although of great importance, is too extensive to enter into in the brief space that could be allotted to it in an elementary treatise such as this. (See, however, p. 172.) An excellent chapter on photomicrography and a bibliography will be found in Gage's *The Microscope* (14th ed.), pp. 206-245.

Pointer Ocular.—See "Demonstration Ocular" (p. 214).

Polariscope.—As used in microscopy the polariscope consists of two parts, each composed of a Nicol prism of Iceland spar; one, the *polarizer*, fits into the substage, and the other, the *analyzer*, is inserted between the objective and the tube of the microscope or, in some forms, just above the ocular. The polariscope is used more in chemical and in geological than in histological work. Some of the uses are as follows: determining whether an object is singly or doubly refractive; detecting the presence of minute crystals; determining the composition of rocks; examining sections of bone, hoof and horn, hairs and fibers of animals and plants, starch, etc., for certain characteristic and striking effects.

Projection Ocular.—An ocular specially designed for projecting a microscopic object on to a screen or for use in photomicrography. While ordinarily used with apochromatic objectives, they may be used with ordinary objectives of large numerical aperture. The eye-lens is movable so that a sharp focus (indicated by a distinct image of the diaphragm) may be obtained at different screen distances.

Resolving Power.—The quality of an objective which enables the observer to make out fine details of structure. It is the most essential property for precision in observation, and determines largely the excellence of an objective. Resolving power depends upon careful correction of aberrations, general accuracy in the mechanical construction of the microscope, and upon the aperture of the objective (see "Angular Aperture," "Numerical Aperture," pp 206, 222). Resolving power is tested by the resolution of fine parallel lines ruled on glass or the striae on the surface of diatoms. The test is to determine how many lines to the inch or centimeter may be distinguished, and whether the objective simply glimpses the markings or whether it resolves them clearly. The wider the angle of aperture the better the resolving power, provided the width is not so great as to interfere with the correction of the lenses. The increased resolution of immersion lenses is due to the fact that the immersion fluid practically widens the angle of aperture (see "Immersion Objective," p. 218).

Tube-Length.—The distance between the places of insertion of ocular and objective into the tube of the microscope. There are two standard tube-lengths; the short standard is 160 mm. ($6\frac{3}{10}$ inches), the long standard, 216 mm. ($8\frac{5}{6}$ inches). Some makers, however, do not adhere to the standards. The optical efficiency of the instrument is the same in either case. The short length is more advantageous in that it is more compact. The lenses must be corrected for the length of tube with which they are to be used. The short standard is in use in most American laboratories.

Ultramicroscopy.—A system of microscopical inspection in which objects are examined by reflected light. The object appears to be self-luminous against a dark field, hence the term *dark-field illumination* is often used as descriptive of the method. *Dark-field microscopy* is more inclusive than *ultra-microscopy*, since in strict sense the latter deals with objects so small that they are not directly visible; one detects their presence by the points of light which they deflect ("Tyndall effect"). Dark-field microscopy, however, also includes many objects which fall within the resolving power of the microscope. Objects to be studied in this way are usually semi-transparent or consist of fine particles such as occur in colloidal suspensions. In lighting, rays of great obliquity are used so that only such traverse the objective as are deflected from some object in the field. The great value of the method lies in the fact that particles may be rendered visible which are wholly invisible with the microscope as ordinarily used.

For *low powers* without a condenser, the diaphragm must be wide open and the mirror so tilted that the object is lighted by oblique rays which cannot get directly into the front lens of the objective. With a condenser, a *central-stop diaphragm* is used which admits only marginal rays. By making an ordinary diaphragm eccentric, somewhat the same effect may be secured. For practice, place a drop of 10 per cent solution of salicylic acid in 95 per cent alcohol on a slide and leave it until the alcohol evaporates. Examine the residue of crystals in the ordinary way and then by dark-ground illumination. By the latter method the crystals should appear brilliantly lighted on a dark background. Add a small drop of the solution to the crystals and watch crystallization under dark-ground illumination.

For *high powers*, according to one system, very wide apertures (greater than 1.00 N.A.; see p. 222) are necessary in the condensers. Some makers (e.g., Leitz, Reichert) use specially modified condensers. Others (e.g., Beck, Siedentopf) substitute a parabolic reflector for the condenser. In the method of Siedentopf and Zsigmondy, the field is lighted from one side, at right angles to the axis of the microscope, by a wedge or cone of bright light. In another method, useful for both high and low powers, an objective of wide aperture and a condenser of moderate aperture are employed. The field is lighted, as in

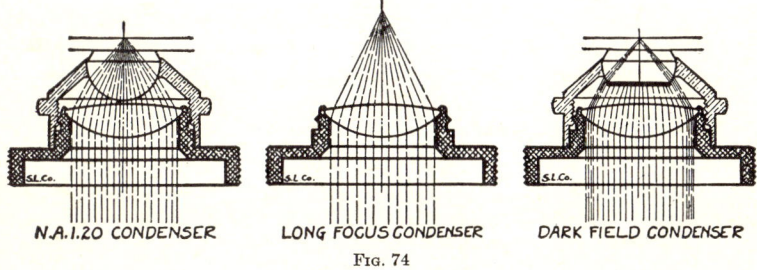

N.A.1.20 CONDENSER   LONG FOCUS CONDENSER   DARK FIELD CONDENSER

Fig. 74

ordinary microscopy, by a cone of light from the condenser, but a diaphragm or stop of the right size to cut out the central rays of light is placed on the back lens of the objective. In this way only those rays which have entered the marginal zones of the objective pass on to form an image, and among these are the rays which have been deflected by objects in the field. A somewhat similar result may be attained by using a stop in the eye-point (p. 216). **Figure 74** represents a divisible condenser which is converted into a long-focus condenser by removing the upper lens and into a dark-field condenser by replacing the upper lens with a special dark-field illuminator.

Dark-field illuminators for all powers have been so perfected in recent years that minute details in living structures can be seen with remarkable clearness. An excellent chapter (pp. 424–69) on "The Dark-Field Microscope and Its Applications" will be found in Gage's *The Microscope* (14th ed.).

Overcorrection and Undercorrection.—In correcting for chromatic aberration, if the concave lens is stronger than is necessary to neutralize the aberration of the convex lens, the blue rays are brought to focus beyond the true principal focus of the objective, and the latter is said to be overcorrected; if the concave lens is **not strong enough**, the result is what is known as undercorrection. In case of **overcorrection**, the object takes on an orange tint if, after focusing, the distance between object and objective is slightly increased; or it becomes of bluish color if the distance is decreased. In case of undercorrection, just the reverse is true. In some instances the objective is purposely undercorrected, and the eyepiece (e.g., compensating ocular) is equally overcorrected.

Working Distance.—The distance between the front lens of the objective and the object when the latter is in focus. With high powers it is very small, so that with some oil-immersion objectives, if a thick cover is used, it is impossible to focus upon the object. For this reason thin cover-glasses (No. 1) should be used on preparations which are to be used with high-power immersion lenses. For examination under high-power dry lenses, however, see remarks under "Correction Collar" (p. 214).

MANIPULATION OF THE COMPOUND MICROSCOPE

1. Always handle the instrument cautiously; it is a delicate mechanism. Lift it by the base or by a handle specially provided, never by the tube.

2. The work-table should be of such a height that the observer can sit at it comfortably without compressing the chest or tiring the neck. Sit as upright as possible. If the instrument is inclined, it should set farther in on the table than if it is in the upright position.

3. With a piece of old linen, a chamois skin, or a bit of lens-paper, carefully clean the eyepiece to be used and put it in place. Always use the low-power eyepiece first.

4. Likewise clean and attach the objective (low power first) after elevating the tube far enough above the stage for this purpose. Guard particularly against screwing the objective in crooked, as this will injure the threads. It is best to swing the objective between the first and second fingers of one hand and bring the screw squarely into contact with the screw of the tube (or nosepiece); with the thumb and forefinger of the other hand it is then screwed into place.

5. Bring the draw-tube to the standard length (see "Tube-Length," p. 224) for which the lenses are corrected. If a nosepiece is used, allowance must be made for its height. In some of the more recently made microscopes, however, the scale on the back of the draw-tube includes the nosepiece. In pushing in or drawing out the draw-tube always grasp the milled head of the **coarse adjustment** also, so that the tube as a whole will not be shifted.

6. Place the slide which bears the object on the stage with the object over the central opening of the latter, and clamp it in place by means of the spring clips. While looking at the object from one side, turn the mirror until a flood of light shines up through the center of the stage.

7. Lower the tube until the objective nearly touches the cover-glass, then look through the eyepiece and slowly raise the tube by means of the coarse adjustment until the specimen to be examined is plainly visible. Focus accurately by means of the fine adjustment. If a high-power objective is being used, since it must come very near the cover, the operator should lower his head to the level of the stage, and look toward the light between objective and cover-glass in order to prevent actual contact. This is of great importance, for otherwise the objective or the object is liable to injury. Remember that in focusing *up* the lowest part of the object comes into view first, the highest part last. It is often easier to locate the object if the preparation is moved about slightly while focusing.

8. The higher the power the more difficult it is to find an object or a particular part of it. For this reason the finding is usually done by means of a low-power objective, or a low-power ocular, or both, and after accurately centering the object in the field, the high power is attached. In case a revolving nosepiece is used, great care should be used in turning in the high power not to strike the slide with the objective. This is very likely to happen if the objectives are not parfocal. When objectives are not parfocal they may usually be made so by putting a paper or bristol-board collar on them.

9. After the object is in focus give any further attention to the illumination that is necessary (see "Illumination" and "Mirror," pp. 216, 222). If intensified illumination is desired, use the concave mirror, or use the substage condenser and the plane mirror. For ordinary purposes the field should be evenly illuminated, although oblique light is frequently useful. Manipulate the diaphragm until the structure to be studied shows with the greatest distinctness. Too much light "drowns" the object, and is hard on the eyes. (To determine the proper distance at which the concave mirror should stand below the stage, let direct sunlight shine upon the mirror, and then adjust the latter so that the apex of the cone of light comes just at the top of the stage where the object will rest.)

If particles of dust or cloudiness appear in the field, determine, by moving the slide and rotating the ocular one after the other, whether slide, objective, or ocular requires further cleaning. A camel's hair brush is often more effective than lens-paper or cloth in removing bits of dust.

10. In using oil-immersion objectives, a small drop of cedar oil (specially prepared by the maker of the lens) is applied to the front lens by means of a small rod or brush. It is very important to keep the oil free from dust, and to see that it does not contain air bubbles when applied to the lens. Carefully lower the tube until the oil on the objective comes in contact with the cover-glass. The operator should lower his head to the level of the stage to observe this properly. Focus up as with a dry objective. For critical work immersion oil should also be placed between the condenser and the lower side of the slide. With a piece of lens-paper or a soft cloth clean the immersion lens immediately after you have finished using it. Likewise remove the oil from the cover-glass. Oil which has hardened on the cover-glass should be removed with lens-paper wet with xylol.

If the immersion oil becomes too dense, as is likely after some months, it may be diluted with pure cedar oil.

11. The range of the fine adjustment is limited. Keep it as near the middle point as possible. If the tube does not respond to the movement of the screw, you have probably gone beyond the range of the fine adjustment.

12. In working with the microscope *keep both eyes open*. The eye which is not in use soon becomes accustomed to ignoring objects in the field of vision. To avoid fatigue it is well to use first one eye and then the other for observation. The eye should be placed at the eye-point (p. **216**) of the lens. This is some distance from the eye-lens in low-power eyepieces, close to it in high-power eyepieces.

While observing, with one hand keep the fine adjustment moving. This relieves the eye of the strain of attempting to focus on different depths of the object.

13. Put the microscope in its case when you have finished using it, or at least cover it with a cloth or cone of paper. For further details regarding the use or care of the microscope consult one of the following books: *The Microscope*, by Gage; *Principles of Microscopy*, by Wright; *The Microscope and Its Revelations* (1,200 pages), by Carpenter and Dallinger.

14. Do not apply alcohol to any part of the instrument. The lenses

may be cleaned ordinarily by breathing upon them and wiping them with a rotary motion on lens-paper or a piece of soft old linen. In case a solvent must be used for balsam or oil, benzene is the one commonly recommended. It must be quickly wiped away so that it will not affect the setting of the lens.

15. Read carefully in the catalogue of the maker of your instrument what is said about its construction.

16. Determine the magnifications of your various combinations of lenses as described under the heading of "Micrometer" (p. 219).

The beginner in microscopy should acquaint himself with various common objects that are liable to get into his preparations in the form of dust, etc., so that he may not mistake them for essential parts of his specimen. Such objects are hairs, fibers of silk, wool, linen, cotton, and the like, and particularly air bubbles. Air bubbles are usually circular with black borders and bright centers; they may show tinges of color. Examine a drop of saliva for examples.

APPENDIX B
SOME STANDARD REAGENTS AND THEIR USES
I. FIXING AND HARDENING AGENTS

1. Acetic Acid.—Acetic acid is more commonly used in mixtures or in diluted form than pure. It is valuable because it tends to produce good optical differentiation and facilitates penetration. When employed alone it causes some tissues to swell and disintegrate. Inasmuch as most fixing agents give the best results when they have an acid reaction, from 1 to 5 per cent of acetic acid is generally added to acidify them in case they are not naturally acid. Any reagent containing a very large proportion of acetic acid should be allowed to act for only a short time. Acetic acid is also of great value in mixtures because it counteracts the shrinking action of certain reagents. It has to be omitted or used sparingly, however, in reagents after which study of cytoplasmic detail is desired, because of its destructive action on mitochondria and other cell inclusions. Ordinary acetic acid is of about 36 per cent strength; glacial acetic, of about 99.5 per cent strength. The latter is meant when mentioned in this book unless otherwise specified.

A strength of from 0.2 to 1 per cent is recommended by Flemming for work on cell nuclei. Strong glacial acetic acid is sometimes used for highly contractile animals, such as Coelenterata, Mollusca, and Vermes. The animal is rapidly flooded with the acid and remains immersed until it is thoroughly penetrated (6 to 10 minutes). It is then washed in repeated changes of 50 or 70 per cent alcohol and left to harden in 70 to 83 per cent alcohol. The pure acid, if allowed to act for more than a few minutes, swells and softens the tissues. Acetic acid should not be used when connective tissue or delicate calcareous structures are to be preserved.

2. Acetic Alcohol.—Carnoy recommends each of the following formulae:

a) Glacial acetic acid.......................... 1 part
    Absolute alcohol........................... 3 parts
b) Glacial acetic acid.......................... 1 part
    Absolute alcohol........................... 6 parts
    Chloroform................................ 3 parts

The chloroform is said to hasten the action of the mixture. Either of these reagents penetrates well and acts rapidly. Solution *b* is especially good for glandular or lymphatic tissue. Almost any stain will follow them. Even such difficult objects as the eggs of Ascaris may be fixed by the second mixture. The reagent should be washed out in absolute or at least in strong alcohol.

Absolute alcohol to which 20 per cent acetic acid has been added was used in Boveri's laboratory, for Ascaris. Material is left overnight in it.

A mixture of absolute alcohol, glacial acetic acid, and chloroform, equal parts, saturated with corrosive sublimate (formula of Carnoy and Lebrun), becomes even more valuable for the fixation of difficult objects. According to Lee, isolated ova of Ascaris are fixed in 30 seconds, entire oviducts in 10 minutes, in this liquid.

3. **Alcohol.**—Alcohol is used especially for gland cells and for preserving the brain and spinal cord for Nissl's method of staining nerve cells. See "Alcohol Fixation," page 34; also reagents 1 and 2, page 7, and memoranda on page 11.

Alcohol and Chloroform.—See 2, *b*.

Altmann's Osmic-Bichromate Mixture.—See 9, page 234.

Bensley's Formol-Bichromate-Sublimate Mixture.—See page 239.

Bichloride of Mercury.—See "Corrosive Sublimate."

4. **Bichromate of Potassium.**—Bichromate of potash is one of the oldest and best known fixing reagents. At present it is more commonly used in mixtures than alone. It is widely used in hardening nervous tissue. Its fixation of nuclei is unsatisfactory unless it is properly corrected through the addition of acetic acid. It acts very slowly, about 3 weeks being necessary to harden properly a sheep's eye, and from 3 to 6 months for a good-sized brain. A weak solution (2 per cent) should be used at first, to be replaced gradually by stronger solutions up to 5 per cent. When hardening is completed the object should be thoroughly washed in running water and then put into alcohol; begin with low percentages of alcohol and gradually increase the strength up to 70 or 80 per cent. Change the alcohol as often as it becomes yellow. After the object has been placed in alcohol, keep it in the dark in order to prevent a precipitate forming on the surface. Either carmine or hematoxylin may be used as a stain after bichromate of potash. In case carmine is used, the staining is best done before the object is placed in alcohol. Tissues which do not stain well should be

placed for 3 hours in acid alcohol and then washed in alcohol before staining.

5. **Bichromate of Potassium and Acetic Acid** (Tellyesnicky's fluid).—

Bichromate of potassium	3 grams
Glacial acetic acid	5 c.c.
Water	100 c.c.

It is best not to add the acetic acid until just before using. This is a good general reagent. It is valuable for embryos. Objects should remain in some 20 volumes of the fluid from 24 to 48 hours, according to size. It is well to change the fluid once after a few hours. After fixation, tissues should be washed thoroughly in running water (6 to 12 hours) and passed through alcohols of increasing strength beginning with 15 per cent.

By adding 10 c.c. of formalin to each 100 c.c. of this fluid just before using a good reagent for fixing frog eggs and frog embryos is obtained. Fix for 24 hours. Wash in running water for 6 hours and preserve in 3 per cent formalin.

Smith (*Turtox News*, X [1932], 203) finds the following modification more suitable than the original formula for frog eggs and tadpoles up to 10 mm. in length:

Potassium bichromate	0.5 gram
Glacial acetic acid	2.5 c.c.
Formalin (commercial)	10.0 c.c.
Water	87.5 c.c.

6. **Bichromate of Potassium and Corrosive Sublimate** (Zenker's fluid).—For formula, see page 8, reagent 7.

Zenker's is a valuable reagent for both histological and embryological material (embryos up to 25 mm.). Several hours are required for fixation: 2 to 4 hours for a 2-day chick; 8 to 10 hours for objects or embryos of 6 to 8 mm.; 24 hours for embryos of 12 to 14 mm., etc. For washing, running water is employed for from 12 to 24 hours. The object is then transferred to gradually increasing strengths of alcohol up to 70 per cent, leaving it according to size from 1 to 3 hours in each alcohol. To remove the excess of corrosive sublimate, see p. 18; also "Caution" 1, p. 237. Almost any stain follows this reagent well. Both nuclear and cytoplasmic structures are properly fixed. See page 29.

7. **Bichromate of Potassium, Corrosive Sublimate, and Formalin** (Zenker formalin mixtures).—

A. *Helly's Fluid:*

Prepare a Zenker's fluid, but, instead of acetic acid, just before using add formalin in the same proportion and in the same way. Good for tissues in which it is desired to examine the granular cytoplasmic contents. Adding double the amount of formalin gives *Maximow's fluid*. Zenker's fluid containing 5 to 10 per cent formalin in addition to the acetic acid (as much formalin as acetic acid) is highly recommended by Meeker and Cook (*Arch. für Ophthal.*, LVII [1928], 185) as an eye fixative.

Such combinations as the foregoing 5, 6, and 7 are inconsistent in that in potassium bichromate and formalin they bring together an oxidizing and a reducing agent in the same mixture.

B. *Danchakoff's Mixture:*

Corrosive sublimate.....................	50 parts
Potassium bichromate...................	25 parts
Sulphate of soda......................	10 or 12 parts
Water................................	1,000 parts

Boil to dissolve. Just before using add sufficient formalin to make the solution contain 5 per cent for soft tissues or 10 per cent for dense tissues. Fix for from 2 to 4 hours, never more than 6 hours. Keep the mixture at about 37° C. during fixation. This fluid is useful for some kinds of cytological work.

8. **Bichromate of Potassium and Cupric Sulphate (Erlicki's fluid).—**

Bichromate of potash..................	5 grams
Sulphate of copper....................	2 grams
Distilled water.......................	220 c.c.

Pulverize the crystals before adding the water.

Erlicki's fluid is a good reagent for general use, and is especially valuable for voluminous objects such as advanced embryos. Its principal drawback is the length of time required properly to harden objects (5 days to 3 weeks). The process may be hastened by keeping the fluid containing the tissue at the temperature of an incubator (39° C.). At the end of this time transfer the object to 35 per cent alcohol, keeping it in the dark for 2 hours to avoid precipitation. The alcohol should be changed occasionally during this time. Repeat the process, using 50 per cent alcohol, and finally preserve the material in 70 per cent alcohol.

9. Bichromate of Potassium and Osmic Acid (Altmann's mixture).—

> Bichromate of potash, 5 per cent aqueous solution................................... 1 part
> Osmic acid, 2 per cent aqueous solution........ 1 part

This is a useful reagent for fixing cell granules such as Altmann's "bioblasts."

10. Bichromate of Potassium, Osmic Acid, and Chromic Acid (Champy's mixture).—

> Bichromate of potash, 3 per cent aqueous solution................................... 7 parts
> Chromic acid, 1 per cent aqueous solution...... 7 parts
> Osmic acid, 2 per cent aqueous solution........ 4 parts

This fluid keeps well and is a useful cytological reagent. Tissues are fixed for from 6 to 24 hours, then washed in running water for about the same length of time. Iron-hematoxylin follows it well as a stain, as does the Champy-Kull triple stain (acid fuchsin, toluidin blue, and aurantia).

11. Bichromate of Potassium and Sodium Sulphate (Müller's fluid).—

> Bichromate of potassium............... 20 to 25 grams
> Sodium sulphate...................... 10 grams
> Water................................ 1,000 c.c.

Müller's fluid was formerly widely used. It is used mainly for the nervous system. It is a hardening rather than a fixing agent. It is doubtful if the sulphate is of any value since the reagent seems to work as well without as with it. Nuclear structures are not well preserved. It acts very slowly. Specimens require immersion in a large quantity of the fluid from 3 to 10 weeks, according to size. The solution should be changed every 2 days for the first 10 days, and later about once a week. If a scum appears at any time, the fluid should be changed. In washing, the tissues are placed in running water for a number of hours and are then treated with gradually increasing strengths of alcohol in the usual manner. For some purposes, however, the tissue is transferred directly from the fluid to 70 per cent alcohol. In any event, the material should always be kept in the dark to prevent precipitation.

A mixture of Müller's fluid 1 part and physiological saline 9 parts makes a good macerating and isolation medium for epithelia. Wash the tissue in normal saline and then in a 1.5 per cent sodium bicarbonate to remove mucin. Subject to the maceration mixture for 36 hours, or until the cells come away freely when the tissue is lightly stroked with a small brush.

Bouin's Picro-Formol.—See pages 9, 30, and 242.

Dr. Pearl E. Claus finds that to secure sharp differential staining it is advisable, after Bouin fixation, to run slides from water into 1 per cent permanganate of potash till the sections turn brown, followed by 1 per cent oxalic acid till the brown disappears, then to water again, followed by the stain desired.

Carnoy's Acetic Alcohol.—See 2.

12. Chloride and Acetate of Copper (liquid of Ripart and Petit).—

Camphor water	75.0	grams
Crystallized acetic acid	1.0	gram
Distilled water	75.0	c.c.
Acetate of copper	0.30	gram
Chloride of copper	0.39	gram

This is a good reagent for cytological work where objects are to be studied in as fresh a condition as possible. Methyl green (reagent 67) should be used for staining. Only aqueous media are employed with such material.

13. Chromic Acid.—Aqueous solutions of from 0.2 to 1 per cent are used. The acid is best kept in the form of a 1 per cent stock solution. Tissues are left in at least fifty times their volume of the acid for from 24 hours for small pieces to one or more weeks for larger ones. The objects are then washed in running water for several hours, after which they are treated with gradually increasing strengths of alcohol. Do the washing and dehydrating in the dark. If sections of chromic-acid material do not stain readily, they should be treated for 3 hours with acid alcohol, washed out with ordinary alcohol, and then stained. Hematoxylin or some of the anilins are the best stains for chromic material. Chromic acid hardens much more rapidly than bichromate of potash. It makes tissues extremely brittle. Since it is an oxidizer, it should not be used in mixtures containing alcohol or formaldehyde.

14. Chrom-Acetic-Osmic Acid (Flemming's solution).—

Chromic acid, 1 per cent aqueous solution	15 parts
Osmic acid, 2 per cent aqueous solution	4 parts
Glacial acetic acid	1 part

This is the so-called "strong" solution of Flemming. The mixture should not be made until immediately before using, because it deteriorates if allowed to stand for any considerable length of time. The fluid is valuable for cytological work, especially for the study of karyokinetic figures. Only small pieces of tissue should be used, as the reagent penetrates poorly. They should remain in the fluid for from 24 to 48 hours and then be washed in running water for from 6 to 24 hours. From water they are transferred to gradually increasing strengths of alcohol. Particles of fat are blackened by the mixture. Sections stain well with safranin or hematoxylin.

Winiwarter's modification:

Osmic acid	4.0 grams
Chromic acid	7.5 grams
Distilled water	950.0 c.c.

At the time of using add 3 or 4 drops of glacial acetic acid, or 6 or 8 drops of trichloracetic acid, to 20 c.c. of the mixture. Fix for from 48 to 72 hours. Wash for 12 to 24 hours in running water. Dehydrate very gradually and clear in cedarwood oil.

D. M. Kaufman (*Anatomical Record*, May, 1929) finds that a Flemming-formol mixture (Flemming's strong solution with 40 per cent formaldehyde in place of acetic acid) followed by iron-hematoxylin gives the best results in the study of megakaryocyte nuclei.

Read the remarks on osmic acid, 26.

15. Chrom-Acetic-Bichromate Mixture (Goldsmith's).—

Chromic acid, 1 per cent solution	15 parts
Bichromate of potassium, 2 per cent solution	4 parts
Acetic acid, concentrated glacial	1 part

Goldsmith recommends this fluid for protozoa, planaria, and embryonic avian tissues. Heidenhain's iron-hematoxylin follows it well. Fixation requires from 2 to 24 hours. For protozoa concentrated in a centrifuge tube Goldsmith dilutes the fixing fluid one-half with water; for the cover-slip method he uses it undiluted.

16. Chrom-Acetic-Formalin Mixture.—

Chromic acid, 1 per cent solution	16 parts
Glacial acetic acid	1 part

Just before using add to two volumes of this mixture one volume of formalin. This is a good fixing fluid for general embryological work.

17. **Corrosive Sublimate** (mercuric chloride, bichloride of mercury).—Corrosive sublimate is ordinarily used as a saturated solution in distilled water (about a 7 per cent solution) or in normal saline. The latter keeps better and contains a greater percentage of the sublimate. Corrosive sublimate is an excellent and rapid fixing fluid for many objects (glands, epithelia, etc.). Objects should remain in the fluid only long enough to become thoroughly fixed; this has been accomplished when they have become opaque throughout. Only a few minutes or even seconds are required to fix very delicate objects, but denser tissues may require from 4 to 24 hours. The value of the fluid is usually enhanced by the addition of 5 per cent of glacial acetic acid. Small pieces of tissue (not over 0.6 cm. in diameter) should be used where practicable. Washing may be done in running water (several hours) or in 50 to 70 per cent alcohol.

CAUTIONS.—(1) With corrosive sublimate or mixtures containing it, the mercuric salt is often not wholly removed in washing. If the tissues are to remain several days or weeks in alcohol, the alcohol will gradually extract it. If they are to be used within a few days, however, it is necessary to remove the excess of sublimate by adding a few drops of a 10 per cent alcoholic solution of iodine to the 70 per cent alcohol. Sufficient of the solution is added to give the alcohol a port-wine color; as often as the color disappears the iodine must be renewed. After from 12 to 48 hours of this treatment, the iodine color persists, and the object should then be transferred to fresh 70 or 80 per cent alcohol, which must be renewed until it no longer extracts iodine from the specimen. Some workers prefer not to treat tissues fixed in a mercuric fixer with the iodized alcohol until they are sectioned and on slides. The treatment then requires only about 30 minutes.

(2) In handling corrosive sublimate, a glass or horn spoon should be used instead of a metal instrument, because it corrodes metal.

(3) Use *distilled* water, not tap water, in making an aqueous solution.

18. **Corrosive Sublimate and Acetic Acid.—**

  Corrosive sublimate, saturated aqueous solution.................................... 100 parts
  Glacial acetic acid....................... 5 to 10 parts

This is an excellent reagent for embryonic tissues and for organs which do not contain a very great amount of connective tissue. See remarks under 17.

19. Corrosive Sublimate and Alcohol (Schaudinn's fluid).—

    Corrosive sublimate, saturated aqueous solution.. 2 parts
    Alcohol, 95 per cent........................... 1 part

This fluid is employed widely for protozoa. Most workers add from 1 to 5 parts of glacial acetic acid. It is no better than Worcester's formol sublimate (25, p. 240) for protozoa. For variants of Schaudinn's fluid and critical comments, see Wenrich and Geiman, *Stain Technology*, VIII (1933), 158.

20. Corrosive Sublimate, Nitric-Acid Mixture (Gilson's mercuronitric mixture).—

    Corrosive sublimate........................ 5 grams
    Nitric acid (approximately 80 per cent)...... 4 c.c.
    Glacial acetic acid......................... 1 c.c.
    Alcohol (70 per cent)...................... 25 c.c.
    Distilled water............................ 220 c.c.
    Filter after three days.

Gilson's is an excellent general reagent and gives a very delicate fixation. Objects should be left in the fluid from 15 to 30 minutes for delicate ones to 6 hours for those which are larger or denser, although many tissues may be left for 24 hours without injury. This is one of the most satisfactory killing and fixing reagents that the beginner can use.

Danchakoff's Mixture.—See 7 B.

Erlicki's Fluid.—See 8.

21. Ether-Alcohol.—Equal parts of sulphuric ether and absolute alcohol.

Flemming's Solution.—See 14.

22. Formalin.—See reagent 6, page 8, and reagent 3, page 30. It should be borne in mind that formalin is a reducing agent and will rapidly decompose such reagents as osmic acid or chromic acid if mixed with them. It preserves fat and myelin, so that they may be stained by the standard methods, and various substances, such as amyloid and hemosiderin, to which it may be desirable to apply chemical tests.

Commercial formalin is always slightly acid. This is not objectionable for ordinary fixation. If neutral formalin is required, add magnesium carbonate to the commercial variety, keeping a deposit of the carbonate on the bottom of the formalin container.

23. **Formalin, Alcohol, and Acetic Acid (Lavdowsky's mixture).—**

Formalin, commercial	10 parts
Alcohol, 95 per cent	50 parts
Glacial acetic acid	2 parts
Distilled water	40 parts

This mixture is recommended in some cases for the treatment of embryos, especially when the nervous system is to be studied. It penetrates well and preserves faithfully; the alcohol counteracts the swelling effects of the acetic acid and the formalin. Material may remain in it without injury for several days. The fluid should be sooner or later be replaced by 70 per cent alcohol. No preliminary washing is necessary.

Dr. Mossman recommends the following modification as preferable for mammalian embryos since it penetrates exceptionally well and decalcifies if lime salts are present:

Formalin, commercial	10 parts
Alcohol, 95 per cent	30 parts
Glacial acetic acid	10 parts
Distilled water	50 parts

Kahle's fluid is much like the foregoing in composition and effect but inasmuch as it is widely used for arthropod tissues, particularly insect larvae, the exact formula is given:

Formalin, commercial	6 parts
Alcohol, 95 per cent	15 parts
Glacial acetic acid	1 part
Distilled water	30 parts

Tissues should be left 12 to 24 hours or longer. Put directly into 50 per cent alcohol, then 70, and keep in 85 per cent alcohol. Larvae are best cut into pieces before fixing, or injected with the fluid. It may be used hot if preferred.

Formalin-Zenker.—See 7.

24. **Formol-Bichromate-Sublimate Mixture (Bensley's).—**

Neutral formalin (freshly distilled)	10	c.c.
Water	90	c.c.
Potassium bichromate	2.5	grams
Mercuric chloride	5	grams

Use the solution soon after making. Bensley (*Biological Bulletin*, XIX [August, 1910], 3) recommends it as a fixing fluid for demonstration of "Holmgren's canals" (Golgi apparatus). It is also one of the best fixing agents for mitochondria.

25. **Formol Sublimate** (Worcester's fluid).—*a*) Make a saturated solution of corrosive sublimate in 10 per cent formalin. This reagent is recommended by Raymond Pearl (*Journal of Applied Microscopy*, VI, 2451) as "extremely satisfactory" for killing and fixing protozoa. Washing may be done in water or 4 per cent formalin. The material may be preserved in 4 per cent formalin or carried up the grades of alcohol to 70 per cent alcohol.

b) If to 9 parts of this formol-sublimate mixture 1 part of glacial acetic acid is added, Worcester's formol-sublimate-acetic mixture is obtained. Pearl recommends this highly for teleost eggs and for embryological material in general. It will not produce coagulations and cloudiness in the gelatinous envelopes of amphibian eggs, if thoroughly washed out after fixing. Preservation is the same as for (*a*). Johnson (*Journal of Applied Microscopy*, VI, 2652) also recommends this reagent very highly for general work except in the case of nervous tissue.

Personally, I have found it advisable not to prepare either of the above mixtures until needed because the formalin, which is a reducing agent, causes much of the mercuric salt to pass over into the insoluble mercurous salt.

Gilson's Mercuro-Nitric Mixture.—See 20.
Helly's Fluid.—See 7 A.
Hermann's Fluid.—See 33.
Kleinenberg's Picro-Sulphuric.—See 32.
Lavdowsky's Mixture.—See 23.
Müller's Fluid.—See 11.

26. **Osmic Acid** (really the tetroxide of osmium OsO_4).—Osmic acid kills quickly and fixes well. It is exceedingly volatile. The chief objections to it, aside from its extremely poisonous nature, are its poor powers of penetration, and the fact that it becomes reduced in the presence of the least amount of dust containing organic particles. Since it is an oxidizer, it should not be mixed with alcohol or formalin. The substance must be handled with the greatest care, as even the vapors are dangerous. It is usually put up in small quantities (0.1 to 1 gram) in hermetically sealed glass tubes. In making up solutions, the wrappings are removed from such a tube, and the tube is dropped into a reagent bottle, where it may then be broken by means of a glass rod.

Aside from its use in mixtures (see reagents 14 and 33), the vapor or a 0.05 to a 1 per cent aqueous solution is commonly used. A stock solution of 1 per cent is usually kept on hand. It *must* be kept free from *dust*. As the most practical way of preventing reduction, Lee recommends that the osmic acid for ordinary work be kept as a solution in chromic acid (a 2 per cent solution of osmic acid in a 1 per cent aqueous solution of chromic acid). This solution may be employed in making up Flemming's solution or for the purpose of fixation by means of osmium vapor. For vapor fixation, however, many workers prefer the vapor from the solid crystals.

To fix by means of the vapor, the tissue is pinned to the lower end of a cork which fits tightly into the bottle containing the osmic acid, or it is suspended by a thread. Objects which will adhere to a slide are fixed by simply inverting the slide over the mouth of the bottle. The time required for such fixation varies from 30 seconds or a few minutes for isolated cells to several hours for thicker objects, such as the retina. For fixing in the solution, 24 hours are required ordinarily. Objects are then washed in running water for the same length of time. Only small or thin pieces can be fixed by means of either the solution or the vapor. The stains which follow osmic acid best are hematoxylin, methyl green (for study in aqueous media), alum-carmine, picro-carmine, and safranin.

27. Picric Acid.—A cold saturated aqueous solution (about 1.2 per cent) of picric acid is commonly used. Small objects are fixed in from a few minutes (infusoria) to 6 hours; objects up to 1 cm. in size in from 24 to 36 hours. They may be left a much longer time, however, without injury. Such cell structures as mitochondria are fixed poorly or not at all in picric acid or its mixtures. Large objects may require weeks for proper fixation. After fixing, tissues should be washed in 70 per cent alcohol until the alcohol is no longer deeply colored by the picric acid. The tissue should not pass, during subsequent treatment (with a few exceptions in case of staining), into an aqueous medium or into an alcohol of less than 70 per cent strength, because such media seem to undo the work of fixation.

28. Picric Alcohol.—Gage recommends a 0.2 per cent solution of picric acid in 50 per cent alcohol as an excellent fixer and hardener for almost any tissue or organ. Time required, 1 to 3 days. Entire objects which have been fixed in picric acid or in picric alcohol stain readily in borax-carmine or paracarmine.

29. Picro-Acetic.—Saturate a 1 per cent aqueous solution of acetic acid with picric acid. This liquid is widely used as a general reagent. and is to be preferred for most purposes to picric acid alone. For washing, etc., see remarks under 27.

30. **Picro-Acetic-Formalin.**—See *Bouin's Picro-Formol*, pages 9, 30.

Smith's modification of Bouin.—Dr. Harvey M. Smith, after several years of experimentation with various picro-formol mixtures in perfecting a technique for embryonic and adult rabbit eyes, informs me that the following mixture fixes just as well as Bouin's and does not harden fibrous tissue so much:

Picric acid, saturated aqueous solution	45 parts
Alcohol, 95 per cent	45 parts
Formalin	5 parts
Acetic acid (glacial)	5 parts

31. **Picro-Sublimate.**—

Rabl's:

Picric acid, saturated aqueous solution	1 vol.
Corrosive sublimate, saturated aqueous solution	1 vol.
Distilled water	2 vols.

This mixture has been especially recommended for embryos. They are left in the fluid for 12 hours, then washed in weak alcohol and transferred to gradually increasing strengths of alcohol.

O. vom Rath's:

Picric acid, cold saturated solution	1 vol.
Corrosive sublimate, hot saturated solution	1 vol.
Glacial acetic acid	0.5 to 1 vol.

After fixing for several hours, transfer the material directly into alcohol.

32. **Picro-Sulphuric** (Kleinenberg's).—

Picric acid, saturated aqueous solution	98 vols.
Sulphuric acid	2 vols.
Water	200 vols.

This is an excellent reagent for embryos, either for entire mounts or for sectioning. Chick embryos of 24 to 48 hours should remain in the liquid for from 2 to 4 hours; older embryos for from 3 to 6 hours. For washing, 70 per cent alcohol is used. It should be changed (frequently at first) until the color ceases to come out of the embryos. Preserve in about 80 per cent alcohol.

Lillie recommends the addition of glacial acetic acid sufficient to make a 5 per cent solution of acetic acid.

33. **Platino-Aceto-Osmic Mixture** (Hermann's fluid).—

Platinum chloride, 1 per cent aqueous solution... 60 c.c.
Osmic acid, 2 per cent aqueous solution......... 8 c.c.
Glacial acetic acid............................ 4 c.c.

Hermann's fluid is one of the most valuable cytological reagents. Only small pieces of tissue should be used. The washing and subsequent treatment are the same as for Flemming's solution (14). For subsequent treatment with pyrogallol, see 77. Read, also, remarks on osmic acid (26).

Rabl's Picro-Sublimate.—See 31.
Rath's (O. vom) Picro-Sublimate.—See 31.
Ripart and Petit, Liquid of.—See 12.
Schaudinn's Fluid.—See 19.
Tellyesnicky's Fluid.—See 5.
Van Gehuchten's Fluid.—Same as 2, *b*).
Worcester's Fluid.—See 25.
Zenker's Fluid.—See 6.

II. STAINS

Read the general statement about stains in chapter ii, pp. 19–22.

Altmann Acid Fuchsin and Picric Acid Stain.—See page 250.

34. **Alum-Cochineal.**—For formula see page 9. Alum-cochineal is one of the best stains for entire objects. It is easy to work with, and does not overstain. The time required for staining is from 24 to 36 hours ordinarily depending upon size and density of the object. After staining, the object should be washed in water for 15 or 20 minutes to extract the alum, which would otherwise crystallize when the preparation is placed in alcohol. Too long an immersion in water may extract the stain to too great an extent. From water the object should be passed upward through the grades of alcohol, remaining about an hour in each. The writer has found alum-cochineal especially valuable for flatworms (tapeworms, flukes, etc.), small roundworms and embryos. If the worms are pigmented, they should be bleached in chlorine vapor or hydrogen peroxide before staining. Alum-carmine differentiates structure as no other carmine will. Embryos for sectioning should not be destained but should be washed in 70 per cent alcohol for 24 hours. If it is desired to use a counterstain with it, Lyon's blue, picric acid orange G, or light green will answer.

35. Alum-Carmine.—

Powdered carmine.......................... 1 gram
Ammonia alum (2.5 per cent aqueous solution) 100 c.c.

Boil for 20 minutes, and filter when cool. The uses and manipulation are the same as for reagent 34. These stains affect calcareous structures injuriously.

36. Anilin Stains.

—Read the general remarks about anilin stains in chapter ii (p. 20). The formulae for some of the most important are given separately in this list in their proper alphabetical position.

The dyes are dissolved in water, in alcohol of any desired strength, or in anilin water, according as they are soluble in these media, or as they meet the needs of the operator. Some workers even use some of them as counterstains dissolved in the clearing fluid. For the study of nuclei, after Hermann's or Flemming's fluid has been used for fixing, the writer has found a weakly alcoholic anilin-water solution to be the most satisfactory. As cytoplasmic contrast stains alcoholic solutions (in 70 to 95 per cent alcohol) have given the best results. Anilin water is made by shaking up 4 c.c. of anilin oil in 90 c.c. of distilled water and filtering the mixture through a wet filter. Enough alcohol may be added to make it a 20 per cent alcohol, if a weakly alcoholic solution is desired.

The length of time which sections should be immersed in the stain varies from a few seconds or minutes for some of the dyes (especially when used for cytoplasm) to 24 to 36 hours for others (especially nuclear). Sections usually overstain, in which case they are differentiated by means of alcohol, either pure or slightly acidulated with hydrochloric acid. The color is thus extracted rapidly; decolorization should be stopped immediately after the color ceases to come from the tissue in clouds (20 seconds to 3 minutes). If acidulated alcohol is employed, it must be in much weaker solution than that used for extracting carmines or hematoxylins. One part of hydrochloric acid to 1,000 of water or alcohol is about the correct proportion. When one desires to study the karyokinetic figures of nuclei, the acid-alcohol differentiation should be employed, but if resting nuclei are to be studied, only neutral alcohol should be used.

37. Anilin Blue, Orange G, and Acid Fuchsin (Mallory's triple connective-tissue stain).—

Solution I:
Acid fuchsin........................... 0.2 gram
Distilled water........................ 100 c.c.

Solution II:
Anilin blue (Grübler's water soluble)..... 0.5 gram
Orange G (Grübler).................... 2.0 grams
Phosphomolybdic acid, 1 per cent aqueous
solution........................... 100 c.c.

The tissue should have been fixed in Bouin's or in Zenker's fluid. Stain either paraffin or celloidin sections in the acid-fuchsin solution for 30 seconds to 5 minutes, depending upon the tissue. Transfer directly to solution II and stain for from 15 to 20 minutes or longer. Wash and dehydrate in two changes of 90 per cent alcohol. Stronger alcohol often causes precipitation. Let stand in the second jar of 90 per cent alcohol until the tinge of blue desired in the tissue is obtained. Pass paraffin sections through absolute alcohol, clear in xylol, and mount in balsam. Clear celloidin sections from 90 per cent alcohol in creosote or other celloidin clearer, or blot and clear in xylol, finally mounting in balsam.

Connective-tissue reticulum, collagen fibrils, mucus, amyloid, and various other hyaline substances stain in different shades of blue; nuclei, cytoplasm, axis-cylinders, neuroglia fibers, fibroglia fibrils, and fibrin stain red; elastic fibers, pale pink or yellow; and red blood corpuscles and myelin sheaths, yellow. If the acid fuchsin is omitted, nuclei and protoplasm stain yellow and the connective-tissue fibrillae and reticulum stand out sharply in deep blue. This is an excellent stain for developing bone, inasmuch as cartilage stains light blue and bone dark blue.

Krichesky's modification (*Stain Technology*, VI [1931], 97) is preferred in our own laboratory as a cytological stain:

Solution 1:
Acid fuchsin (dye content 62 per cent)... 0.25 gram
Distilled water........................ 100.00 c.c.

Solution 2:
A. Anilin blue........................ 2.00 grams
Distilled water..................... 100.00 c.c.
B. Orange G (dye content 83 per cent) . 1.00 gram
Distilled water..................... 100.00 c.c.
C. Phosphomolybdic acid (1 per cent).. 100.00 c.c.

Use dyes certified by the Stain Commission. Keep solutions A, B, and C in separate bottles and, just before using, mix in equal parts to make solution 2. (1) Pass sections through a descending series of alcohols and, after 5 minutes in water, transfer them to solution 1 for from 1 to 3 minutes (time determined by trial). (2) Wash by dipping the slide into water until surface stain is removed; then transfer to solution 2 for from 3 to 5 minutes or longer. (3) Wash by dipping the slide into water 1 to 3 times only; then transfer rapidly through 70, 80, and 95 per cent alcohols, dipping hurriedly 3 to 5 times in each. Since anilin blue is readily soluble in water, the depth of this stain can be controlled by the amount of washing. With very thin sections we find that, to hold the stain, we have to transfer directly from solution 2 to 95 per cent alcohol without the intervening washings. (4) Dehydrate in absolute alcohol for several minutes, clear in xylol, and mount in balsam. Dr. Pearl E. Claus finds that for tissues that take anilin blue deeply, such as thyroid or uterus, the amount of anilin blue may well be reduced.

Bensley's Acid Fuchsin-Methyl Green Method.—See page 163.

Bensley's Copper Chrome Hematoxylin Method.—See page 164.

38. Bismarck's Brown.—Boil 1 gram of the stain in 100 c.c. of water, filter, and add 30 c.c. of strong alcohol. Bismarck brown is a nuclear stain which does not overstain, although it acts rapidly. After staining, wash in 95 per cent or absolute alcohol. This stain is also used in aqueous solution for *intra-vitam* staining; the nucleus of the living cell may thus be colored. It has been used as an *intra-vitam* stain mostly in the study of infusoria. The stain may be fixed by means of a 0.2 per cent chromic-acid solution, but this, of course, destroys the life of the cells.

39. Borax-Carmine (Grenacher's).—

Borax (4 per cent aqueous solution)	100 c.c.
Carmine	1 gram

Boil until the carmine dissolves, then add 100 c.c. of 70 per cent alcohol. Filter after 24 hours.

This is a stain much used in the past for staining in bulk. Objects must be left in it for from 24 hours to several days. They are then transferred, without washing, to acid alcohol and left until the color no longer comes away in clouds. Objects should become bright scarlet in color. Finally they should be washed and hardened in neutral alcohol.

40. Bordeaux Red.—

Bordeaux red	1 gram
Distilled water	100 c.c.

This is a good plasma stain. Recommended by Heidenhain as a contrast stain for iron-hematoxylin in the demonstration of centrosomes (p. 168). Stain for 12 to 24 hours.

41. Carmalum (Mayer's).—

Carminic acid	1 gram
Alum	10 grams
Distilled water	200 c.c.

Dissolve with heat and filter the solution when cold. Add a few crystals of thymol or a little salicylic acid to prevent the formation of mold. Carmalum is one of the best stains for staining objects in bulk and will follow almost any fixing reagent, even osmic acid. If the object has an alkaline reaction it does not stain so well. Washing is done in water.

42. Carmine (Beale's).—

Powdered carmine	1 gram
Ammonia	3 c.c.
Pure glycerin	96 c.c.
Distilled water	96 c.c.
Alcohol, 95 per cent	24 c.c.

The ammonia and part of the water are first mixed and the carmine dissolved in the mixture. The remaining water is added and the solution is left in an open dish until the ammonia has almost evaporated. The alcohol and glycerin are then added. For staining, equal parts of the stain and glycerin are used. The staining is carried on for 24 hours under a bell-jar in an uncovered dish. A second open dish containing acetic acid is placed under the bell-jar. After staining, the sections are washed in water, then in weak hydrochloric acid (1 to 500 of water), and again in water. Minot recommends this stain and method of treatment especially for the placenta and for the central nervous system of embryos.

43. Carmine, Picric Acid, and Indigo Carmine (Calleja's staining fluid).—

Solution I:

Carmine	2	grams
Lithium carbonate, saturated aqueous solution	100	c.c.

Solution II:

Indigo-carmine	0.25	gram
Picric acid, saturated aqueous solution	100	c.c.

Place sections in solution I for from 5 to 10 minutes, then into acid alcohol until they become pale red (20 to 30 seconds); wash well in water. Next place the sections in solution II for 5 to 10 minutes, then into acetic acid (0.2 to 0.5 per cent) for a few seconds, and wash well in water. Dehydrate rapidly and clear in xylol. The method is useful for epithelial cells and connective tissue.

44. Carmine, Acid (Schneider's).—Add carmine to boiling acetic acid of 45 per cent strength until no more will dissolve. Filter the solution when cool. This is a valuable reagent for the study of the nuclei of fresh cells. It is very penetrating and gives a brilliant stain. The strong acetic acid ultimately destroys the cell. See comments under memorandum 2, p. 173.

Champy-Kull Triple Stain.—See page 250.

45. Congo Red.—For formula, see page 10. This is a good counterstain for hematoxylin, especially when applied to fetal and young tissues. Its solutions become blue in the presence of free acid, hence it is useful in determining the existence of free acid in tissues.

46. Cyanin (Chinolin Blue; Quinoline Blue).—Dissolve 1 gram of cyanin (prepared by H. A. Metz & Co., of New York) in 100 c.c. of 95 per cent alcohol, and add 100 c.c. of distilled water. This is a good cytological stain. Sections are stained for 5 to 10 minutes. Chromosomes stain a deep blue. I have found cyanin followed by erythrosin (0.5 per cent alcoholic solution) especially valuable for spermatozoa.

47. Ehrlich-Biondi Triple Stain (Heidenhain).—The ingredients should be obtained from K. Hollborn & Sons (successors to G. Grübler & Co.), Leipzig, Germany, or from their agents (e.g., Pfaltz & Bauer, 300 Pearl Street, New York City).

> Acid fuchsin, saturated aqueous solution......... 4 parts
> Orange G   "   "   "   .... 7 parts
> Methyl green (Methylgrün OO) saturated aqueous
>   solution.................................... 8 parts

The solution of orange should be prepared first, and the solutions of fuchsin and methyl green added to it with continual stirring. Each solution must be thoroughly saturated; it takes several days for this to occur. The above-mentioned mixture constitutes a stock solution which should be diluted with about 50 or 100 times its volume of water before using. According to Lee (*Microtomist's Vade-Mecum*), "if a drop be placed on blotting paper it should form a spot bluish green in the center, orange at the periphery. If the orange zone is surrounded by a broader red zone, the mixture contains too much fuchsin." For use with this method, tissues should be fixed in pure corrosive-sublimate solution. Sections should be thin (3 to 5 microns) and must remain in the stain from 12 to 24 hours. They should then be rapidly washed in 95 per cent alcohol, placed for a short time in absolute alcohol, and cleared in xylol. If the sections remain in the alcohols any considerable length of time, the methyl green will be extracted. The stain is very uncertain in its action, but when it is successfully applied the results are excellent. It is used chiefly in cytological studies, especially in connection with gland cells. Grübler prepares a dry powder for this three-color mixture, but the results are usually not as satisfactory as when the mixture is properly made fresh. To prepare the stain from the powder, a 0.4 per cent solution of the latter in distilled water is made, and to 100 c.c. of this solution 7 c.c. of a 0.5 per cent aqueous solution of acid fuchsin is added.

48. **Ehrlich's Triple Stain.**—For blood films Ehrlich's so-called triacid mixture is a serviceable stain which is widely used.

Orange G, saturated aqueous solution	14 c.c.
Acid fuchsin, saturated aqueous solution	7 c.c.
Distilled water	15 c.c.
Absolute alcohol	25 c.c.
Methyl green, saturated aqueous solution	12 c.c.
Glycerin	10 c.c.

Each solution must be thoroughly saturated (several days). Add the ingredients in the order named, shaking the mixture well before each addition. It is best for the stain to stand several weeks before it is used. Neutrophil granules stain violet, oxyphil granules a brownish red. The mixture stains in from 5 to 15 minutes.

49. **Eosin.**—See reagent 16, page 10. The term as commonly used refers to Eosin Y, sometimes called "water-soluble eosin," or "yellowish eosin." This is the dye used in the various blood stains (Giemsa stain, Jenner stain, Romanovsky stain, tetrachrome stain, Wright stain). An ethyl ester of eosin Y, ethyl eosin, is often sold as alcohol-soluble eosin. This acid dye of the xanthene group is often used after hema**toxyl**in as a contrast stain. It is specific for certain granules of leucocytes and for red blood corpuscles, giving to the latter a very characteristic coppery-red tinge. Some workers prefer to dissolve it in water or in some cases in the clearer.

50. **Erythrosin.**—An eosin; properties and manipulation much the same as ordinary eosin (see reagent 49).

51. Fuchsin, Acid (Rubin S, Acid Magenta, Magenta S).—

Acid fuchsin............................ 0.5 gram
Distilled water......................... 100 c.c.

This is an excellent anilin stain for cytoplasmic structures. It is also used in some instances as a specific stain for nerve tissue. Acid fuchsin should not be confounded with *basic fuchsin*, which is a nuclear stain. It too is used in aqueous solution. When fuchsin alone is mentioned by writers, without specifying whether it is acid or basic, the basic fuchsin is ordinarily meant.

52. Fuchsin-Methyl-Green Stain (Auerbach's).—Keep in separate bottles 0.1 per cent aqueous solutions of acid fuchsin and methyl green respectively. When ready to use, mix 2 parts of the acid-fuchsin solution with 3 or 4 parts of the methyl green, after acidulating every 50 c.c. of the former with 1 drop of a 10 per cent solution of acetic acid.

This stain works best after a sublimate fixer. Chromosomes stain green, linin and plasmosomes red. Sections should not be over 3 or 4 microns thick. Stain for 15 minutes and transfer directly to 95 per cent alcohol. As soon as the green stain ceases to leave the sections in clouds, pass the slides rapidly through absolute alcohol and xylol and mount in balsam.

53. Fuchsin (Acid) and Picric Acid (Van Giesen's stain).—

Acid fuchsin, 1 per cent aqueous solution........ 10 c.c.
Picric acid, saturated aqueous solution.......... 90 c.c.

This stain is frequently used in conjunction with hematoxylin in the study of fibrous or of nerve tissue. Small bits of tissue should be fixed in corrosive sublimate or its mixtures. Sections are slightly overstained with hematoxylin, rinsed in water, and then stained 5 minutes in the picro-fuchsin mixture. To avoid extracting too much of the yellow color in dehydrating and clearing, the alcohols and clearer should each have a few crystals of picric acid added to them. The result should be: nuclei and epithelia brown; white fibrous connective tissue red; elastic tissue and muscle yellow.

54. Fuchsin (Acid), Toluidin Blue, and Aurantia (Champy-Kull triple stain).—(1) Stain first in Altmann's acid-fuchsin anilin-oil mixture (5 to 10 grams of acid fuchsin in 100 c.c. of anilin-water) and heat until steaming. (2) Cool the slide for 6 minutes, pour off the stain and rinse in distilled water. (3) Counterstain in a 0.5 per cent solution of toluidin blue (a saturated solution of thionin in distilled water may be used instead) for 1 to 2 minutes. (4) Wash in distilled water and differentiate in 0.5 per cent solution of aurantia in 70 per cent alcohol for 20 to 40 seconds (watching extraction of red stain under the microscope that it may not go too far). (5) Wash in 95 per cent alcohol, then absolute alcohol, followed by xylol, and mount in balsam. The mitochondria (and sometimes Golgi apparatus) should be red; chromatin, blue; and the cytoplasmic background, brownish-yellow to green. It follows best Champy's fluid, but also is satisfactory after Altmann's and Flemming's (without acetic acid).

If after step 1 the preparation is cooled for 6 minutes, the dye poured off, the sections then washed in a dilute picric-acid solution (1 part saturated aqueous solution of picric acid diluted with 2 parts of distilled water), heated, blotted, dehydrated, cleared in xylol and mounted in balsam, the Altmann acid-fuchsin and picric-acid stain for mitochondria ("bioblasts") will be obtained.

55. Gentian Violet.—This is one of the best of the nuclear anilin stains. It is best made up in anilin water and weak alcohol (see reagent 30).

Gentian violet	1 gram
Anilin water	80 c.c.
Alcohol, 95 per cent	20 c.c.

The stain works well with thin sections. It is also widely used in the study of bacteria. For differentiation, Gram's method is used.

Gram's solution:

Iodine	1 gram
Iodide of potassium	2 grams
Water	300 c.c.

After staining, the sections are placed in this solution until they are black (2 to 3 minutes) and are then decolorized in absolute alcohol until they appear gray. See also reagent 80.

NOTE.—Considerable confusion exists concerning this dye. Gentian violet, as referred to by microscopists, is a brand sold by Grübler and Hollborn. It is a mixture of crystal violet and one of the higher methyl violets. Crystal violet is often sold in the United States under the trade name of gentian violet. The Commission on Standardization of Biological Stains (see p. 21) is recognizing two grades of gentian violet: (1) a *bluish,* which is mainly or wholly crystal violet; (2) the other, a *reddish,* which is seemingly a mixture of crystal violet with one of the deeper methyl violets. For staining bacteria and for most histological and cytological work, the bluish is usually regarded as preferable.

56. Gold Chloride.—The gold-chloride method is used chiefly in the study of nerve-fiber terminations, both motor and sensory, although it is sometimes used for the coloration of other tissue elements (capsules of cartilage, etc.). The process is really an impregnation; through the agency of sunlight and of certain reagents (acetic, citric, formic, or oxalic acid) the gold is deposited in the tissues in the form of very fine particles. There are numerous modifications of the method, one of which is given in chapter x. For use as a bleach in the study of mitochondria see page 176.

57. Golgi's Chrome-Silver Method.—See chapter x (p. 82).

Hemalum, Mayer's, see page 67.

Hematoxylin.—For general statement see chapter ii (p. 20) and the remarks under 12 (p. 9).

58. Hematoxylin, Conklin's Picro.—

Delafield's hematoxylin	1 part
Water	4 parts

Add one drop of Kleinenberg's picro-sulphuric (23) to each cubic centimeter of the solution. This is a good stain (1 to 3 hours) for embryos which are to be mounted entire. If the embryos are to be sectioned they should be stained for 12 hours.

59. Hematoxylin, Delafield's.—See reagent 12, page 9.

Harris' modification.—One gram of hematoxylin is dissolved in 10 c.c. of absolute alcohol and this is added to a warm solution of 20 grams of alum in 200 c.c. of water. The whole is brought to boiling and 0.5 gram of mercuric oxide is added. After a minute more of boiling the flask is plunged into cold water and cooled rapidly under the faucet. Just before using, 4 per cent of acetic acid is added. The mercuric oxide ripens the stain, hence it is ready for immediate use. It keeps well. Several workers in my laboratory have adopted this stain as superior to Delafield's for routine work.

60. Hematoxylin, Ehrlich's Acid.—

Hematoxylin	2 grams
Absolute alcohol	100 c.c.
Glacial acetic acid	10 c.c.
Glycerin	100 c.c.
Distilled water	100 c.c.
Potassium alum	10 grams

Dissolve the hematoxylin in the acetic acid with 25 c.c. of the alcohol; then add the glycerin and the remaining alcohol. Dissolve the alum in the water by the aid of heat and slowly pour the warm solution while stirring into the solution of hematoxylin. The solution must be exposed to light and air at least 3 weeks to ripen. It is not ready for use until it acquires a deep red color. This solution is an excellent nuclear stain and will keep for years.

Mann's acid hematein is the same as this, except that hematein (Grübler's) is substituted for the hematoxylin. This solution should stain without having to ripen.

61. Hematoxylin, Heidenhain's Iron-.—See reagent 17, page 10, for formula; page 54 for method; chapter xviii for cytological uses. This stain is much used in the study of cell structures such as centrosomes, chromosomes, etc. Tissues are best fixed in Bouin's (or modifications of it), in some of the sublimate solutions, or in acetic alcohol, although it will follow liquid of Flemming, Hermann, or Champy. Sections not over 6 microns thick give best results. The ferric solution must be renewed occasionally as it soon spoils.

62. Hematoxylin, Mallory's Phosphotungstic Acid.—

Hematein ammonium	0.1	gram
Water	100	c.c.
Phosphotungstic acid crystals (Merck)	2	grams

Dissolve the hematein by heating it in a little water. When cool add it to the rest of the mixture. If the stain is too weak at first, it may be ripened by adding 5 c.c. of a 0.25 per cent aqueous solution of permanganate of potassium or by allowing it to stand 2 or 3 weeks. Hematoxylin may be used instead of hematein ammonium if 10 c.c. of the potassium-permanganate solution is added to ripen it. Stain sections 2 to 24 hours.

The stain is recommended for the demonstration of neuroglia, myoglia, and fibroglia fibrils, for fibrin, muscle, connective tissue, and for centrosomes and spindles of mitotic figures.

Janus Green.—See pages 165, 167.

63. Light Green (Lichtgrün S.F.).—This is a beautiful cytoplasmic anilin stain which is frequently used after safranin as a counter-stain. Not more than than 0.5 per cent solution should be used as it stains very rapidly and very deeply. It may be used either as an aqueous or as an alcoholic solution. The writer has found a 0.5 per cent solution in 95 per cent alcohol very satisfactory. Sections should remain in it only a few seconds. Do not confuse it with methyl green, which is sometimes called light green by dealers.

64. Lugol's Solution.—A solution of iodine in water containing iodide of potassium. It is used in various strengths. One of the commonest formulae is that of Gram (p. 251), although some prefer a solution with only one-third the amount of water used by him.

65. Lyons Blue (Bleu de Lyon, Spirit Blue).—This is one of the numerous anilin blues. It is a good contrast stain when used after such nuclear stains as safranin and carmine, particularly with embryonic tissues. It brings out growing nerve fibers well. See reagent 15, p. 10; also p. 54.

Magenta, Acid.—See 51.

Mallory's Triple Connective-Tissue Stain.—See 37.

66. Methylen Blue.—This reagent is an extremely useful one; it is of great value in the study of the nervous system, and it can be made to give results with intercellular cement substance, lymph spaces, etc., as satisfactory and with greater certainty than impregnations obtained with gold chloride or silver nitrate. It is also serviceable as an *intra-vitam* stain. Furthermore, methylen blue (saturated solution in 70 per cent alcohol) followed by eosin is sometimes used for the double staining of blood corpuscles. Methylen blue should not be confounded with *methyl blue*.

Ordinary commercial methylen blue usually contains, in addition to the blue dye, a small quantity of a reddish-violet dye. Such methylen blue is termed *polychromatic* and is especially serviceable in staining certain cell granules. Only the pure methylen blue (so-called medicinal grade), however, should be used for nerve staining and other *intra-vitam* work.

a) **"Intra-Vitam" Stain for Small, Comparatively Transparent Aquatic Organisms.**—Add sufficient methylen blue to the water containing the organisms to tinge it a light blue. Different tissues will take up the color after different intervals of time. A given tissue after having attained a maximum degree of coloration will rapidly lose its color again. It is necessary, therefore, to watch the organisms closely for the maximum of color in the tissue desired. If the observer wishes, the stain may be fixed for more prolonged study by following the processes indicated under (*b*). The order in which various tissues take the stain seems to vary in different organisms. Usually gland cells stain first, then with more or less deviation, other epithelial cells, fat cells, blood and lymph cells, elastic fibers, smooth muscle, and striated muscle. Nerve cells and nerve fibers do not ordinarily take the stain when the entire animal is immersed.

b) **Ehrlich's Method for Nerve-Terminations and the Relations of Nerve Cells and Fibers to the Central Nervous System.**—The stain should be Grübler's methylen blue (*rectificiert nach Ehrlich*). A 1 per cent solution in normal saline is used. Warm the solution till it steams, stir it thoroughly, and, when cool, filter. The tissue must be perfectly fresh. Chloroform the animal and immediately inject the stain into the main artery of the part to be investigated. If the animal is small, the entire body may be injected. The vessels should be filled full, but care must be taken not to rupture them. The part should become decidedly blue in color. It is well after 10 or 15 minutes to inject more stain. At the expiration of half an hour after the second injection remove small pieces of tissue containing the nerve elements desired, and *expose them freely to the air* on a slide wet with normal saline. Examine every two minutes under the microscope (without cover-glass) until the particular element to be investigated (cell, axone, termination) has developed a well-marked blue color. It is important to catch the color at the proper stage and fix it because it soon begins to fade.

Fixing the Stain.—When the desired element has developed a satisfactory blue color, the tissue is transferred immediately to a saturated aqueous solution of *Ammonium picrate* (Dogiel's method) and left for from 6 to 24 hours. For final mounting the tissue should be teased out sufficiently to show the proper elements and then mounted in a few drops of a mixture of pure glycerin (free from acid) and ammonium picrate (saturated aqueous solution), equal parts. It is well to let the tissue stand in 20 to 30 volumes of this glycerin-picrate mixture for a day or two before mounting it. If the preparation is to be kept the cover-glass should be sealed (p. 108).

Sections.—If it is desired to make paraffin sections and mount them in balsam, after treatment with the ammonium picrate (10 to 15 minutes), the tissue must be placed into 20 or 30 volumes of Bethe's fluid, which renders the color insoluble in alcohol.

Bethe's fluid:

Molybdate of ammonia	1 gram
Chromic acid, 2 per cent aqueous solution	10 c.c.
Hydrochloric acid, concentrated c.p.	1 drop
Distilled water	10 c.c.

The tissue is left in this mixture for from 45 to 60 minutes (for small objects) and then washed 1 to 2 hours in distilled water. Dehydrate directly in absolute alcohol; follow this with xylol, imbed in paraffin, and section in the ordinary manner. Sections may be counterstained in alum-carmine or alum-cochineal.

c) **Immersion Method.**—Material which cannot be readily injected or which has failed to stain may be stained by immersion. A 0.1 per cent solution of the stain is used (dilute 1 volume of the solution used for injection with 9 volumes of normal saline). To small pieces (2 to 3 mm. thick) of the tissue add a few drops of the stain at intervals of about three minutes. The tissue should always be moist, but never covered sufficiently by the solution to exclude air. Examine the preparation from time to time under the microscope and when the nerve elements are well stained, fix in ammonium picrate and proceed as in (*b*). In case of the central nervous system, fairly good results may sometimes be obtained by dusting the methylen-blue powder over the freshly cut surface of the part to be studied. The development and fixing of the color is the same as in (*b*).

Cole (*Stain Technology*, IX [1934], 89) prefers the following ammonium molybdate solution for fixing:

Distilled water	50 c.c.
Glycerin	50 c.c.
Ammonium molybdate (hepta) crystals	in excess
Concentrated HCl	15 drops

Shake the solution from time to time until it is saturated with the molybdate.

d) **Nissl's Method of Staining Basophil (Tigroid) Substance in Nerve Cells.**—

Methylen blue	3.75	grams
Venetian soap (white castile soap)	1.75	grams
Water	1,000	c.c.

It is best to keep the stain for some months before using.

Ganglia should be fixed in alcohol, formalin, or corrosive sublimate and sectioned in paraffin. Fix the sections to the slide, dissolve out the paraffin with xylol, and run the preparation down to the aqueous stain in the ordinary way. In a test-tube heat a few cubic centimeters of the stain until it steams, then apply it while still warm to the sections on the slide, which has been placed flat on the desk. It takes about 6 minutes for the stain to act. Pour off the surplus stain and rinse the slide in distilled water. Lay it flat on the desk again and flood the sections with *anilin-alcohol* (95 per cent alcohol, 9 parts; anilin oil,

1 part). Let the sections decolorize (20 to 30 seconds) until they are a pale blue; then drain off the anilin-alcohol and transfer the preparation to absolute alcohol. Clear in xylol and mount in balsam. The basophil granules should appear deep blue in color. They are arranged for the most part concentrically around the nucleus.

e) **Unna's Method of Staining Unstriated Muscle in Sections.**—Stain in a 1 per cent aqueous solution of polychromatic methylen blue, rinse in water, and then leave for 10 minutes in a 1 per cent aqueous solution of potassium ferricyanide. Transfer to acid alcohol until sufficiently decolorized, then complete the dehydration, and mount in the usual way.

f) **For Ordinary Section Staining** where a nuclear stain is desired methylen blue answers very well. It is usually used (2 to 24 hours) in aqueous solution. The treatment is the same as for safranin.

g) **Impregnation of Epithelia, etc.**—Place the fresh tissue, preferably a thin membrane, into a 4 per cent solution of methylen blue in normal saline. To demonstrate the outline of cells, leave the tissue in the stain not longer than 10 minutes. To get a negative image of lymph spaces, canals, etc., in contrast to the ground substance which becomes deeply impregnated, leave the tissue in the stain 20 to 30 minutes. For this purpose it is advisable to remove any membranous covering which invests the organ. In either case, after staining, fix the tissue for 30 to 40 minutes in a saturated aqueous solution of ammonium picrate, changing it once or twice, and examine in dilute glycerin. To preserve the preparation permanently, proceed as in (*b*). To do away with the macerating action of the ammonium picrate, add 2 per cent of a 1 per cent osmic-acid solution to the fixing bath.

67. **Methyl Green.**—This is one of the best of the nuclear anilin stains. It is particularly valuable because it often instantly stains the chromatin of nuclei in fresh tissues. Use in strong aqueous solutions, acidulated to about 1 per cent with acetic acid. It does not give a satisfactory chromatin stain if the tissue has been fixed in acetic acid or mixtures containing it. It follows pure corrosive-sublimate solution admirably.

68. **Methyl Violet.**—This stain is commonly used in 0.5 to 2 per cent aqueous solutions for staining bacteria, nuclei, and amyloid. It may often be substituted for gentian violet. Hanstein's so-called "rosanilin-violet" consists of methyl violet, 0.5 gram; basic fuchsin, 0.5 gram; **70 per cent** alcohol, 99 c.c.

69. Muci-Carmine (Mayer).—

Carmine	1.0 gram
Aluminium chloride	0.5 gram
Distilled water	2 c.c.
Alcohol, 50 per cent	100 c.c.

Mix in the order given; heat gently till the fluid darkens (about 2 minutes); filter after 24 hours. To use, dilute with 5 to 10 volumes of water. The stain (3 to 10 minutes) is specific for mucus-containing cells.

70. Muci-Hematein (Mayer).—

Hematein	0.2 gram
Glycerin	40 c.c.
Aluminium chloride	0.1 gram
Distilled water	60 c.c.

Rub up the hematein in a mortar with the glycerin and the aluminium chloride, then add the water. It stains in from 3 to 10 minutes, mucin appearing blue. If a drop or two of nitric acid is added, its nuclear staining capacity is enhanced.

71. Neutral Red.—Neutral red is used widely as an *intra-vitam* stain. It is a good stain for cytoplasmic granules, and in some cases for mucus-cells. For *intra-vitam* staining it may be used in the same way as methylen blue (with the omission of fixation). For staining fixed material, a 1 per cent or stronger aqueous solution is employed. Granules are stained orange red (bright red in acid medium, yellow in alkaline medium). Rosin finds that in nerve cells stained in neutral red (followed by water, acid-free alcohols, xylol, and balsam) nucleoli and Nissl's granules are stained red, the rest of the cell yellow.

72. Orange G.—This is an excellent cytoplasmic stain and is often used on sections as a contrast to carmine, hematoxylin, and safranin. It is especially good as a counterstain in tissues of vertebrate embryos. Grübler's orange G is the most reliable. It should be used in saturated aqueous solution. The solution does not keep very well.

73. Orcein (Unna's method for elastic fibers).—

Orcein (Grübler's)	1 gram
Hydrochloric acid	1 c.c.
Absolute alcohol	100 c.c.

Sections are stained in a watch-glass or porcelain dish. The dish is warmed over a flame or in an oven until the stain becomes thick through the evaporation of the alcohol. Rinse the stained sections thoroughly in 70 per cent alcohol, wash in water, run up through the alcohols, clear in xylol, and mount in balsam. Elastic fibers should appear dark brown, connective tissues a pale brown. Nuclei may be brought out by staining in Unna's polychrome methylen-blue solution after washing the sections in water.

74. Paracarmine (Mayer's).—

Carminic acid	1.0 gram
Aluminium chloride	0.5 gram
Calcium chloride	4.0 grams
Alcohol, 70 per cent	100 c.c.

Paracarmine is an excellent stain for large objects. It does not overstain ordinarily. The stained tissue is washed in 70 per cent alcohol. In case overstaining occurs add 2.5 per cent glacial acetic acid or 0.5 per cent aluminium chloride to the alcohol used for washing. Objects to be stained should not have an alkaline reaction nor contain limy materials.

75. Picro-Anilin Blue.—

Anilin blue (water sol.)	2 grams
Picric acid	1 gram
Alcohol (80 per cent)	1,000 c.c.

Used as a test for chromatin in fresh tissue, following Congo red.

Picric acid alone is widely used as a contrast stain with carmine, hematoxylin, etc. It is best manipulated as a stain by adding a little to each of the alcohols used in dehydrating, after application of the nuclear stain. However, if acid alcohol is to be used, the picric acid should be used only in the grades above the acid alcohol. It may be employed in staining entire objects as well as sections. See also remarks on washing under 27.

76. Picro-Carmine.—

Ammonium hydrate	5 c.c.
Distilled water	50 c.c.
Carmine	1 gram
When dissolved, add picric acid (saturated aqueous solution)	50 c.c.

Expose to air and light for 2 days, then filter. A few crystals of picric acid should be added to the alcohols used for dehydration after staining.

Picro-Fuchsin.—See 53.

Pyridine-Silver Method (Ranson's Cajal).—See page 86.

77. **Pyrogallol.**—Tissues which have been fixed in Hermann's or in Flemming's fluid for 24 to 36 hours may be treated (without previous washing) with a weak solution of pyrogallol or with crude pyroligneous acid. Lee (*Microtomist's Vade-Mecum*) recommends the pyrogallol as much preferable. Tissues should remain in the fluid from 1 to 24 hours, depending upon size. The result is a black stain which colors both nucleus and cytoplasm. If desired, an additional chromatin stain may be employed. Safranin (79) for 24 hours is recommended; decolorize slightly with very dilute acid alcohol. The stain is excellent for cytological work (for "sphere," etc.).

78. **Resorcin-Fuchsin** (Weigert's elastic tissue stain).—

> Basic fuchsin, 1 per cent aqueous solution....... 100 c.c.
> Resorcin, 2 per cent aqueous solution........... 100 c.c.

Heat the mixture in a porcelain dish and while boiling add 30 c.c. of a 30 per cent solution of ferric chloride; stir and keep boiling for 2 to 5 minutes. Cool and filter. Throw away the liquid. Dry the precipitate which remains on the filter-paper thoroughly in a porcelain dish over a water or sand bath. Return the dried precipitate together with the filter-paper to the first porcelain dish, add 200 c.c. of 95 per cent alcohol, and boil. Remove the paper when the precipitate is dissolved off. Cool, filter, and replace the alcohol lost through evaporation, up to 200 c.c. Add 4 c.c. of hydrochloric acid. The stain works best after formalin-fixed material.

Stain the section 20 minutes to an hour in this solution. Wash in alcohol, dehydrate in absolute alcohol, clear, and mount in balsam. Elastic fibers should appear dark blue on a clear background. If desired, the sections may also be stained, either before or after the staining of the elastic fibers, with one of the carmine or hematoxylin stains.

French (*Stain Technology*, IV [1929], 11) finds that the addition of a trace of dextrin, and the substitution of crystal violet for (or its addition to) the basic fuchsin, makes more certain a satisfactory preparation.

79. **Safranin.**—Safranin is one of the most important of the basic anilin dyes. Read carefully the remarks on anilin stains under 36.

> Safranin..................................... 1 gram
> Anilin water (see 36)........................ 90 c.c.
> Alcohol, 95 per cent......................... 10 c.c

Filter before using. Grübler's "Safranin O" is the most reliable dye. Sections of tissue fixed in Hermann's or Flemming's solution are left in the stains for from 24 to 48 hours. Decolorize as directed under 36.

80. Safranin, Gentian Violet, and Orange G has been a favorite combination ever since the days of Flemming. It is almost indispensable in the study of cell problems, especially spermatogenesis. Tissues are best fixed in Flemming's or Hermann's solutions. The following modification by Foley (*Anatomical Record*, XLIII [1929], 171–85), is one of the best techniques. (1) Unless the sections are of material which has been fixed in a fluid that contained osmic acid, they should be mordanted for 30 to 60 minutes in 0.5 per cent aqueous solution of osmic acid. (2) Rinse thoroughly in water; then stain 1 to 3 hours or more in 1 per cent aqueous safranin. (3) Rinse in water and differentiate in a 0.025 N solution of hydrochloric acid until nuclear outline and chromatin granules are just visible. (4) Rinse in water; then stain in a 0.27 to 0.3 per cent solution of crystal violet in 7 per cent alcohol for about 20 minutes. (5) Rinse in water and again differentiate in the acid alcohol, stopping when the nuclei show definite outlines. (6) Rinse in water and place in Gram's solution (55) or until the sections turn black. (7) Wash off excess iodine thoroughly and immerse in 1 per cent solution of bichloride of mercury until sections turn bright blue (1 to 3 minutes). (8) Wash in water; blot excess, but do not let sections dry. (9) Immerse in 95 per cent alcohol for 4 to 6 seconds and transfer to carbol-xylol while the violet is still coming out in clouds. Leave in carbol-xylol until proper differentiation has occurred (15 seconds to several minutes). (10) Wash in xylol and stain in 1 per cent solution of orange G in clove oil for from a few seconds to one minute. (11) Wash with pure clove oil and xylol, and mount in balsam.

Dyes recommended: Safranin and the clove-oil solution of orange G, dispensed by the Oceanic Company; crystal violet, by the National Anilin and Chemical Company.

Solution 1:
Safranin (dye content not stated) .... 1.0 gram
Distilled water .... 100.0 c.c.

Solution 2:
Crystal violet (91 per cent representing
 0.273 gram dye) .... 0.3 gram
Ethyl alcohol (95 per cent) .... 7.0 c.c.
Distilled water .... 88.0 c.c.

Solution 3:

Orange G (dye content not stated)	0.5 gram
Clove oil	50.0 c.c.

81. Scharlach R.—A saturated solution of the dye in equal parts of 70 per cent alcohol and acetone is used. This is a specific stain for fat. For example, cover-slip preparations are fixed in formalin vapor for 5 or 10 minutes, stained 5 minutes in the Scharlach R solution, rinsed in 70 per cent alcohol, washed in water, counterstained with alum-hematoxylin or methylen blue, washed in water, and mounted in glycerin-jelly. Frozen sections of formalin-fixed material may be treated in much the same way.

With any evaporation of the alcohol a precipitate forms, hence staining should be done in a tightly closed vessel. After staining with alum-hematoxylin, if the sections are put into a 1 per cent aqueous solution of acetic acid for 3 minutes, the contrast is sharper.

82. Silver Nitrate.—The nitrate-of-silver method is used largely as an impregnation method for work on nerve tissue and for demonstrating intercellular substances and outlining boundaries of cells in the epithelial coverings of membranes, etc. Wash the fresh tissue in distilled water, then place it for 2 to 5 minutes in 0.5 to 1 per cent aqueous solution of silver nitrate. Rinse in distilled water, then expose the tissue to bright sunlight in water or glycerin (or in 70 per cent alcohol, if it be mounted in balsam) until a brown coloration appears. Temporary mounts should be made in glycerin. For applications see pages 82, 85.

83. Sudan III.—This is a specific stain for fat. See remarks on page 169. A saturated alcoholic solution is used (5 to 10 minutes). Wash rapidly in alcohol. Since alcohol is a solvent of fat, too long an immersion will destroy the preparation. Mount in glycerin or glycerin-jelly. The tissue should have been fixed previously in Müller's fluid (11) or other medium which does not dissolve fat.

84. Thionin.—This is an excellent stain for chromosomes when used in saturated aqueous solution for about 5 minutes. After corrosive-sublimate fixation it is, when used dilute for 10 to 15 minutes, a specific stain for mucin (mucin red, everything else blue).

Van Giesen's Stain.—See 53.

85. Weigert-Pal Stain for Medullated Nerve Fibers.—

Solution I:

Hematoxylin	1	gram
Absolute alcohol	10	c.c.
Distilled water	90	c.c.
Lithium carbonate (1 part of a saturated aqueous solution to 80 of distilled water)	1	c.c.

Solution II:

Potassium permanganate	0.25	gram
Distilled water	100	c.c.

Solution III; mix just prior to using:

Oxalic acid, 1 per cent aqueous solution	50	c.c.
Potassium sulphate, 1 per cent aqueous solution	50	c.c.

Tissues should have been fixed in Müller's fluid or in 10 per cent formalin, hardened in alcohol, and sectioned. Stain sections until black (6 to 24 hours) in solution I. Wash well in water to which a few drops of lithium carbonate have been added. Transfer to solution II and leave until the gray matter of the nervous tissue becomes brown ($\frac{1}{4}$ to 2 minutes). Rinse in water and decolorize in solution III until the gray matter of the tissue becomes a light brown and the white matter a steel blue ($\frac{1}{4}$ to 1 minute). Each section must be carefully watched to get a satisfactory result. Wash in running water or in several changes of water, dehydrate, clear, and mount in balsam. If desired, a counterstain of alum-cochineal may be given before the final dehydration.

86. Wright's Stain (for blood and for the malarial parasite).—
See memoranda 5 and 6, pages 128–29.

III. NORMAL OR INDIFFERENT FLUIDS
(For Fresh Tissues)

87. Aqueous Humor.—
Obtained by puncturing the cornea of a freshly excised beef's eye. A small amount may readily be obtained by means of a capillary pipette from the eye of a freshly killed frog. Amniotic fluid from pig or cow fetuses is a serviceable fluid for the examination of fresh tissues.

88. Blood Serum.—
Blood is allowed to clot and after 24 hours the serum is poured off. If necessary it may be further freed of blood cells by means of a centrifuge. The serum will keep for only a day

or two. *Schultze's iodized serum* made by saturating blood serum with iodine is sometimes classed as an indifferent fluid, but it is really a dissociating fluid.

89. Locke's Solution.—See page 155.

90. Normal Saline.—

Sodium chloride	0.7 to 0.9 gram
Distilled water	100 c.c.

91. Ringer's Solution.—

Sodium chloride	0.80	part
Calcium chloride (anhydrous)	0.02	part
Potassium chloride	0.02	part
Sodium bicarbonate	0.02	part
Distilled water	100	parts
Dextrose (may be left out)	0.10	part

The following formula is probably better adjusted to tissues of warm-blooded animals:

Sodium chloride	0.900	part
Calcium chloride (anhydrous)	0.024	part
Potassium chloride	0.042	part
Potassium bicarbonate	0.020	part
Distilled water	100	parts

Ringer's solution corresponds more nearly to normal blood serum than does normal saline and is therefore less likely to produce distortions in tissue elements.

IV. DISSOCIATING FLUIDS

92. Bichromate of Potassium.—A 0.2 per cent aqueous solution is commonly used. Nerve cells of the spinal cord and also various epithelia dissociate well in it (2 to 3 days).

93. Caustic Potash.—A solution of 35 parts in 100 parts of water is often used for isolating fibers of smooth muscle or heart fibers. It acts by rapidly destroying the connective tissue (20 to 30 minutes). Examination of the tissue is made by mounting it in the dissociating fluid. If water is added, the tissue will be destroyed. Usually only temporary preparations are made in this fluid, but tissues may be made permanent by neutralizing the alkali by means of acetic acid.

Digestion Method.—See pages 90, 302.

94. Gage's Formaldehyde Dissociator.—

Formalin	0.5 c.c.
Normal saline solution	250 c.c.

Good for epithelia and for nerve cells.

95. Hertwig's Macerating Fluid.—See page 90.

96. MacCallum's Macerating Fluid.—

Nitric acid	1 part
Glycerin	2 parts
Water	2 parts

This fluid is recommended for heart muscle of adults or embryos. Hearts should remain in it from 8 hours to 3 days, according to size. The method is valuable for showing the arrangement of cardiac muscle fibers.

97. Ranvier's One-Third Alcohol.—This is one of the commonest as well as one of the best macerating fluids. It is simply a 30 per cent alcohol. Epithelia will macerate in it sufficiently in 24 hours. A still weaker alcohol (20 to 25 per cent) is used for isolating the nerve fibers of the retina.

98. Sodium Chloride.—A 10 per cent solution of sodium chloride is excellent for tendon, etc. It dissolves the cement substance of epithelial cells and of connective tissue. As a stain, a saturated aqueous solution of picric acid (stain for 24 to 36 hours) followed, after thorough washing in water, by a dilute alcoholic solution of acid fuchsin gives excellent results.

V. DECALCIFYING FLUIDS

Tissues are fixed in Zenker's or other fluid, thoroughly washed, and hardened for at least 24 hours in alcohol before decalcification.

99. Chromic Acid.—Chromic acid diluted to 1 per cent or in combination with other fluids is frequently used for decalcification. Chromic acid, 1 gram; water, 200 c.c.; nitric acid, 2 c.c., is a mixture widely used. It decalcifies well but acts more slowly than the 10 per cent nitric-acid mixture. Bone should first be hardened in Müller's fluid (11).

100. Nitric Acid.—A 10 per cent solution of nitric acid in 70 per cent alcohol may be used. If nitric acid is used for young or fetal bones, it is advisable to use only 1 part of the acid to 99 parts of the alcohol. After washing out in 70 per cent alcohol, the decalcified bone may be kept in 95 per cent alcohol.

101. Phloroglucin Method.—This is a rapid method. Young bones may be decalcified in half an hour and old and hard ones in a few hours. Teeth require a somewhat longer time. Phloroglucin itself does not decalcify, but protects the tissue from the action of the strong nitric acid. One gram of phloroglucin is dissolved in 10 c.c. of pure non-fuming nitric acid with the aid of gentle heat. Ten c.c. of nitric acid in 100 c.c. of water is added to the mixture. Wash thoroughly and stain in Delafield's hematoxylin. After staining leave sections in tap water for 12 hours.

102. **Picric Acid.**—A solution kept fully saturated is useful for delicate bones. It stains and decalcifies the tissue at the same time. Wash in 70 per cent alcohol.

103. **Von Ebner's Fluid.**—

Alcohol, 95 per cent	500 c.c.
Water	100 c.c.
Sodium chloride	2.5 grams
Hydrochloric acid	5.5 c.c.

This is an excellent fluid for bone because in it the ground substance of the bone does not swell up. Sections are best examined in a 10 per cent solution of sodium chloride. See also memorandum 4, page 95.

APPENDIX C

TABLE OF TISSUES AND ORGANS WITH METHODS OF PREPARATION

Numbers refer to reagents described in Appendix B. Unless otherwise directed, preparations are to be dehydrated and mounted in balsam. Abbreviations: do. is used to indicate that the same reagent or process is to be used as the one immediately above; p.c. = per cent; chap. = chapter; mem. = memorandum; sat. = saturated; sol. = solution; alc. = alcohol; aq. = aqueous; g. = gram.

Object or Element	Animal, Organ, or Tissue Recommended for Demonstration	Fixing and Hardening, or Other Preliminary Treatment	Section or Isolation Method P.=Paraffin C.=Celloidin F.=Freezing H.=Free Hand	Staining, etc.	General Remarks
I. BLOOD AND BLOOD-FORMING ORGANS					
Blood, cover-glass preparation	Frog Pigeon Man	Dry		Wright's stain, mem. 5, p. 128.	See also II, p. 122
Crystals		Corrosive sublimate (17) for 15 minutes		Hematoxylin, eosin (59, 49) or Ehrlich (48)	The preparation should be washed with water or alcohol before staining
Erythrocytes (red corpuscles)					See I, e, p. 121
Fibrin					Technique same as for cover-glass preparation
Fresh blood	Frog; pigeon; man	Examine fresh			See I, d, p. 121 See I, b, p. 121; for effects of reagents, etc., on blood
Granules, acidophil (eosinophil, oxyphil)	Leucocytes of man	Ether-alcohol (21) 1 to 2 hours		Ehrlich's hematoxylin (60). To every 100 c.c. add 0.5 g. eosin. Time 2 to 24 hours	
Granules, amphophil (indulinophil)	Leucocytes of guinea-pig, rabbit, or pigeon. Not in man	Do		Equal parts, sat. glycerin sol. of indulin, naphthylamin yellow, and eosin. Stain 4 to 6 hours	Rinse with water, blot, dry, and mount in balsam. Nuclei, red; amph. granules, black; acido. granules, red; hemoglobin, yellow
Granules, basophil	Mononuclear leucocytes of man	Do		Sat. aq. sol. of methylen blue (66f.)	Rinse in water, blot, dry, and mount in balsam. Baso. gran., blue

Granules, mast cells..	Small numbers in normal tissues and normal blood. Large numbers in leukemic blood	Do, for blood For tissues use strong alcohol (3)	P. Thin sections. (Mucous membrane of mouth, intestine, etc.)	Stain for 24 hours in alum-carmine dahlia (dahlia, 1 g.; abs. alc., 25 c.c.; pure glycerin, 12 c.c.; glacial acetic, 5 c.c. In to this mixture pour 25 c.c. of alum-carmine; see 35) Ehrlich's triple (48)	Differentiate in abs. alc. for 24 hours and finally mount in balsam. Nuclei, red; mast granules dark blue
Granules, neutrophil.	Polynuclear leucocytes of man, some transitional cells, pus cells, and myelocytes	Ether-alcohol (21) 1 to 2 hours			Rinse in water, blot, dry, and mount in balsam. Nuclei, green; neut. granules, violet; acido. granules, brownish red
Leucocytes (white corpuscle)............					Technique same as for cover-glass preparation. See II, chap. xv, and particularly mem. 5, p. 128
Lymph glands........	Mesenteric glands of kitten or dog	Acetic alcohol (2a)	P. or C.	Hematoxylin, eosin (59, 49)	Longitudinal sections passing through the hilum are the best See mem. 6, chap. xv, p. 129
Malarial parasite......	Cover-glass preparation from rib of guinea-pig, rat, or young rabbit	Ether-alcohol (21) 1 to 2 hours		Ehrlich-Biondi (47) 24 hours, or hematoxylin, eosin (59, 49)	After 47 rinse quickly in strong alcohol, clear in clove oil followed by xylol, and mount in balsam
Marrow, red..........					
Platelets............					
Spleen, elements of.....	Scrape the cut surface of a fresh spleen				See I, c, chap. xv, p. 121 Examine in normal saline (90)
Spleen, section........	Cat	Gilson (20) or Zenker (6)	P. or C.	Hematoxylin, eosin (59, 49) or, better, hematoxylin and Van Giesen's (59, 53)	Fix the spleen entire and later cut segments of the proper size for sectioning
Thymus..............	New-born infant; young rat	Bouin (30)	P. or C.	Hematoxylin, eosin (59, 49)	
II. CIRCULATORY SYSTEM					
Aorta................	Man; dog; cat	Absolute alcohol (3) or Gilson (20)	C.	Hematoxylin, acid fuchsin (59, 51)	Make both transverse and longitudinal sections
Artery (medium size)..	Do	Absolute alcohol (3) or Zenker (6)	C.	Do	Do
Capillaries and small vessels..............	Pieces of pia mater from base of brain	Zenker (6) for 1 hour	Teased preparation	Carmalum (41)	Mount in glycerin or dehydrate and mount in balsam. For distribution of capillaries see injection method, chap. xiii, p. 99
Elastic elements......	Aorta, artery, etc.	Absolute alcohol (3)	C.	Orcein (73) or Resorcin-Fuchsin (78)	
Endothelium of blood vessels..............	Vessels from the mesentery			Silver nitrate (82) 1 p.c. solution injected into the blood vessels	Capillaries may be examined entire; larger vessels should be slit open and spread out

Table of Tissues and Organs

TABLE OF TISSUES AND ORGANS WITH METHODS OF PREPARATION—*Continued*

Object or Element	Animal, Organ, or Tissue Recommended for Demonstration	Fixing and Hardening, or Other Preliminary Treatment	Section or Isolation Method P.=Paraffin C.=Celloidin F.=Freezing H.=Free Hand	Staining, etc.	General Remarks
Heart	Dog; cat; man	Bouin (30)	C. or P.	Hematoxylin, acid fuchsin (59, 51), or borax-carmine, Lyons blue (39, 65)	For muscle fibers, see under "Muscular tissues," IX
Hemolymph gland	In the fat around heart or lungs of calf	Zenker (6)	P. Thin sections	Methylen blue (66), eosin (49)	Red bodies looking almost like clots of blood
Intercalated disks in heart muscle	Mammalian heart	Zenker (6) or Bouin (30)	P. Thin sections	Mallory's phos. ac. hematoxylin (62)	Overstain (6 to 24 hours) and decolorize with potassium permanganate (3 to 15 min.). Wash quickly and place in a 1 p.c. aq. sol. of acid potassium sulphite until steel-gray in color
Lymph capillaries					Injection by puncture, see mem. 4, p. 101
Purkinje fibers	Heart of a sheep	MacCallum (96) until macerated	Teased preparation	Methylen blue (66)	Mount in glycerin
Spleen, blood vessels of		Alcohol (3)	C.	Ehrlich's hematoxylin (60)	First inject the animal with a carmine mass (chap. xiii). For elements of spleen, see "Spleen," under I
Valves of heart		Bouin (30)	C. or P.	Hematoxylin, acid fuchsin (59, 51)	Longitudinal sections
Vein	Man; cat; dog	Do	C.	Do	
III. CONNECTIVE OR SUPPORTING TISSUES					
Adipose tissue	Subcutaneous tissue	Formalin (22) or Müller (11)	F.	Scharlach R (81)	See remarks under reagent 81; also pp. 163, 169
Areolar tissue	Strips of intermuscular connective tissue. Subcutaneous tissue		Spread out the film with needles	Picro-carmine (76)	Mount in glycerin or examine in normal saline without staining
Bone corpuscles and their processes		Fix in a mixture of Müller (11) 9 parts and formalin (22) 1 part. Decalcify (see next)	C.	Stain for 10 to 15 min. in: thionin (50 p.c. alc. sol.) 1 part; carbolic acid (1 p.c. aq. sol.) 9 parts. Rinse in water and stain 2 to 3 min. in sat. aq. sol. of picric acid (75)	Sections are brought from water into the stain. Cells, reddish-violet; ground substance, yellow; bone corpuscles and processes, dark brown

Bone, decalcified	Fix as above and decalcify in von Ebner's fluid (103) or nitric acid (100)	C. or F.		Picro-carmine (76)	See also p. 95, mem. 5
Bone, development of (endochondral)	Extremities of fetal pigs, cats, or human fetuses	Zenker (6)	C.	Mallory's stain (37)	In advanced fetuses the bone should be decalcified (see p. 95, mem. 4, 5)
Bone, development of (intra-membranous)	Parietal, frontal, or lower jaw of embryos	Do	C.	Do	
Bone, fibers of Sharpey	Vault of a fowl's skull	Decalcify in von Ebner's fluid (103)	H. Thick sections		Tease lamellae apart; fine tapering fibers and apertures from which other fibers have been withdrawn should be visible
Bone, Haversian canals and lamellae. Bone, isolation of corpuscles	Thin fragments of bone	Strong nitric acid for several hours	Teased preparation		See method for grinding bone, chap. xii, p. 95. Cover on a slide and drum upon the cover-glass with the handle of a dissecting needle
Bone, isolation of lamellae		Fix in a mixture of Müller (11) 9 parts and formalin (22) 1 part. Decalcify in von Ebner's fluid (103)			Examine in 10 p.c. aq. sol. of sodium chloride
Cartilage (in general)		Zenker (6) or corrosive sublimate (17)	H. or C.	Hematoxylin, picric acid (59, 75)	Thin sections of costal cartilage may be cut and examined readily without any previous preparation. Finer details of structure vanish in xylol-balsam mounts. Try euparal.
Cartilage, capsule of	Small pieces	Fresh	H.	Treat with 1 p.c. aq. sol. of gold chloride (56)	
Cartilage, connective tissue and elastic fibers in		Fresh, or corrosive sublimate (17)	H. or C.	Picro-carmine (76)	Examine in glycerin. Elastic fibers, yellow; connective-tissue fibers, pink
Cartilage, elastic (yellow fibro-)	Epiglottis; vocal process of arytenoid cartilage (man, beef); external ear (man, rabbit)	Zenker (6) or corrosive sublimate (17)	H. or C.	Hematoxylin, eosin (59, 49)	Thin sections of fresh cartilage should be examined in normal saline (90)
Cartilage, glycogen in		Fresh	H.	Lugol's iodo-iodide of potassium (64)	Glycogen, if present, is stained a mahogany brown; elastic fibers are stained a different shade of brown. See p. 169
Cartilage hyaline	Head of femur; costal cartilage; ensiform process of frog	Zenker (6) or corrosive sublimate (17)	H. or C.	Hematoxylin, picric acid (59, 75)	Thin sections of the fresh cartilage also should be examined in normal saline (90)
Cartilage, white fibro-	Intervertebral disks	Do	H. or C.	Do	Do

TABLE OF TISSUES AND ORGANS WITH METHODS OF PREPARATION—*Continued*

Object or Element	Animal, Organ, or Tissue Recommended for Demonstration	Fixing and Hardening, or Other Preliminary Treatment	Section or Isolation Method P. = Paraffin C. = Celloidin F. = Freezing H. = Free Hand	Staining, etc.	General Remarks
Cells of fibrillar connective tissue	Intermuscular connective tissue		Spread out the film with needles	Wright's stain (mem. 5, chap. xv)	
Connective tissue (general)	Sections of any organ	Zenker (6) or corrosive sublimate (17)	C. or P.	Mallory (37), or Calleja (43)	
Elastic fibers (fine)	Intermuscular or subcutaneous connective tissue	Irrigate with acetic acid	Tease		The white fibrous tissue swells; the elastic fibers remain unaltered and stand out distinctly. See reagents 73 and 78 for special stains
Elastic fibers (coarse)	Ligamentum nuchae of a beef	Fresh	Tease	Picro-carmine (76)	Tease in normal saline (90); mount in glycerin
Encircling fibers	See preparation for fine elastic fibers				
Fat cells	Fatty areas of mesentery	Müller (11) or formalin (22)	H.	Scharlach R (81)	The constrictions seen at intervals along the swollen connective-tissue bundles are caused by encircling fibers
Fenestrated membrane	Basilar artery, cut open lengthwise; endocardium of sheep's heart	Strong caustic potash solution, 6 hours		Acid fuchsin (51)	Do not subject to any treatment with alcohol except as directed under the stain. Osmic acid (0.5 to 1 p.c. sol.) is also used as a test for fat; it turns fat black. See p. 169 Wash well in water before staining
Fibrillar (white fibrous) connective tissue	Tendon from tail of rat or mouse; intermuscular connective tissue	Fresh	Tease		Examine in normal saline (90). See also "Tendon" (p. 272)
Jelly of Wharton	Umbilical cord of young (2 to 3 months) human embryo, or of a 3 to 5 cm. pig embryo	Zenker (6) or Lavdowsky (23)	H. or P.	Hematoxylin, eosin (59, 49)	
Ligament	Ligamentum teres	Zenker (6)	H. or C.	Hematoxylin, Van Gieson (59, 53)	Cross and longitudinal sections
Ligamentum nuchae	Beef	Zenker (6)	Do	Do	Do
Mucoid connective tissue	See "Jelly of Wharton"				
Reticular (adenoid) connective tissue	Spleen or lymph gland	Fresh	F. or H. Thin sections		See p. 90

272   Animal Micrology

Synovial villi	From capsular ligament near border of patella	Fresh	H.	Methylen blue (66 f.)	With low power and without a cover-glass, examine inner surface of a strip n normal saline (90). See also "Fibrillar connective tissue". Cross-section
Tendon		Dry	Tease	Picro-carmine (76)	Examine in glycerin
Tendon cells	Tendon from tail of rat or mouse	Fresh	C.	Stain in bulk in borax-carmine (39) and after sectioning, in picric acid (75)	
Tendon to muscle	Tendo-Achillis of frog	Zenker (6) or Gilson (20)			Longitudinal sections
IV. DIGESTIVE ORGANS					
Blood vessels of	Cat or dog				Carmine injection method, chap. xiii
Cells of Paneth	Ileum of man or rodents. Not present in carnivora	A mixture of potassium bichromate (3.5 p.c. aq. sol.) 80 c.c.; strong formalin 20 c.c. After 24 hours transfer to 3.5 p.c. potassium bichromate sol. for 3 days	P. Thin sections	Iron-hematoxylin (61)	
Crescents of Gianuzzi (demilunes of Heidenhain)	Submaxillary gland (man, dog)	Zenker (6) or Flemming (14)	P.	Hematoxylin, eosin (59, 49)	
Duodenum	Dog	Zenker (6) or Gilson (20)	P. or C.	Hematoxylin (59, mucicarmine (69)	
Epithelium of mouth	Man or dog	Do	P.	Do	For isolation see 92, 94 or 97
Epithelium of stomach	Do	Do	P.	Do	Do
Epithelium of small intestine and villi	Dog or cat	Acetic alcohol (2)	P.	Do	
Esophagus	Man or dog	Zenker (6)	P. or C.	Do	The tissue must be perfectly fresh, otherwise the alkaline bile renders it unfit for microscopical examination
Gall bladder	Do	Do			
Gall duct	Do	Acetic alcohol (2)	C.	Do	Do
Gastric glands (sections)	Stomach of cat or dog, mucous membrane only		P.	Stain in dilute hematoxylin (59) 3 to 5 min.; wash in water; 0.03 p.c. aq. sol. Congo red, 3 to 6 minutes	The animal should have fasted for 1 or 2 days. Chief cells pale blue; parietal cells, red
Gastric glands (fresh)	Rabbit	Fresh. Dissect off mucous membrane and tease		Picro-carmine (76)	Tease in 0.5 p.c. solution of sodium chloride
Goblet cells	Large intestine	Corrosive sublimate (17)	P.	Mallory (37)	

TABLE OF TISSUES AND ORGANS WITH METHODS OF PREPARATION—*Continued*

Object or Element	Animal, Organ, or Tissue Recommended for Demonstration	Fixing and Hardening, or Other Preliminary Treatment	Section or Isolation Method P.=Paraffin C.=Celloidin F.=Freezing H.=Free Hand	Staining, etc.	General Remarks
Granules of salivary glands and pancreas..		Bensley's formol-bi-chromate-sub. (24)	P.	Iron-hematoxylin (61) *et al.*	Alpha and beta cells are differentiated by neutral gentian
Intestinal absorption of fat................	Frog; rat	Osmic acid, 1 p.c. (26)	P. or tease in normal saline (90)	Safranin (79)	Feed the animal on fat bacon for a couple of days before killing
Large intestine........		Zenker (6) or Gilson (20)	P. or C.	Hematoxylin eosin (59, 49); or Mallory (37)	
Lip....................	Lower lip of man or dog	Zenker (6) or Tellyesnicky (5)	C.	Hematoxylin, eosin (59, 49)	Vertical sections including both skin and mucous membrane should be made
Liver, amyloid infiltration of............ Liver, bile capillaries.	Guinea-pig	Alcohol (3) Do	P. C. or P.	Mallory (37)	Inject a concentrated aq. sol. of Berlin blue through the bile duct, after clamping the cystic duct. Avoid too great pressure
Liver, blood vessels of Liver cells...........	Dog; cat Man, pig, or dog	Do Zenker (6) or Gilson (20)	C. P. Thin sections	Acid hematoxylin (60) Hematoxylin, acid fuchsin (59, 51)	First inject with carmine mass, see chap. xiii
Liver, hepatic lobules.	Pig	Acetic alcohol (2)	P. or C.	Do	Make sections both parallel and vertical to the surface
Liver, interlobular connective tissue.......		Do	F. or H.	Do	After staining, shake out cells by shaking sections in a test-tube with some water
Mucin.................	Glands (sections and thin membranes)	Müller (11) or Zenker (6)	P.	Thionin (2 to 3 drops of sat. aq. sol. to 5 c.c. of water), 15 minutes	Tissue blue, mucin reddish. See also "Goblet cell"; also reagents 69 and 70
Nerve plexuses (alimentary canal)......	Stomach or duodenum of guinea-pig	Corrosive sublimate (17), 5 p.c, aq. sol. for 2 to 8 hours	Tease	Gold chloride (56)	
Pancreas.............	Man or dog	Macerate in 0.25 p.c. acetic acid Bouin (30)	P.	Hematoxylin, Congo red (59, 45). Also stain pieces in borax-carmine and section (39)	The granules have frequently been extracted by water in mounted specimens. To see granules tease fresh pancreas in normal saline (90)
Parotid gland........ Peyer's patches (agminated nodules)........		Do Do	P. P. or C.	Do Hematoxylin, eosin (59, 49)	
Small intestine.......	Small intestine at its point of opening into the large intestine (cat or dog) Kitten or puppy	Do	P. or C.	Do. Also stain pieces in borax-carmine before sectioning (39)	Keep without food 1 to 3 days. Kill by blow on the head or by decapitation

274   Animal Micrology

Stomach, cardiac end.	Dog	Acetic alcohol (2)	P. or C.	Hematoxylin, eosin (59, 49), or borax-carmine, Lyons blue (39, 65)	Cut some longitudinal sections to show transition from stomach to intestine
Stomach, fundus		Do	P. or C.	Do	Technique same as for "Pancreas"
Stomach, pyloric end.		Do	P. or C.	Do	Do
Sublingual glands					
Submaxillary glands					
Taste-buds	Foliate papilla of rabbit (at sides near base of tongue)	Flemming (14)	P. Thin sections	Safranin, gentian-violet (80)	Orient the buds carefully for sectioning. Both longitudinal and cross sections are instructive
Tongue (general)		Bouin (30)	P. or C.	Hematoxylin, eosin (59, 49)	Cross-sections
Tongue, papillae and folliculi linguales	Mucous membrane of upper surface (rabbit, man)	Do	P. or C.	Do	For fungiform papillae, tip of tongue. For filiform papillae, middle of tongue. For circumvallate papillae, root of tongue. For folliculi, root of tongue
Tonsil	Cat or rabbit	Acetic alcohol (2) or Tellyesnicky (5)	P. or C.	Do	Technique same as for decalcified bone. See also chap. xii, p. 94
Tooth, decalcified	Canine tooth of dog, cat, or man	Fix in Tellyesnicky (5); decalcify in 1 p.c. nitric acid (100)	P. or F. Serial sections	Borax-carmine, picric acid (39, 75)	First stage, 7 cm. embryo. Second stage, 10 cm. embryo. Later stages, use new-born kitten
Tooth, development	Sheep or pig embryo				
Tooth, enamel prisms					Tease out bits of enamel from teeth prepared as for "Odontoblasts"
Tooth, ground					See chap. xii, p. 94
Tooth odontoblasts	New-born child or other young animal	Müller (11), 6 to 8 days	Withdraw pulp and tease a bit of its surface	Picro-carmine (76)	Seen best in teeth which have not cut the gum. Examine in slightly acidulated glycerin
V. EAR					
Ceruminous glands		Alcohol (3)	C. or P. Cut from base to apex	Iron-hematoxylin (61)	
Cochlea	Guinea-pig or rabbit. Cut away the lower jaw and reach the inner ear by dissecting through the bulla	Fix in Flemming (14) and decalcify in 2 p.c. chromic acid (13)		Do	In the guinea-pig decalcification is accomplished in about a week. Before fixing make a hole or two in the turns of the cochlea for the reagent to enter; do this under the liquid so that air will not get in. Sections should be transverse and should include the cartilage
Eustachian tube		Zenker (6) or Gilson (20)	C.	Hematoxylin, eosin (59, 49)	
External ear		Do	P. or C.	Do	
Middle ear		Zenker (6)	P. or C.	Do	
Nerve fibers and nerve endings of cochlea	Young fetuses; new-born mouse	Golgi method (57), or methylen blue (66, b)			

Table of Tissues and Organs 275

TABLE OF TISSUES AND ORGANS WITH METHODS OF PREPARATION—*Continued*

Object or Element	Animal, Organ, or Tissue Recommended for Demonstration	Fixing and Hardening, or Other Preliminary Treatment	Section or Isolation Method P.=Paraffin C.=Celloidin F.=Freezing H.=Free Hand	Staining, etc.	General Remarks
Otoliths					
Semicircular canals	Young dog; skate	Flemming (14). Also decalcify in case of dog; as for "cochlea."	P.	Safranin (79)	Chisel into the sacculus, remove bits of the macula, and examine in dilute glycerin
VI. EPITHELIAL TISSUES					
Ciliated epithelium (living)	Roof of frog's mouth; gill plate of mussel, clam, or oyster				Scrape the surface with a scalpel and examine the material thus obtained in normal saline (90)
Ciliated columnar epithelium	Trachea; epididymus	Gilson (20) or Zenker (6)	P. or C.	Hematoxylin, eosin (50, 49)	
Columnar and glandular epithelium	Bronchus; Intestine; stomach	Do	P. or C.	Do	
Cubical epithelium	Smaller bronchioles; uriniferous tubules	Do	P.	Do	
Endothelial cells	Slit open a medium-sized vein or artery from a freshly killed cat or dog	Subject the inner surface to 1 p.c. silver nitrate (82) until opaque (10 to 15 min.)	To be mounted flat	Expose to sunlight until a brownish-red color is visible, then stain lightly in hematoxylin (59)	The vessel should be pinned out flat, endothelium uppermost, before applying the nitrate. Another method is to inject the solution into the vein or artery
Intercellular bridges	Epidermis of larval salamander	Flemming (14)	P. Thin	Iron-hematoxylin, acid fuchsin (61, 51)	Also mount bits of epidermis flat without sectioning
Isolation of epithelial tissues	Fresh epithelia	Formaldehyde dissociator (94) or alcohol (97)	Tease, or shake in a vial	Picro-carmine (76)	Examine in the dissociating fluid or in dilute glycerin
Mesothelial cells	Central tendon of diaphragm, peritoneum, or pericardium	1 p.c. silver nitrate (82) until opaque (10 to 15 min.)	To be mounted flat	Expose to sunlight until a brownish-red color is visible, then stain lightly in hematoxylin (59)	Rinse in distilled water before placing in the silver nitrate and again upon removal from it. Mount in glycerin, or dehydrate and mount in balsam
Pigmented epithelium	Pigment layer of eye	Formalin preserved material (22)		Carnalum (41)	

276 Animal Micrology

Stratified epithelium. Squamous or pavement epithelium	Sloughed off epidermis of amphibian				See "Cornea," "Skin," "Œsophagus," etc.
Terminal bars	Small pieces of intestine				Examine also scrapings from inside your cheek
Transitional epithelium	Bladder of frog, guinea-pig, or rabbit	30 p.c. alcohol (97) for 24 to 36 hours. Distend the bladder with some of the alcohol	To be mounted flat. P. Thin	Picro-carmine (69). Stain in toto for 24 hours	Scrape off some of the mucous membrane and examine in glycerin
VII. EYE					
Blood vessels of eyeball	Injected eye (chap. xiii) of albino rabbit or rat	Gilson (20) for 20 min.		Hematoxylin (52)	The halves of the rat's eyeball may be dehydrated and mounted whole in balsam
Choroid	Cat	Corrosive sublimate (17) in normal saline, or Flemming (14)	Bisect into anterior and posterior halves	Iron-hematoxylin (54)	
Cornea		Alcohol (3) or Müller (11)	Tease C. Thin	Picro-carmine (69). Stain in toto for 24 hours	Examine in glycerin. For sections see "Eyeball"
Corneal corpuscles and nerves		Flemming (14)		Iron-hematoxylin (54)	
Corneal spaces and canaliculi	Fresh eye	Use the gold-chloride method (p. 88), or the methylen-blue method (64, b)	Tangential sections with a razor	Place in water for 2 to 3 days	Spaces and canaliculi show dark upon a light field. See also methylen blue (66g)
Eyeball (general)	Use whole eyeball	Bouin (Smith) 30	C.	Borax-carmine, picric acid (39, 75)	After fixing, the part desired may be removed when the preparation has reached 95 per cent alcohol. Imbed, section, and mount as usual.
Eyelid	Eyelid of an infant	Tellyesnicky (5)	P. or C.	Hematoxylin, eosin (59, 49)	
Harder's glands	Rabbit (median angle of eye) See "Eyeball"	Zenker (6) or Gilson (20)	P.	Do	Do not confuse with lachrymal glands
Iris					Meridional sections of the anterior half of the eye will show it
Lachrymal gland	Man or rabbit	Zenker (6) or Gilson (20)	P.	Hematoxylin, eosin (59, 49)	In case of the rabbit do not confuse with Harder's glands
Lens	Beef or sheep	Müller (11)	C.	Borax-carmine, picric acid (39, 75)	Make antero-posterior sections, also other sections, at right angles to these
Lens, capsule, and epithelium	Do	Flemming (14)	P.	Hematoxylin, eosin (59, 49)	Fix the lens entire, then peel off the anterior capsule for sectioning

Table of Tissues and Organs

TABLE OF TISSUES AND ORGANS WITH METHODS OF PREPARATION—Continued

Object or Element	Animal, Organ, or Tissue Recommended for Demonstration	Fixing and Hardening, or Other Preliminary Treatment	Section or Isolation Method P.=Paraffin C.=Celloidin F.=Freezing H.=Free Hand	Staining, etc.	General Remarks
Lens fibers	Beef or sheep	0.2 p.c. sol. potassium bichromate for a week, or 30 p.c. alc. for 2 to 3 days	Tease		Junctional lines of lens fibers will be seen. Peel off lamina and tease
Membrane of Descemet	Rabbit; fowl	Draw off aqueous humor and inject 1 p.c. silver nitrate	Cut out cornea with a razor	Expose under surface to sunlight until it becomes brownish red	Wash in distilled water and examine in glycerin or dehydrate and mount in balsam. See also "Cornea"
Retina	Both mammal (human if possible) and amphibian. Cone visual cells are particularly well seen in the eyes of fish	Flemming (14). Bisect the eye into anterior and posterior halves	P.	Hematoxylin, orange G (59, 72)	Some sections should pass through the fovea centralis. Also tease bits of retina from eyes which have been prepared in Müller (11). Try Golgi (57) or methylen-blue method (66, b)
Sclera	See "Eyeball"				
Visual purple in rods	Eye of freshly killed frog	Rapidly cut out a segment consisting of a third of the posterior wall of the eyeball. Quickly peel out the retina			Examine in normal saline (90). If the preparation has been made quickly enough, the retina will appear purplish red. Some rods may be green. The color soon fades. Note also the mosaic formed by the ends of rods and cones

VIII. MUSCULAR TISSUE

Object or Element	Animal, Organ, or Tissue Recommended for Demonstration	Fixing and Hardening, or Other Preliminary Treatment	Section or Isolation Method	Staining, etc.	General Remarks
Areas of Cohnheim	Man; lingual muscle of rabbit	Flemming (14)	P. Thin	Hematoxylin (59)	
Branched striated fibers	Tongue of frog	Macerate in 20 p.c. nitric acid (2 to 3 days), or in 96	Tease, or shake in a vial of water P. or C.	Picro-carmine (76)	Examine in glycerin and mount in glycerin-jelly, or dehydrate and mount in water
Cardiac muscle	Man; dog; sheep	Corrosive sublimate (17)		Acid hematoxylin (60)	Longitudinal and transverse sections
Cardiac muscle, isolated fibers	Adult or embryo	MacCallum (96)			
Ends of striated fibers	Gastrocnemius of frog	Dissociate in 35 p.c. caustic potash for 15 min. (93)			Tease on a slide and examine in the dissociating fluid

Animal Micrology

Fibrillae in striated muscle	Frog	Macerate in 0.1 p.c. chromic acid (24 to 36 hours)	Tease in water		Ends of teased fibers show fibrillae. To make a permanent preparation, wash thoroughly, stain in hematoxylin, dehydrate Tease on a slide and examine in the dissociating fluid
Muscle to tendon	Small muscle with its tendon (e.g., sartorius of frog)	Dissociate in 35 p.c. caustic potash (93) for 15 min.			
Non-striated fibers (sections)	Bladder; intestine	Flemming (14)	P.	Iron-hematoxylin, acid fuchsin (61, 51)	
Non-striated fibers (isolated)	Intestine, stomach, or bladder of frog	Macerate in 20 p.c. nitric acid (2 to 3 days)	Tease, or shake in a vial containing water	Picro-carmine (76)	
Purkinje fibers	See "Purkinje fibers" under III				
Sarcolemma	Fresh striated muscle	Cold saturated solution of ammonium carbonate	Tease		Within 5 or 10 minutes the sarcolemma separates from the muscle substance. Examine in normal saline (90)
Striated fibers (fresh)	Frog; mammal; wing (thoracic muscle of insect)	Isolate by teasing in normal saline (90)			One of the thin muscles from the leg of Hydrophilus (a water beetle) is the classic example of insect muscle. Examine without adding fluid; waves of contraction are observable
Striated fibers (fixed)	Do	Stretch the muscle by extending, an extremity; inject 0.5 to 1 c.c. of 1 p.c. osmic acid hypodermically	Tease	Hematoxylin (59). Need not be stained, however	Cut out small pieces of the muscle from the Injected area and wash in distilled water before teasing
IX. NERVOUS SYSTEM					
Axone (axis cylinder)	Small nerve (lay it along a toothpick or small splinter without stretching, and tie it fast)	0.5 p.c. osmic acid for 4 hours; then wash in water 4 hours, followed by 90 c.c. alcohol for 24 hours	P. Very thin sections	Sat. aq. sol. of acid fuchsin (51) for 24 hours, followed by absolute alcohol (3 days)	Make longitudinal and transverse sections. In imbedding use paraffin method for delicate objects, only substitute toluol for chloroform (pp. 57, 86)
Brain cells	Small pieces of brain	Macerate in Gage's fluid (94)	Tease	Beale's carmine (42) diluted with 1 volume of the macerated fluid	See also "Cerebellar cortex," "Cerebral cortex," etc.
Brain sand	Pineal body		Tease		Tease in normal saline (90) and examine under low power
Central nervous system (general topography)	Brain and cord	Müller (11)	C.	Borax-carmine (39), or Beale's carmine (42)	
Cerebellar cortex		Müller (11) or Erlicki (8)	P. or C.	Borax-carmine, picric acid (39, 75)	Also try Golgi method (57). See chap. x. Failures with the Golgi method are more frequent than in the case of cerebrum or cord
Cerebral cortex		Do	P. or C.	Do	Also try Golgi method (57). See chap. x.

TABLE OF TISSUES AND ORGANS WITH METHODS OF PREPARATION—Continued

Object or Element	Animal, Organ, or Tissue Recommended for Demonstration	Fixing and Hardening, or Other Preliminary Treatment	Section or Isolation Method P. = Paraffin C. = Celloidin F. = Freezing H. = Free Hand	Staining, etc.	General Remarks
Choroid plexus........	Spread out pieces under the microscope and choose one which shows blood vessels well	Flood with acetic alcohol (2) for 30 min., then wash well in 50 p.c. alcohol	To be mounted flat	Hematoxylin, eosin (59, 49)	
Cylindrical end bulbs.	Fresh scleral conjunctiva (e.g., of calf). Oral mucous membrane			Methylen blue 66, b or c)	The end bulbs lie close beneath the epithelium of the conjunctiva and may be torn off if the tissue is not handled cautiously. This is a difficult preparation to make
Degenerating fibers... Ganglion canaliculi... Ganglion cells........	See "Marchi's method" Spinal ganglion of cat Gasserian or spinal ganglia; cerebral and cerebellar cortex; spinal cord	Picro-sublimate (31) Macerate in Gage's fluid (94)	P. Thin Tease	Iron-hematoxylin (61) Beale's carmine (42) diluted with 1 vol. of the macerating fluid or picro-carmine (76)	See also 24 Most of the cell processes will be torn off. See also sections of brain, cord, and ganglia
Grandry's corpuscles.	Waxy skin from lateral edges of duck's or goose's upper beak; tongue of woodpecker	Treat thin pieces with 1 p.c. osmic acid (26) for 18 to 24 hours	H. In hardened liver	Mount unstained or stain in carmalum (41)	In sectioning, cut from corium toward epithelium
Herbst's corpuscles..	Do	Do	H. In hardened liver	Do	Do
Hypophysis...........		Bouin (30)	P. Thin	Hematoxylin, eosin (59, 49)	
Intra-epithelial nerve fibers...............	Skin from freshly amputated toe or finger (volar surface)		C. Very thin sections	Gold chloride (56)	Remove all fat before treating the tissue with any of the solutions. Look for tactile corpuscles near the excretory ducts of the sweat glands
Marchi's method for degenerating fibers...	Nerve or cord, 2 to 4 weeks after the lesion has been made	Müller (11) for 8 days; then in a mixture of Müller 2 parts and 1 p.c. osmic acid 1 part for 6 days			Degenerated parts appear black, others yellowish or gray
Medulla oblongata ..		Müller (11) or Erlicki (8)	P. or C.	Borax-carmine, picric acid (39, 75)	Also try Golgi method (57)
Medullary sheath (myelin).............	Small fresh nerve	Press an end of the nerve firmly against the slide. To prevent drying, breathe on the preparation occasionally	Tease loose end		Finally examine in a drop of normal saline (90). Look for exuded drops of myelin. See also "Medullated fiber" and Weigert-Pal method (85)

Structure	Material	Fixation	Method	Stain	Remarks
Medullated fibers of cord and brain		Müller (11) or Erlicki (8)		Weigert's hematoxylin (85)	Mallory's Phos. hem. (62) is especially good for nerve fibers.
Medullated nerve fiber	Small nerve	1 p.c. osmic acid (26)	Tease		Dehydrate. On a slide tease and at the same time clear small pieces in clove oil. Mount in balsam
Meissner's corpuscles	Papillae of corium on volar surface of hand, finger-tip, or foot			Gold chloride (56), or methylen blue (66, b or c)	For simple demonstration, boil the fresh skin for 15 minutes, strip off the epidermis, slice off some papillae from the cutis and examine them in 3 p.c. acetic acid
Neurokeratin	Frog; rabbit	Osmic acid (1 p.c. for frog; 2 p.c. for rabbit)		Dehydrate. Clear with bergamot oil (48 hours)	The oil dissolves out the myelin and thus renders visible the neurokeratin
Nerve-endings (in general)		The gold chloride (56), the methylen blue (66, b or c), or the chrome silver (57) method			The methods are enumerated in order of their simplicity. For nerve-endings in striated muscle see chap. x, p. 88
Nerve-fiber bundles (transverse sections)	Sciatic	Zenker (6) or Gilson (20)	C. or P.	Carmalum, Lyons blue (41, 65)	
Neuroglia	Spinal cord	Müller (11) or Erlicki (8)	P.	Safranin (79) for 24 hours	Differentiate in absolute alcohol. See also Golgi method (57)
Nissl's granules (tigroid granules)	Lumbar enlargement of spinal cord	Corrosive sublimate (17), formalin (22), or alcohol (3)	P.	Methylen blue (66, d)	See also neutral red (71)
Nodes of Ranvier	Small nerve	0.5 p.c. osmic acid (26)	Tease	Silver nitrate (82)	
Non-medullated (Remak's) fibers	Sympathetic nerve; vagus nerve	Spread out in normal saline and isolate from fat	Tease	Picro-carmine (76) 24 to 48 hours	Medullated fibers (myelin) are stained black; non-medullated show no black
Pacinian corpuscle	Mesentery of cat			Treat with 1 p.c. osmic acid (26) until the core appears brownish	To find the corpuscles, look for minute oval bodies between the strands of fat
Purkinje cells	Cerebellum of young kitten or new-born guinea-pig				See "Cerebellar cortex," Golgi method (57)
Spinal cord (general; transverse sections)	Cat	Corrosive sublimate (17)	C. or P.	Hematoxylin, eosin (59, 49)	Thin paraffin sections may be stained in Ehrlich-Biondi (47)
Spinal cord; axones and cells (transverse sections)	Cat or dog	Müller (11) for 4 weeks. Transfer to stain without washing	C. Stain before sectioning	1 p.c. aq. sol. sodium carminate (3 days)	Before dehydrating wash the stained pieces for 24 hours in running water. Try also Golgi method (57)

Table of Tissues and Organs

TABLE OF TISSUES AND ORGANS WITH METHODS OF PREPARATION—*Continued*

Object or Element	Animal, Organ, or Tissue Recommended for Demonstration	Fixing and Hardening, or Other Preliminary Treatment	Section or Isolation Method P. = Paraffin C. = Celloidin F. = Freezing H. = Free Hand	Staining, etc.	General Remarks
Spinal ganglia (sections)	Cat or dog	Zenker (6)	P.	Hematoxylin, eosin (59, 49)	Spinal ganglia of higher vertebrates are difficult to dissect out. The Gasserian ganglia may usually be substituted for them
Sympathetic ganglia (sections)	Frog, mammal (first thoracic or superior cervical)	Do	P.	Do	
Tactile corpuscles	See "Intra-epithelial nerve fibers"				
Tactile menisci	Snout of pig or of the mole	0.33 p.c. aq. sol. of chromic acid (13)	P.	Overstain with hematoxylin (59)	Differentiate in alcoholic sol. of potassium ferricyanide
X. NOSE					
Isolated olfactory cells	Small pieces of olfactory mucous membrane	30 p.c. alcohol (97) 24 hours, followed by 1 p.c. osmic acid for 5 minutes	P.	Picro-carmine (76)	Scrape off a little of the epithelial covering and examine in glycerin
Mucous membrane	Rabbit (divide head longitudinally; nasal mucous membrane is of brownish color)	30 p.c. alcohol 2 hours, followed by 1 p.c. osmic acid for 24 hours. Harden in alcohol	P. Make section perpendicular to surface of membrane	Iron-hematoxylin (61)	
Nerve processes of olfactory cells	Young animals and fetuses	Golgi method (57), or methylen-blue method (66, b)			
XI. REPRODUCTIVE ORGANS					
Clitoris (transverse sections)	Human; monkey	Gilson (20) or Zenker (6)	P. or C.	Hematoxylin, eosin (59, 49)	
Corpus luteum	Rabbit	Do	P. or C.	Iron-hematoxylin (61)	
Epididymus	Rabbit	Do	P.	Do	
Graafian follicle (ripe)	Rabbit	Acetic alcohol (2)	C.	Hematoxylin, eosin (59, 49)	
Fallopian tube (transverse section)	Child, dog, cat, or rabbit	Gilson (20)	C. or P.	Do	
Oögenesis	Ascaris (see mem. 20, p. 154); mouse (see p. 148, mem. 8)				

Ova	Ripe Graafian follicle of rabbit; ovary of cow	Normal saline (90)	Tease	Methyl green (67)	
Ovary	Mammal	Bouin (30)	C. or P.	Hematoxylin, eosin (59, 49)	
Oviduct		Gilson (20)	P.	Iron-hematoxylin (61)	
Penis (cross-section of different regions)	Man; monkey	Acetic alcohol (2)	C.	Hematoxylin, eosin (59, 49)	
Pflüger's egg tubes	New-born kitten	Gilson (20)	C. or P.	Borax-carmine, picric acid (39, 75)	
Placenta	Human	Technique same as for "Uterus"		Hematoxylin, eosin (59, 49)	
Prostate	Man	Acetic alcohol (2)	C. or P.	Iron-hematoxylin, acid fuchsin (61, 51)	
Seminal vesicle	Man; dog	Do	P.		
Seminiferous tubules	See "Spermatogenesis"				
Spermatogenesis	Rat; guinea-pig; salamander	Bouin (30) or Hermann (33)	P. Thin	Iron-hematoxylin, Bordeaux red (61, 40)	After Herrmann, the safranin-gentian-violet method (80) gives beautiful results. Tubules should also be macerated in Hertwig's fluid (95), teased, and stained in acid carmine (44)
Spermatozoa	Smear preparations from vas deferens or testis. Also, examine some alive in normal saline (90)	Allen's B-15 (p. 170)		Cyanin (46) followed by erythrosin (50)	For structure of tail let spermatozoa macerate in normal saline (90), then stain with acid fuchsin (51)
		Osmic-acid vapor (26)		Safranin, gentian violet (80)	
Testis (general)	New-born child; cat or dog; bisect the organ longitudinally	Allen's B-15 (p. 170)	C.	Hematoxylin, eosin (59, 49)	In the cat or dog the mediastinum is in the interior instead of at the margin of the testis
Umbilicus	Human or pig embryo	Zenker (6) or Bouin (30)	C.	Do	
Urethra	See "Urethra" under XIV. For male urethra see sections of penis				
Uterus	Human, if possible (fresh after surgical operation)	Fix in a mixture of formalin (22) 1 part, Müller (11) 9 parts, for 4 days, then to pure Müller (11)	C.	Do. Or Beale's carmine (42)	
Vagina	Child; dog	Bouin (30)	C. or P.	Hematoxylin, eosin (59, 49)	
Vas deferens (transverse section)		Flemming (14) or Gilson (20)	P.	Iron-hematoxylin (61)	

TABLE OF TISSUES AND ORGANS WITH METHODS OF PREPARATION—*Continued*

Object or Element	Animal, Organ, or Tissue Recommended for Demonstration	Fixing and Hardening, or Other Preliminary Treatment	Section or Isolation Method P.=Paraffin C.=Celloidin F.=Freezing H.=Free Hand	Staining, etc.	General Remarks
XII. RESPIRATORY ORGANS					
Bronchi	Rabbit	Flemming (14)	C. or P.	Safranin (79)	Before sectioning wash the preparation in distilled water and transfer it to 70 p.c. alcohol
Epithelium of lung	Kitten	Fill the lung with 0.5 p.c. silver-nitrate solution and place it in a similar solution for several hours	H.		
Fetal lung		Gilson (20) or Zenker (6)	C. or P.	Hematoxylin, eosin (59, 49)	
Larynx		Technique same as for "Trachea"			
Lung, blood vessels of					Carmine or Berlin-blue injection through pulmonary artery. See chap. xiii, p. 100
Lung, elastic tissue of alveoli		Alcohol (3)	P.	Orcein (73), **or resorcin** (78)	Also, treat pieces of fresh lung with about 15 p.c. potassium hydrate
Lung (sections; general topography)	Small mammal	Zenker (6)	P. or C.	Hematoxylin, eosin (59, 49)	Fill the lung with the fixing fluid as well as immersing it in the fluid. Kill by blow on the head
Thymus gland	New-born child; young rat	Bouin (30)	C. or P.	Do	The stain should differentiate the chief from the colloid cells. Try dioxan technique
Thyroid gland	Young rat	Flemming (14)	C. or P.	Ehrlich-Biondi (47)	
Trachea	Child; cat	Acetic alcohol (2) or Flemming (14)	C. or P.	Hematoxylin, eosin (59, 49)	If very accurate results are desired the mucous membrane should be removed and sectioned alone
XIII. SKIN AND ITS APPENDAGES					
Epidermis under-surface view	Volar surface finger or toe	Macerate in 0.25 p.c. acetic acid until epidermis may be readily separated from dermis	Mount flat with under surface upward	Acid hematoxylin (60)	Many of the sweat glands withdraw from the dermis and remain attached to the epidermis
Hair		Examine in water (under cover-glass) or mount dry in balsam			The hair of the mouse and of most bats is peculiar

284 Animal Micrology

Hair, elements......	Warm in sulphuric acid until the hair begins to curl				Tease if necessary and examine under a cover-glass
Hair, development...	Skin from forehead (not scalp) of human fetus of 5 or 6 months	Zenker (6) or Müller (11)	C.	Hematoxylin, eosin (59, 49)	
Hair follicle........	Preferably the upper lip of man	Zenker (6)	C.	Do	The orientation should be precise, so that exactly longitudinal or cross sections result
Hair, renewal.......	Eyelid of new-born child. If this is not available, try scalp of adult	Tellyesnicky (5)	C.	Do	
Mammary gland (general)............	Preferably human. Nipple and portion of gland	Bouin (30)	C.	Do	Make vertical sections through nipple and gland. Use dioxan technique, chap. vii
Mammary gland (special)............	Mammal during gestation or lactation	Flemming (14)	C. or P.	Safranin (79)	
Milk and colostrum..	Obtain colostrum from a pregnant animal shortly before parturition	Examine milk in normal saline (90) without further treatment			Add a drop of picro-carmine (76) and examine colostrum in normal saline (90). Avoid pressure of the cover-glass
Nail (sections).......	First finger-joint of little finger	Müller (11); then decalcify (99); harden a second time in alcohol (3)	H. or C.	Alum-carmine (35)	Make longitudinal sections of the entire piece of finger
Nail, elements......		Heat in strong caustic potash (93), until the latter boils	Scrape	Unstained	Transfer to a slide without staining and examine in the dissociating fluid
Prickle cells........	Stratum Malpighii	Flemming (14)	P. Very thin sections	Hematoxylin, eosin (59, 49)	To section in paraffin, take only small pieces. Examine in glycerin
Sebaceous glands....	See sections of hair follicle				
Skin (general).......	Volar surface of finger or toe	Bouin (30)	C.	Unstained	
Skin, blood vessels..	Hand of child. Forefoot of cat or dog	Müller (11) after injection	C.		Inject with Berlin blue or carmine mass through ulnar artery. See chap. xiii, p. 100
Sweat glands........	See sections of skin, also in the under-surface view of epidermis				
XIV. URINARY ORGANS					
Bladder.............		Distend the bladder in Gilson (20), tie it shut and suspend it in more of the fluid	P. Thin	Iron-hematoxylin (61)	For epithelial cells see "Transitional epithelium"
Kidney, blood vessels	Rabbit or cat	Alcohol (3) or Zenker (6)	C.	Acid hematoxylin (60)	First inject the fresh kidney (through renal artery) with carmine mass. See chap. xiii, p. 101

TABLE OF TISSUES AND ORGANS WITH METHODS OF PREPARATION—*Continued*

Object or Element	Animal, Organ, or Tissue Recommended for Demonstration	Fixing and Hardening, or Other Preliminary Treatment	Section or Isolation Method P. = Paraffin C. = Celloidin F. = Freezing H. = Free Hand	Staining, etc.	General Remarks
Kidney, cortex, and medulla	Small mammal	Formol-sublimate (25) or Gilson (20)	C.	Hematoxylin, eosin (59, 49)	Make radial horizontal sections embracing the whole organ
Kidney, epithelial cells of uriniferous tubules		5 p.c. aq. sol. of neutral ammonium chromate	Tease		
Kidney; glomerulus and its capsule		Zenker (6) or Gilson (20)	P. Thin	Iron-hematoxylin (61)	
Kidney; isolation of uriniferous tubules	A freshly killed animal. Inject strong (75 p.c.) HCl into blood vessels of kidney under moderately high pressure	After 15 min. remove, cut into several pieces and leave in more of the HCl for 3 hours. Wash thoroughly		Stain 24 hours in hematoxylin (59)	Examine in glycerin, teasing further if necessary
Kidney; medullary rays (vertical sections)	Young animals	Treat as for "Cortex and medulla"			
Kidney, nerves of		Golgi method (57), or methylen-blue method (66)			
Suprarenal gland (Adrenal gland)		Bouin (30)	P.	Hematoxylin, eosin (59, 49)	Bisect gland with very sharp razor. Fix immediately after death
Ureter		Gilson (20)	P.	Iron-hematoxylin (61)	
Urethra (female)	Human; dog	Do	P. or C.	Do	For male urethra see "Penis" under XI

APPENDIX D

PREPARATION OF MICROSCOPICAL MATERIAL FOR A GENERAL COURSE IN ZOÖLOGY

(In addition to the methods enumerated here, see also II, chap. xi, and chap. xiv.)

PROTOZOA

A good general discussion of methods for securing and cultivating protozoa will be found in *Transactions of the American Microscopical Society*, October, 1925, by L. H. Hyman.

Amebae, etc., may usually be obtained in quantities sufficient for class use by the following modification of the well-known method of H. S. Jennings (*Journal of Applied Microscopy*, VI, No. 7, 2406). The directions are those of Bertram G. Smith:

Loosely pack a deep battery jar of 1-gallon capacity with timothy hay and add enough tap water to cover. On top of this place a layer 2 inches deep of the pond-weed *Ceratophyllum*. Add more water, if necessary, to bring its surface to within half an inch of the top of the jar. Keep the jar covered with a glass plate, and place it in a warm, well-lighted room, avoiding direct sunlight. Several such cultures should be provided at intervals of about 2 weeks, beginning at least 2 months before the amebae are needed. Cultures that become overgrown with mold, or from which the brown scum disappears, should be discarded. In favorable cultures, Ameba reaches its maximum in about 1 or 2 months, and may endure for 6 months or even longer. The amebae reach a large size, and may become remarkably abundant. Keep a number of such cultures going in the laboratory throughout the year and it will always be possible to find amebae. Besides Ameba, many other protozoa (Arcella, Difflugia, Carchesium, Stentor, etc.) and rotifers appear in the cultures.

Make smaller cultures in glass dishes about 4 inches in diameter and 3 inches deep, covered with an inverted moist chamber. Fill these dishes with chopped timothy hay, sterilized by dry heat to minimize danger from mold, and inoculate with a little material known to contain amebae (add water of course). These smaller cultures serve to keep the amebae going in seasons when Ceratophyllum cannot be obtained, but are not always necessary.

For a valuable and full account of culturing amebae see Dawson, *The American Naturalist*, LXII (1928), 453–66. For effects of media of different densities on the shape of amebae, see Hogue, *Journal of Experimental Zoölogy*, XXII (1917), 565–72.

Chalkley's Fluid is a popular culture medium for ameba:

NaCl	0.1	gram
KCl	0.004	gram
CaCl$_2$	0.006	gram
Water (glass distilled)	1,000	c.c.

To each 250 c.c. of this fluid 1 grain of unboiled rice is added.

In our own laboratory the "wheat culture" method has proved satisfactory. A handful of wheat is boiled for 5 minutes in a little water. A culture is prepared by placing about 32 grains of boiled wheat and 1 liter of water into a battery jar and then inoculating it with amebae and with unicellular green algae. Several of such cultures should be started. Although amebae may appear in numbers at the end of 7 to 10 days, they usually disappear again for a time. If in the meantime the green algae have formed a luxuriant growth, the amebae reappear as rich permanent cultures at the end of about 1 month from the time of inoculation. Search for amebae on the bottoms of such cultures. If, from time to time, some of the water is removed and fresh water containing a little freshly boiled wheat is added, the life of the culture may be prolonged for months.

Chao-Fa Wu renews ameba cultures as follows: cut alfalfa hay (stalks only) into 5-cm. pieces; boil (50 pieces to each 100 c.c. of water) for 5 minutes; filter; rinse boiled hay in clean water; put hay back into filtrate and expose for 2 days; put 5 pieces of hay with 10 c.c. of filtrate into a petri dish (prepare at least half a dozen) half full of tap water; inoculate with amebae from old culture.

Pieces of frog or mussel allowed to decay for about 10 days in pond water will usually afford an abundance of a small species of ameba.

Barker keeps amebae and paramecia from dying out by adding a sheet of fish food whenever the culture begins to be depleted. For a pure culture method see Kofoid, *Transactions of the American Microscopical Society*, XXXIV, No. 4 (October, 1915). Hance, *ibid.;* April, 1916.

To culture amebae, M. B. Sheib (*Science*, LXXXII [July 5, 1935]) dissolves 1.5 grams of agar in 100 c.c. of hot water and pours the hot solution through an absorbent cotton filter into dry, clean finger-bowls, putting enough in each bowl to form a layer 0.2 cm. thick. Several grains of ordinary rice are dropped on each layer before it sets. When the agar has hardened, a few cubic centimeters of water containing a number of amebae is poured into each bowl and an equal quantity of distilled water is added. Each day 5 c.c. of distilled water is carefully added until the bowl contains about 50 c.c. of liquid. The amebae tend to gather around the grains of rice. The appearance of large numbers of Chilomas in the cultures is favorable, as they are food for amebae. The bowls should be covered to minimize evaporation and may be kept in a water-bath with flowing tap water circulating around them to keep the temperature down to 17° or 18° C. Such cultures should develop thousands of amebae to each bowl in from 2 to 4 weeks, and may last for several months without subculturing. Fresh grains of rice should be added from time to time.

Chilomonas and Infusoria usually appear in a few days in cultures of hornwort and partly decayed water-lily leaves packed in bacteria dishes as for amebae, but with proportionately more water.

Permanent stock cultures of **Colpoda** may, according to Bodine (*Science*, January 28, 1921), be secured as follows:

"From a young soil culture active Colpoda are isolated, transferred to syracuse watch glasses and ordinary hay infusion added. After one or two days the culture fluid in the watch glass is allowed to evaporate slowly by leaving exposed to the air. During this slow evaporation the animals encyst. The dried-up culture is left exposed for 1 or 2 days, when new hay infusion is added. The animals, having divided within the cysts, revive and are found in greatly increased numbers. This drying-up process can be repeated until a more or less concentrated culture of the organisms is obtained. The concentrated culture of organisms is then pipetted into a petri dish in which a piece of ordinary filtered paper, cut so as to exactly cover the bottom of the dish and moistened with hay infusion, is placed. The petri dish is then left uncovered to slowly evaporate. The filter paper, with the encysted organisms on it, when thoroughly dry can be cut into small pieces and kept indefinitely.

"To start fresh cultures, pieces of the filter paper are put into watch glasses or other containers and hay infusion added. In a short time the animals revive and new cultures of the original are thus obtained."

Paramecium may often be kept from dying out by keeping bits of stale bread in cultures. A culture of pond water and bread will usually develop large numbers of paramecia in from a week to 10 days. See also last paragraph under *Euglena*.

To see *food vacuoles*, irrigate the slide carrying paramecia with India ink or add a bit of powdered carmine. To *slow movements of the pulsating vacuoles* add drops of a 0.25 per cent aqueous solution of sodium chloride. To demonstrate trichocysts, (1) place a small drop of ink on a cover-glass and drop the glass, ink side down, on the paramecia, or (2) run a little of a 1 per cent aqueous solution of tannic acid under the edge of the cover-glass on a fresh preparation.

The following directions for culturing, inducing conjugation, and for fixing are by Chao-Fa Wu, formerly of our own laboratory:

a) Culturing.—*Paramecium caudatum* can be raised in large numbers in a water-lily-pad medium. Boil a small handful of dried water-lily pads and leaves in a liter of water for 10 to 15 minutes. Skim off floating parts. Let the solution stand overnight and then inoculate with paramecia. Cover the culture dish (a large crystallization-dish or a battery-jar) with a glass plate, leaving ample space for air to pass through, and set it in a place of moderate light but out of direct sunlight. The culture will be ready for class work in a week or so.

b) Inducing conjugation.—At the height of the culture, concentrate a mass of the paramecia by centrifuging or by any other means. Dilute the concentrated medium with 10 volumes of ordinary tap water and place it in open petri dishes in the dark. After 6 to 12 hours, as high as 10 to 30 per cent of the paramecia will be found pairing. Examine at intervals and select the stages of conjugation desired.

c) Fixation.—Again concentrate the paramecia. Put them in vials, with the culture medium filling one-fifth of the container. Fill the vial quickly with Schaudinn's fluid and let the mixture stand for 10 to 30 minutes. Shake the vial a little from time to time. When the paramecia have finally settled on the bottom, decant the fixative and refill the vial with 70 per cent alcohol. After the paramecia have again settled, wash them with fresh 70 per cent alcohol once more and then treat them (6 to 12 hours) with 70 per cent alcohol which contains enough tincture of iodine to give it a port-wine color. Wash them in fresh 70 per cent alcohol, and store. Staining may also be carried out in vials.

Euglena will be found in some of the cultures, but usually not in any quantities before the end of 4 or 5 weeks. They appear along the side of the dish toward the light. Stephenson finds that a few grams of pulverized rice covered with pond water provides an abundance of *Euglena* in from 10 days to 2 weeks.

Turner boils 20 grams of dry quince seed for half an hour in $1\frac{1}{2}$ liters of distilled water, then passes the thick exudate which is given off, together with the water, through a wire sieve (Eimer and Amend, No. 80). He then makes up the volume to $2\frac{1}{2}$ liters with distilled water, sterilizes it, and places it in a stoppered bottle. The medium will keep for months. Cultures made by inoculating tubes or flasks of the medium with *Euglena* will keep for a year, and specimens can be obtained in considerable numbers at any time after 4 weeks. Cultures will keep longer in a thicker medium but will not reproduce rapidly.

Standard masses for use in experimental work may be prepared by evaporating the exudate to dryness and making up solutions with distilled water. A 0.2 per cent solution of the dried exudate seems to furnish the optimum density. Mold frequently invades the cultures, but will not grow in a density of 0.2 per cent or less. Mold growths are less likely to occur in cultures which have been rendered slightly alkaline. Cultures should be kept at room temperature and in a moderately lighted place.

Paramecia and other infusoria will live and reproduce for about 2 months in the medium, feeding upon *Euglena* and bacteria.

Carchesium and **Vorticella** are frequently found on decaying duckweed (*Lemna*) and hornwort (*Ceratophyllum*). To secure a culture, have a more plentiful supply of water than for amebae. Professor Walton tells me that he always finds a supply of *Epistylis* on the shells of fresh-water snails. To kill Vorticella extended, narcotize with a few crystals of menthol in a small dish of water, then fix rapidly in 0.25 osmic acid.

Didinium, a form which feeds largely on paramecia, is highly recommended by Mast (*Science*, December 20, 1912, p. 871) as of great value in biological study. It is easy to obtain (in paramecia cultures), shows the phenomena of fission and encystment with particular clearness, and has a remarkable method of feeding. Didinia can be kept indefinitely in the encysted state, and when wanted for study will appear within 24 hours to a few days in active form if introduced into a vigorous culture of paramecia.

Opalina may be obtained readily by killing a frog with chloroform and slitting open the large intestine. Examine scrapings of the epithelial wall in normal saline (reagent 90, Appendix B).

Sporozoa.—Gregarina may be found in the alimentary canal of the meal-worm or the cockroach and Monocystis in the seminal vesicles of the earthworm. They are best studied in normal saline. If it is desired to stain and mount specimens, they may be fixed in corrosive-acetic (reagent 18, Appendix B) for 5 minutes, washed thoroughly in 35 per cent alcohol, to which a little tincture of iodine has been added, and stained with Ehrlich's triple stain (reagent 48), or hematoxylin and acid fuchsin (reagents 59 and 51).

Herpetomonas may be obtained from the intestine of the fly and of the squash bug, and Trypanosoma from the blood of the rat.

Volvox.—*Volvox globator*, the form commonly described in text-books, is found in the early spring, often in great abundance, in small permanent pools which contain duckweed and *Riccia*. A smaller and less desirable form, *Volvox aureus*, may be found later in the spring and throughout the summer in the same pools. When water from such pools, together with a small amount of the water plants, is placed in bacteria dishes, so arranged that one side is strongly exposed to light, the volvox present will collect after a few hours at the edge of the water on the lighted side of the dish. If the contents of the vessels which contain volvox are kept in as near the natural condition of the pond as possible, the organisms may be kept alive for some weeks in the laboratory. Tap water is injurious to them. Avoid having too much decaying material in the water, although some is essential. Keep in glass-covered dishes near windows (out of direct sunlight) in as cool a place as possible. Any considerable rise in temperature beyond that of the original pond will result in their death. Small crustacea feed upon volvox and will, if present in any considerable numbers, soon exterminate them. The sexual stages are more likely to be found in the ooze at the bottom of cultures.

Because of the uncertainty of obtaining living volvox at any stated time, it is well to have an abundance of the material preserved in 5 per cent formalin. Such preserved specimens show the flagella more distinctly than do living ones. See Smith, *The American Naturalist*, XLI, No. 481 (1907), 31. Glycerin is an excellent mounting medium for volvox; the green color is retained. Fresh material is passed through 10, 25, 50, and 75 per cent strengths to prevent distortion and finally into pure glycerin. A margin of melted paraffin should be run around the edge of the cover, and the preparation should then be sealed with duco.

General Infusorial Technique.—

a) *Quieting infusoria.*—1. Let sufficient water evaporate from under the cover to permit the latter to press lightly upon the animals. Guard against too great evaporation of water or the infusoria will be crushed.

2. Entanglement in fibers of cotton, etc., sometimes proves efficacious. Chen (*Science*, June 14, 1929) places a drop of the infusion to be examined on a piece of lens-paper smaller than the cover-glass, covers, and examines.

3. A small amount of gelatin, or, better, cherry-tree gum or quince-seed solution, dissolved in water makes a viscous mass which is often useful in retarding their motions. A bit of white of egg may be used in the same way.

4. Animals may be narcotized by means of a small drop of very dilute alcohol (preferably methyl alcohol) or chloretone (about 1 drop of a 1 per cent solution to 10 drops of water). (Chloretone is manufactured by Parke, Davis & Co., of Detroit, Mich. For its use as an anaesthetic in biological work see *Journal of Applied Microscopy*, V, 2051.)

b) *Feeding.*—Place finely pulverized carmine or indigo under the cover-glass. The colored powder rapidly accumulates in the food vacuoles. In such a preparation the action of the cilia of infusoria is also indicated by the rapid movement of the particles in the vicinity of the animal. See also memorandum 4, page 127.

c) *Staining.*—For *intra-vitam* staining see reagents 66*a*, 38, and 71, Appendix B.

To see *cilia* of infusoria treat the animal with very dilute iodine solution or a drop of a dilute solution of tannin. See Noland's method, *f*), page 293.

To see the *macronucleus* and the *micronucleus* use a drop of a 2 per cent solution of acetic acid, or better, methyl green (reagent 67, Appendix B).

d) *Permanent mounted preparations.*—Benedict's method is as follows:

"Smear a glass slide with albumen fixative, as in preparing for the mounting of paraffin sections. Then place on the surface of the film of fixative a drop or two of water containing the form which it is desired to stain. Let nearly all the water evaporate by exposure to the air of the room until only the film of fixative remains moist. The slide can now be immersed in Gilson or any other fixing reagent, and then passed through the alcohols, stains, etc., in the same way that mounted sections are handled.

"I have had no difficulty in getting preparations of paramecium by this method, with very little distortion of the body and any kind of staining desired. By this method students can prepare in ten minutes very satisfactory preparations of protozoa for demonstration of nuclei, etc."—*Journal of Applied Microscopy*, VI, 2647.

For fixation of protozoa Calkins (*Journal of Experimental Zoölogy*, I, No. 3 [1904]) uses a saturated aqueous solution of corrosive sublimate to which 10 per cent of glacial acetic acid is added. See also reagent 25 (formol sublimate), page 240. Barker fixes, washes, stains, destains, dehydrates, clears, and mixes protozoa with balsam, all in homeopathic vials. After each operation the material is allowed to settle well. Reagents are pipetted off. See Wu's method, page 289.

To concentrate and fix free-living protozoa on a cover-glass, Baldwin (*Science*, LXXIX [1934], 143) smears it thinly with egg albumen and places it face up in a paper box about 20 mm. deep. The paper should be about as porous as ordinary mimeograph paper. The box is then stood on blotting paper and filled to a depth of 4 mm. or more with the fixing fluid to be used. Into this fluid an equal or less amount of water containing protozoa is pipetted. Seepage of the fluid through the paper is uniform and relatively slow, hence the organisms are left evenly distributed and fastened to the cover-slip. When only a very thin film of the fluid remains over the animals, the cover-glass is removed and dropped face down in a dish of the fixing fluid. The usual procedure of staining and mounting is then followed.

e) Noland's combined fixative and stain for demonstrating flagella and cilia in temporary mounts.—

Phenol, saturated aqueous solution	80 c.c.
Formalin	20 c.c.
Glycerin	4 c.c.
Gentian violet	20 mgs.

Moisten the dye with 1 c.c. of water before adding the other ingredients.

"Mix a drop of the reagent with a drop of the culture or infusion containing the organisms to be studied. The flagella and cilia stain clearly, while the cell-body remains quite natural in shape and sufficiently transparent to observe the nucleus and the other cytoplasmic structures, such as granules, pharyngeal rods, chloroplasts, pyrenoids, paramylum bodies and the like. The background remains practically colorless, the dye concentrating itself in the organisms. The depth of the stain can be regulated by varying the proportions of reagent to infusion.

"The reagent promises to be extremely useful in demonstrating flagella to elementary classes and in identifying the minute flagellates in protozoölogy courses. It has been used with surprising success on Oicomonas, Tetramitus, Menoidium, Peranema, Euglena, Astasia, Chilomonas, Polytoma, Naegleria and others.

"The cilia, cirri, membranelles and undulating membranes of the ciliates are stained by it in approximately natural form, permitting an accurate determination of the number of the ciliary rows, and the arrangement and number of the cirri, membranelles and membranes. To any one who has tried to work out the exact arrangement of the locomotor organelles of a small hypotrich the advantage of such a reagent is obvious.

"For staining internal protozoan parasites it is advisable to use more glycerin and dye (approximately 8 c.c. glycerin and 25 mgms. gentian violet have given good results). The presence of mucus interferes with the staining process. It is consequently advisable to mix the material to be examined with three or four times its volume of normal salt solution before using the reagent. With these precautions the method has been used with success on Trichomonas, Chilomastix and Balantidium.

"Unfortunately the reagent does not work well with Paramecium, since the discharge of the trichocysts tends to tear away the cilia, but satisfactory preparations have in some cases been obtained in spite of this difficulty. With smaller ciliates, such as Cyclidium, Colpidium, Urotricha, Colpoda and Aspidisca, the method works beautifully; and in larger forms without a heavy trichocyst layer very satisfactory results have been obtained, for example with Stylonychia, Ophryoglena and Chilodon. The cilia stand out as clear blue, individual threads."

f) Demonstration of the fibrillar system of ciliates (Silver line system of Klein) may be readily made by either of the following methods, which are modifications from Klein by Merlin L. Hayes of our laboratory.

A. DRY TECHNIQUE

1. Place a small drop of the culture containing the ciliates on a clean slide and evaporate the water as rapidly as possible without heat (preferably with the aid of an electric fan).

2. Immerse in 2 per cent silver nitrate solution (6 to 8 minutes for Paramecium).

3. Wash in distilled water (three changes are usually sufficient to remove excess silver). To test, remove slides and add a saturated aqueous solution of sodium chloride to the last wash water. If a white precipitate forms, more washing is necessary.

4. Place the washed slides in distilled water over a white ground and expose to diffuse light from the sky. Direct sunlight is to be avoided. Examine from time to time under a microscope.

5. When the basal granules appear connected by lines, stop the reduction by running the preparation through the alcohols, clear in xylol, and mount in balsam. Hayes finds that more uniform results can be obtained by using a 0.5 per cent solution of hydroquinone for reduction. After washing, the slides are dipped into this reagent for a few seconds, washed quickly, and examined. If the desired intensity has not been obtained, the process is repeated. Care must be taken to prevent carrying the reduction too far and thus rendering the preparation completely black.

B. WET TECHNIQUE

1. For animals that do not withstand drying, fix for 5 to 10 minutes in the following solution:

Saturated aqueous corrosive sublimate	90 c.c.
Formalin (commercial)	5 c.c.
Glycerin	5 c.c.

2. Wash in distilled water several times, concentrating by means of a centrifuge.

3. Place in 1 to 2 per cent silver nitrate solution for 3 or 4 minutes (time depends on the animal).

4. Wash in distilled water as in A, 3.

5. Expose to sunlight in a flat dish of distilled water over a white ground, or place in 0.1 per cent hydroquinone until desired results are obtained.

6. Run through the alcohols, clear in xylol, and mount in balsam.

NOTE.—The animals may be fastened to the slide with a 1 or a 2 per cent agar-agar solution before exposing to the action of silver. The agar is dissolved in water with the aid of heat, filtered, and allowed to cool, but not to solidify. A slide is dipped into the liquid agar; and before the film has set, a drop of water containing the fixed animals is placed on it. After the agar has set, the preparation is run through silver, as before (see A).

Successful preparations by either of these techniques should show basal granules as distinct black or dark brown dots connected by fine dark lines, against a light background. For further details, see E. Chatton et A. Levoff, *Comptes Rendus de la Société de Biologie*, CIV (1930), II; Everett E. Lund, *University of California Publications in Zoölogy*, XXXIX, No. 2 (1933), 35–75; Bruno M. Klein, *Zoologischer Anzeiger*, LXVI–LXVII (1920).

g) *Plankton*, in general, are well fixed if passed directly into fresh Zenker's fluid to which a few drops of a 1 per cent solution of osmic acid have been added. Under such treatment ciliates remain expanded.

Limnological apparatus such as plankton nets, traps, pumps, centrifuges, counting cells, thermometers, etc., will be found fully discussed by Chancey Juday in *Transactions of the Wisconsin Academy of Sciences, Arts and Letters*, XVIII, Part II (August, 1916). *An efficient pipette* for smaller organisms is described by Earl H. Myers in *Science*, LXXVII (1933), 609.

SPONGES

To isolate the spicules of calcareous sponges, boil a bit of the sponge in a 5 per cent solution of caustic potash for a few minutes.

Fairly thick transverse, longitudinal, and tangential sections of *Grantia* showing spicules in the tissues are useful. Make these with an old razor or sharp scalpel. To hold the object while sectioning, place it between two pieces of pith or cork. For a careful study of the relations of the two systems of canals in the body wall, thinner sections are necessary. To prepare these it is best to decalcify (2 per cent chromic acid, 24 to 36 hours) the sponge and cut celloidin or paraffin sections on the microtome, although fairly good sections may be made by hand. They should be dehydrated and mounted in balsam if permanent preparations are desired; if not, they may be examined in glycerin.

To color the collar cells use an aqueous solution of anilin blue.

Spicules of *siliceous* sponges are isolated by treating bits of the sponge with strong nitric acid or a mixture of nitric and hydrochloric acid.

COELENTERATES

Hydra should be sought for in spring-fed pools. In the autumn they are found most frequently on *smooth* dead leaves which are completely submerged. Material should be collected and placed in battery jars or larger glass jars, which are then filled with fresh clear water, and placed in a fairly light place, but not in direct sunlight. Put a small amount of hornwort or Chara in each jar. In a few hours (12 to 36) the hydra will be found attached to the sides of the vessel and to the plants. They may readily be kept in the laboratory throughout the winter if glass plates are placed over the jars to prevent excessive evaporation and the temperature is not allowed to go below freezing. Fresh water should be added from time to time to make up for evaporation. In case their supply of food (*Cyclops, Daphnia,* and other small crustacea) is exhausted it should be renewed by skimming out from other aquaria the small forms upon which the animal feeds and putting them in the hydra jars. If a few Daphnia are pipetted into a watch-glass containing several hydra in a little water, the latter react usually by gradually paralyzing and ingesting the crustaceans.

Keeping hydra in the dark at somewhat lower temperature for several days favors the formation of spermaries and ovaries.

For staining and mounting entire see page 109 or Wu's technique (below). Kill in the same way for sectioning. The most instructive sections are (1) transverse sections, (2) longitudinal sections through the mouth and a bud, and (3) sections showing the sexual organs. Stain in bulk with hematoxylin (reagent 59, Appendix B), imbed in paraffin, using the method for delicate objects (p. 57), and after the paraffin has been removed from the sections, stain them for a few seconds in acid fuchsin. Dehydrate and mount in the usual way.

The sections are much more satisfactory if the hydra have been placed in small stender dishes filled with filtered water (not distilled) and kept from food for a week or 10 days before killing. This eliminates the metabolic products and oil globules which ordinarily obscure the details of structure.

Chao-Fa Wu, formerly of this laboratory, proceeds as follows:

Select the parts of plants bearing the most hydras and immerse them in a 0.1 per cent chloretone water. After 3 to 5 minutes, the animals will react only slightly to the touch of a needle. Quickly transfer plants and hydras from the chloretone water to a large dish of formol-acetic-alcohol (reagent 23). The hydras may be left in this fluid for from 1 hour to 3 or 4 months. This fixative does not render the tentacles brittle. The hydras may then be transferred and preserved in 5 per cent formalin with the plants, or they may be shaken off and preserved alone. By this method one can easily kill, in 1 hour, a few thousand hydras in fairly extended condition and suitable for general class work. For more careful work such as for sectioning, a number of hydras may be put in a small amount of chloretone water in a dish. After complete anaesthesia, add quickly 4 to 5 volumes of the fixative.

To Stain the Nematocysts of Living Hydra, place several of the animals in a small stender dish of water which has been tinted a sky blue through the addition of methylen-blue solution made up as follows:

Methylen blue............................	1.0 gram
Castile soap.............................	0.5 gram
Water..................................	300 c.c.

After 2 hours the hydra may be transferred to fresh water; the nematocyst cells are stained a deep blue. (Method of Little, *Journal of Applied Microscopy*, VI, 2216.)

To Discharge Nematocysts, drum on the cover-glass gently with a pencil. By using a very small opening to the diaphragm they are usually sufficiently distinct without staining.

For Other Polypoid Forms the methods given for hydra will answer in most cases.

For Collecting Free-Swimming Medusoid Forms full directions will be found in Brook's *Invertebrate Zoölogy*.

Compound Hydrozoa should be placed alive into the cells which they are to occupy when mounted; 1 per cent formic acid is then added drop by drop to the sea water. After the animals have been killed, the fluid is replaced by glycerin-jelly and the cover-glass is put in place. Another method is to kill the animals slowly by adding a few crystals of chloral hydrate or menthol from time to time to the small vessel of sea water containing them.

Colonial hydroids fixed in sublimate-acetic and preserved in alcohol may be stained in alum-carmine or Delafield's hematoxylin and mounted in balsam. To avoid crushing, the cover should be supported by bits of broken slide. Nigrosin (1 per cent aq. sol.) is recommended as a superior stain for Obelia hydroids.

Small Jellyfish may be fixed and hardened in 1 per cent osmic acid and, stained or unstained, mounted in cells.

Anemones, Medusae, and other delicate marine forms may usually be killed in the expanded condition by means of magnesium sulphate. Success lies in securing a quick diffusion of a quantity of the sulphate through the water without causing mechanical disturbance of the animal to be anesthetized. Griffin accomplishes this by tying a considerable quantity of the magnesium sulphate in a piece of cheesecloth and hanging it over the dish of sea water containing the animals in such a way that the bottom of the bag barely dips into the water. Mayer's method of anesthetizing medusae by carbon dioxide is also often applicable to other sensitive contractile forms.

PLANARIA

Look for planarians on the under sides of stones in small streams of running water. They are usually examined alive. To see them thrust out the proboscis, keep them from food for a few days and then feed them on dead flies. Planaria which have been kept in the laboratory for months display the internal organs much more clearly than freshly captured ones.

If it is desired to study stained specimens, for preparation see p. 110.

Chao-Fa Wu, who has collected and mounted many excellent specimens for large classes in our own laboratories, proceeds as follows:

Collecting.—Lower a piece of fresh beef about 3 or 4 cubic inches in size at the end of a string into a freshwater stream among the plants where Planaria are known to occur. After half an hour the entire surface of the meat will be covered with Planaria. They may be shaken off into the collecting jar and the meat may be used over and over again. With half a dozen such baits, one can easily collect a few thousand Planaria in 2 hours.

Killing.—Drop the worms, by means of a pipette, into a 1 per cent chromic-acid solution. No matter in what condition they are at the time, they will die in an extended, though somewhat coiled, condition. As chromic acid is a strong hardening fluid, they should be transferred into 5 per cent formalin as soon as they become motionless. Or, the worms may be put between slides immediately after killing and immersed in formalin for 24 hours. They may then be preserved either in formalin or in 70 per cent alcohol. For more careful work, such as for sectioning or for demonstration of whole mounts, put a number of Planaria in a drop of water on a slide under a large cover-glass. Run in some 1 per cent chromic acid by dropping the solution at the edge of the cover-glass on one side and drawing the water out on the opposite side with a piece of blotter. After a few minutes, immerse the worms (on the slide with cover-glass still in place) in Bouin's fluid or formol-acetic-alcohol for 24 hours. They may then be stored in either alcohol or formalin.

Staining.—Before staining, the worms should be bleached either by means of chlorine vapor or by means of hydrogen peroxide. Wash thoroughly in water and stain in alum-cochineal for 6 to 24 hours. Destain in acid alcohol. Dehydrate, clear, and mount in balsam.

To Kill Planaria with Pharynx Protruded, Cole (*Journal of Applied Microscopy*, VI, 2125) recommends covering them in a watch-glass with a 1 per cent aqueous solution of chloretone until they are immobilized and then rapidly transferring them to 5 per cent formalin. Other fixing agents than formalin can be used.

TREMATODES

See memorandum 9, page 115. The most easily obtained forms are those found in the lungs, intestine, or bladder of frogs. A good form for study is occasionally found in the liver of the cat. Search for it in the bile passages. Fix trematodes in corrosive sublimate, wash out with alcohol to which tincture of iodine has been added, and stain for 24 hours in alum-cochineal (reagent 34, Appendix B) or carmalum (reagent 41). As with planaria, they should be compressed between two glass slides (see p. 111). To kill trematodes in a distended condition, Barker flattens out each individual on a glass slide or in a watch-glass with a camel's hair brush and floods it with the killing agent.

If the large liver fluke of the sheep (*Fasciola hepatica*) can be obtained, both the alimentary canal and the excretory system may be injected with India ink or with finely powdered carmine in water. For injection a very fine-pointed cannula with rubber cap is used, or the manipulator may operate the cannula by simply blowing through it. The excretory system is injected through an incision made with a sharp-pointed scalpel in the median line near the hinder end of the animal. For the alimentary canal the incision should be made about 1 mm. to one side of the median line. When the injection is completed, flatten the animal somewhat between two slides (see p. 111), harden in 95 per cent alcohol for 12 to 24 hours, then dehydrate, clear, and mount in balsam.

Larval Stages may frequently be found in the so-called "liver" of pond snails.

CESTODES

See memorandum 9, page 115. Ample supplies can ordinarily be obtained from dogs, or, less frequently, from cats. Near large cities an unlimited supply of the sheep tapeworm (*Monieza*) can usually be secured from slaughterhouses. Tapeworms can be kept alive for a considerable length of time in tepid water. The most instructive portions to mount are scolex, sexually mature, and terminal proglottids. For fixing and staining use the same methods as for distomes. La Rue finds that carmine stains are better for trematodes and hematoxylins for cestodes. The scolex should not be compressed. To kill cestodes in an extended condition, Barker wraps the living worm spirally around a glass slide; then immerses the slide in the killing reagent. The worm is removed as soon as killed.

To Find Cysticerci, open the body cavity of a rabbit and look for large whitish bodies imbedded in the peritoneum or liver (the cysticercus of *T. serrata*). Likewise, the cysticercus of *T. crassicollis* may be found in the liver of the mouse. If a cysticercus is found, its outer wall should be slit open in order to show the reversed scolex. *To evert scolex* place a cysticercus on a slide and cover with *filter* paper. Gently roll a quarter-inch glass rod over the filter paper. The scolex will be everted, while the paper absorbs the moisture.

NEMATODES

See memorandum 10, page 115, also 20, page 154. Nematodes occur frequently in the intestines of pigs, dogs, cats, and rabbits.

Trichinella.—The simplest way to obtain it is to apply for infected pork to the government inspector whose headquarters are to be found near all large slaughterhouses in cities. Bits of the infected muscle

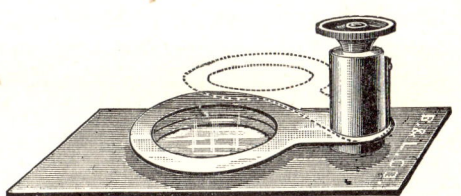

Fig. 75.—Compressor

should be teased and flattened out in a compressor (Figs. 75 and 76) until a favorable area has been found. The flattened tissue may then be dehydrated and mounted unstained or it may be stained in hematoxylin (reagent 59, Appendix B). Better results will be obtained if the material is fixed for from 4 to 6 hours in Carnoy's fluid (reagent 2) before dehydrating or staining. If desired, the tissue may be sectioned in celloidin or paraffin.

Goldsmith prepares sections and whole mounts of *Trichinella spiralis* larvae in striated muscle as follows:

Fix fresh diaphragm containing parasites in Bouin's fluid overnight; rinse in water but do not remove all color; stain in bulk in Delafield's hematoxylin overnight; destain in acid alcohol until all color is out of cytoplasm; make blue in tapwater or alkaline alcohol; section by

FIG. 76.—Compressor Used by the Government Bureaus for Meat Inspection

celloidin method. For whole mounts, flatten fresh material in a press, fix and stain as for sectioning. Result: muscle cytoplasm yellow with striations very clear; nuclei of muscle cells and cells of cysts purple; larvae of parasite sharply demarked, showing all embryonic parts.

To Demonstrate Living Trichinellae, Barnes (*American Monthly Microscopical Journal*, XIV, 104) subjects small bits of trichinized muscle to a mixture of 3 grains of pepsin, 2 drams of water, and 2 minims of hydrochloric acid for about 3 hours at body temperature with occasional shaking. When the flesh and cysts are dissolved, the liquid is poured into a narrow glass vessel and allowed to settle. The live trichinae may be withdrawn with a pipette from the bottom of the fluid and examined on a warm stage.

For details of technique concerned with helminthology the reader is referred to such volumes as Stitt, *Practical Bacteriology, Blood Work and Animal Parasitology* (8th ed., 1927), and Faust, *Human Helminthology* (1929).

ROTIFERS

Rotifers will usually be found in abundance in some of the laboratory aquaria on the lighted side of the vessel. For ordinary class work they are best studied alive. They are difficult to preserve properly. Full directions for killing and preserving will be found in Jenning's paper, "Rotatoria of the United States," *U.S. Fish Commission Bulletin* (1902), page 277.

To Quiet Rotifers, Cole (*Journal of Applied Microscopy*, VI, 2179) anaesthetizes them by adding from time to time a drop of 1 per cent aqueous solution of chloretone to the water on the slide in which the animals are being examined.

BRYOZOA

They may be treated in the same way as compound hydrozoa. *Plumatella* may frequently be found in shallow fresh-water streams on the under side of flat rocks; *Pectinatella*, in rivers and streams on the upper surface of mussel shells, etc.

STARFISH

Barker's technique for *Pedicellaria* is as follows: Boil the aboral part of a ray from a formalin-preserved starfish in 5 per cent caustic soda for from 3 to 5 minutes. Wash quickly and thoroughly in water, stain in water-soluble eosin, wash in acid (acetic) alcohol, dehydrate, and mount.

EARTHWORMS

Earthworms are best collected on warm, rainy nights when they may be found extended on the surface of the ground near their burrows. They are most plentiful in old gardens or rich lawns. A lantern and a pail are the only implements necessary. Earthworms may frequently be found, however, in large numbers on the surface of the ground on cloudy days immediately after prolonged hard rain. In winter living ones can nearly always be found under manure piles.

To Prepare Earthworms for Class Work, secure good-sized specimens, wash them in water, and place them in a vessel containing moist filter paper. Put only a few worms in each dish and adjust the cover so as to admit a little air. After 12 to 24 hours it is well to remove any dead or injured specimens and to change the filter paper. The dish should be kept from direct sunlight in a cool place. After 2 or 3 days the grit and dirt in the alimentary canal will have been passed out and its place taken by paper which the worms have eaten. They are then ready to kill and preserve or section.

Place the worms in a flat vessel and pour on sufficient water to cover them. During the next 2 hours add a little alcohol from time to time until the strength of the liquid is increased to about 8 or 10 per cent. Then wash all mucus from the body of the worms and replace them in 10 per cent alcohol until they no longer respond to pricking or pinching with forceps. Transfer them to 50 per cent alcohol for several hours, keeping them straightened out as much as possible; then to 70 per cent alcohol for 12 hours, followed by 95 per cent alcohol for 24 hours. Preserve finally in 70 per cent alcohol.

Chromic-Acid Method.—Although requiring considerable more work in preparation, specimens hardened in chromic acid are so much superior to alcoholic ones for general dissection purposes that the extra trouble is well worth while. The worms are anaesthetized as in the preceding method, but from 10 per cent alcohol they are injected with a 1 per cent aqueous solution of chromic acid and then immersed in it for 4 hours. While working with the chromic acid the hands and wrists should be coated with vaseline.

Keeping the worms extended and submerged in 1 per cent chromic acid in a large shallow dish, inject the acid into the body cavity *slowly*, about half an inch behind the clitellum, and again near the posterior end of the body. Avoid piercing the alimentary canal. The injection is not complete until the worm is turgid along its entire length. The worms must be kept straight and untwisted while in the chromic acid. Remove them at the end of 4 hours (a longer time in the acid will make them brittle) and wash thoroughly in running water until the yellow color is gone (12 to 16 hours). Remove them to 50 per cent alcohol for 2 days, then to 70 per cent alcohol for 2 or 3 days, and finally preserve in fresh 70 per cent alcohol.

For injection a water-pressure apparatus (Fig. 34) is best. The reservoir *A* should be placed about 4 feet above the compression chamber *B*. The cannula should be made of a piece of quarter-inch glass tubing with one end drawn out to a very fine bore and so broken as to leave a sharp point and edge for piercing the body wall of the worm.

For sectioning, the preliminary steps are the same as in the alcohol method, but from 10 per cent alcohol the worms should be placed into Zenker's fluid (reagent 6, Appendix B) for 4 to 6 hours. For washing, etc., follow the directions given in the discussion of the reagent. To facilitate penetration of the fluid, it is well to slit open the body cavity of the worm in places that are not to be sectioned. The most instructive sections are cross-sections of the middle of the body and sagittal sections of the anterior end which include the pharynx. The worms may be stained in bulk (24 to 36 hours) in borax-carmine (reagent 39) or hematoxylin (reagent 59) before sectioning.

Entire nephridia, together with a small part of the septum which they traverse, should be carefully dissected out, stained in borax-carmine (reagent 39), dehydrated, cleared, and mounted in balsam.

An ovary should be removed entire, stained with borax-carmine, dehydrated, cleared, and mounted in balsam.

A testis should be treated in the same way as an ovary. Tease it in the balsam before adding the cover-glass.

To Keep Earthworms Alive in Winter, Jennings (*Journal of Applied Microscopy*, VI, 2412) places them, immediately after collection, into bacteria dishes (9 in. in diameter by 3 in. deep) between folds of muslin which is kept damp but not dripping wet. Not more than a dozen worms should be placed in one dish, and the cloth should be changed or washed at least every 2 weeks. The worms may be fed on on leaves, etc., from time to time.

The small white annelid, *Enchytraeus albus*, obtainable from aquarium supply houses, can easily be propagated in the laboratory if kept at a temperature of about 60° F. in rich black earth in shallow wooden boxes. They may be fed from time to time by burying bits of bread soaked in water (in summer) or milk (in winter) about an inch in the soil. Overfeeding should be avoided, since unused food will cause the soil to become sour. Enchytraeus is excellent food for aquarium animals such as small amphibians and fishes. When immobilized by chloretone and observed in a little water under a binocular microscope, the ciliated openings of the nephridia and the septa are beautifully displayed.

To Immobilize Earthworms or Other Worms for study of circulation of the blood under the microscope or projection lantern, Cole (*Journal of Applied Microscopy*, VI, 2125) places them in a 0.2 per cent aqueous solution of chloretone for 3 or 4 minutes. Such worms may be slightly compressed between two slides.

To Examine Corpuscles of the Coelomic Fluid, expose the worms for a minute or two to the vapor of chloroform. The coelomic fluid exudes through dorsal pores. Touch a cover-glass to the fluid and mount.

The Setae Can Be Isolated by boiling a bit of the tissue containing them in a solution of caustic potash. When isolated, dry them and mount in balsam.

LEECH

Leeches are obtained from fresh-water pools, streams, and marshes, but to get sufficient numbers for class use it is usually necessary to purchase them from dealers. Live leeches intended for dissection may be killed with chloroform, or, better, by placing a few crystals of menthol on the surface of a small amount of water in which the animals have been placed; leave them there until they are completely immobile, then fix in Bouin's fluid. Sections stained with Mallory's triple stain or with Delafield's hematoxylin are satisfactory. Cross-sections prepared in the same way as for earthworms are very instructive.

ARTHROPODS

In the Crayfish Eye, to see crystalline cones, mount sections in 5 per cent potassium hydrate; to see rods, tease out in 1 per cent osmic acid.

Barnacles may be killed expanded in 0.1 per cent chloral hydrate.

For Mounting Small Crustacea see III, A, chapter xv, page 110.

To Quiet Small Crustacea for Microscopical Examination (Cole, *Journal of Applied Microscopy*, VI, 2180) place them in a watch-glass containing 2 parts of 1 per cent chloretone and 5 parts of water. The same treatment is useful for the larvae of insects. Some, such as the nymph of the dragon fly, will require more chloretone.

Daphnia is easily reared in battery-jar cultures. It constitutes an excellent food supply for such aquarian forms as Hydra and small fish. It may be obtained from stagnant pools which contain an abundance of vegetation. For cultures use pond water and keep a bit of decaying vegetable or animal matter present in the water at all times. If the water becomes cloudy, too much food is being added; it is fermenting so rapidly that the Daphnia will be killed. Euglena is also excellent food for Daphnia. Keep the cultures out of direct sunlight and in a temperature which does not go above 75° F.

For Culturing Artemia and Daphnia, Bond (*Science*, LXXIX [January 19, 1934]), as a modification of an earlier practice of Martin, recommends ordinary Fleischmann's yeast. A quarter of a fresh yeast cake is mixed into uniform suspension in from 50 to 100 c.c. of water and poured into the aquarium containing the Daphnia in some 60 to 70 liters of water. A stream of air must be kept bubbling through the culture constantly. The feeding is repeated every 5 or 6 days. See also Banta, *ibid.*, LIII (1921), 557; and Chipman, *ibid.*, LXXIX (1934), 59.

For Various Dissections and Parts of Insects see especially pages 91, 92, 108, 110–14.

For Mounting Insects Entire (beetles, mosquitoes, gnats, aphids, larvae, etc.) as microscopic preparations, and for mounting muscle, wings, heads, legs, scales, antennae, etc., see chapter xiii.

Live nymphs of the dragon fly are especially valuable for study under the compound microscope because they show very clearly the valvular action of the heart, the tracheal gills and tracheae, and the brain and its relation to the eyes. The heart is located well toward the posterior end of the abdomen between the main tracheal trunks. Cole (*Journal of Applied Microscopy*, VI, 2274) recommends that the animals be anaesthetized by subjecting them to a 1 per cent aqueous solution of chloretone.

A Culture of Mealworms not only affords an easily maintained supply of larvae, pupae, and adults of the beetle *Tenebrio*, but also constitutes a valuable source of food for such laboratory animals as toads, frogs, salamanders, and lizards. The larvae can be secured from bird stores or dealers in aquarium supplies. Place them in a battery jar nearly filled with fresh bran, cover the jar with a glass plate, and let nature take its course. As the adults appear, they mate; and a new generation is soon available.

Small Fresh-Water and Marine Organisms may be slowly killed in the expanded state by spraying menthol on the surface of the water in the vessel containing them. Test the degree of narcotization by touching the animal from time to time with a glass rod.

MOLLUSKS

Gills of the Fresh-Water Mussel may be fixed in corrosive sublimate (reagent 17, Appendix B) for from 20 to 30 minutes, washed out in water and then in dilute alcohol to which tincture of iodine has been added. Make cross-sections in paraffin, stain in dilute hematoxylin (reagent 59), and mount in the ordinary way. See remarks on page 28.

Cross-Sections of the Entire Mussel are valuable to show the relations of the gills, kidneys, and heart. Wedge the valves apart slightly and immerse the animal for 24 hours in 1 per cent chromic acid (reagent 13). Wash out thoroughly in running water and transfer the specimens to 70 per cent alcohol for 2 or 3 days or until needed. To section, remove both valves, place the animal on a board, and with a razor cut transverse sections. These are to be examined with the naked eye or with a dissecting lens.

Glochidia are best treated in quantities in a specimen tube. Borax-carmine is a satisfactory stain. Clear in cedar-wood oil. See LeFevre and Curtis, *Journal of Experimental Zoölogy*, IX (1910), 1.

To Kill Snails, Whelks, etc., in an Expanded Condition, put them into a vessel of cold water, then run a layer of hot water on to the surface of the cold water. See that the vessel is full of water and cover it with a glass plate to exclude the air. Or, run soda water from a syphon down the inner wall of the vessel of water containing the animals.

For Lamellibranch Leucocytes, see page 127, memorandum 3.

AMPHIOXUS

Specimens must ordinarily be secured from dealers. The animals should be stained entire in borax-carmine (reagent 39, Appendix B) and sectioned in celloidin. The most instructive sections are cross-sections of a female with well developed-gonads and longitudinal sections of small individuals. Mounts of entire small specimens should also be made.

VERTEBRATA

For any of the tissues of vertebrates which teachers may desire to prepare, ample directions are given in Appendix C.

For Demonstration of Circulation of the Blood in the frog, see chapter xv, page 126.

To Secure Living Frog Eggs Out of Season, implant frog pituitary anterior lobes into mature females. The number of successive implants required to induce ovulation varies somewhat, although a healthy female implanted with a single fresh anterior lobe for 3 or 4 successive days will usually ovulate. Male frogs also become activated following from two to five daily implants, so that they develop the clasp reflex and shed sperm over the eggs as they are laid. Normal development follows.

The technique of making the implants is simple. Have all instruments sterile. With sharp scissors, cut off the head of the frog from which the anterior lobe material is to be secured, expose the brain, and remove the anterior lobe of the pituitary from the floor of the cranium. Insert it by means of a hypodermic syringe into the lateral or femoral lymph sinus of the recipient. To do this successfully, first draw a small amount of sterile water into a 5 c.c. syringe, and then suck up the piece of pituitary gland into the injecting needle. Next make a very small incision in the skin on the side of the frog. Insert the point of the needle through this just under the skin and inject a little (not over 0.1 c.c.) of the sterilized water, which thus forces the anterior lobe material far into the sinus.

APPENDIX E

TABLE OF EQUIVALENT WEIGHTS AND MEASURES

WEIGHTS, METRIC AND AVOIRDUPOIS

1 kilo = 1,000 grams = 1 liter of water at its maximum density = 2.2 pounds.
1 gram = 1 cubic centimeter of water at its maximum density = 15.43 grains
   = 0.035 ounce.
1 pound = 453.59 grams.
1 ounce = 28.35 grams.
1 grain (Troy) = 0.065 gram.
1 dram = 1.77 grams.

WEIGHTS, METRIC AND APOTHECARY'S

1 kilo = 1,000 grams.
1 gram = 15.43 grains = 0.032 ounce.
1 pound = 373.24 grams.
1 ounce = 31.10 grams.
1 dram = 3.89 grams.
1 scruple = 1.30 grams.
1 grain = 0.065 gram.

MEASURES OF LENGTH, METRIC AND ENGLISH

1 meter = 1,000 millimeters = 39.37 inches.
1 centimeter = 0.394 inch.
1 millimeter = 0.039 inch.
1 micron = 1 mu (μ) = 0.001 millimeter.
1 yard = 0.914 meter.
1 foot = 30.48 centimeters.
1 inch = 2.54 centimeters = 25.40 millimeters.

LIQUID MEASURES, METRIC AND APOTHECARY'S

1 liter = 1,000 cubic centimeters = 2.11 pints.
1 cubic centimeter = 0.034 fluid ounce = 16.23 minims.
1 gallon = 128 ounces = 3.79 liters.
1 pint = 16 ounces = 473.18 cubic centimeters.
1 fluid ounce = 8 fluid drams = 29.57 cubic centimeters.
1 fluid dram = 60 minims = 3.70 cubic centimeters.

THERMOMETERS

To reduce degrees Fahrenheit to degrees Centigrade use the formula $C = 5/9(F - 32)$. For example, if the number of degrees Fahrenheit is 77, then $C = 5/9 (77 - 32) = 25$ degrees. Or, for instance, to reduce -31 degrees Fahrenheit to Centigrade, $C = 5/9 (-31 - 32) = 5/9 \times -63 = -35$ degrees.

To reduce degrees of Centigrade to degrees of Fahrenheit use the formula $F = 9/5\, C + 32$. For example, if the number of degrees Centigrade is 25, then $F = (9/5 \times 25) + 32 = 77$ degrees. Or, to reduce -35 degrees Centigrade to Fahrenheit, $F = (9/5 \times -35) + 32 = -31$ degrees.

APPENDIX F
REFERENCES

Only a very limited bibliography is given. A full one will be found in Gage's *The Microscope*. Above all, for desirable special methods the student is advised to look through the articles in current journals which cover the field of his own researches.

Of American periodicals the *Anatomical Record, Stain Technology*, and the *Transactions of the American Microscopical Society* make a special point of technique. In England the *Journal of the Royal Microscopical Society* does the same. Other periodicals which give prominence to technique are: the *Zeitschrift für wissenschaftliche Mikroskopie und für mikroskopische Technik*, the *Zoologischer Anzeiger, Anatomischer Anzeiger*, and the *Biologisches Centralblatt*.

The following books should be consulted for detailed information on points of zoölogical and histological technique:

Chamberlain, *Methods in Plant Histology*. University of Chicago Press, Chicago, 5th rev. ed., 1932.

Cole, *A Manual of Biological Projection and Anesthesia of Animals*. Neeves Stationery Co., Chicago, 1907.

Conn, *Biological Stains*, 2d ed. Commission on Standardization of Biological Stains, Geneva, N.Y., 1929.

Ehrlich, Krause, Moore, Rosin, and Weigert, *Encyclopädie der mikroskopischen Technik*. Urban und Schwarzenberg, Berlin, 1926.

Gage, *The Microscope*. Comstock Publishing Co., Ithaca, N.Y., 1925.

Hardesty, *Neurological Technique*. University of Chicago Press, Chicago, 1902.

Lee, *The Microtomist's Vade-Mecum*. Blakiston, Philadelphia, 1928.

Mallory and Wright, *Pathological Technique*. Saunders, Philadelphia, 1924.

Mann, *Physiological Histology; Methods and Theory*. Clarendon Press, Oxford, 1903.

McClung, *Microscopical Technique*. Paul B. Hoeber, Inc., New York City, 1929.

Wright, *Principles of Microscopy*. Macmillan Co., New York City, 1907.

Any of the recent textbooks on histology.

INDEX

INDEX

Abbe, 213, 222

Abbot's method for staining spores of bacteria, 136

Aberration: chromatic, 202; correction of chromatic, 204; correction of spherical, 202; spherical, 202

Absolute alcohol: preparing, 7; testing for water, 59

Accessory chromosomes, 173

Acetic acid, 16, 230; for contractile animals, 230

Acetic alcohol, 230

Aceto-carmine, 167, 248

Achromatic objective, 206

Achromatism, 206

Acid carmine, 167, 248

Acid fuchsin, 250; Altman's, 164; methyl-green method for mitochondria, 163; methyl-green stain, Auerbach's, 250; orange G, methyl-green mixture, Ehrlich's, 249

Acid magenta, 250

Acidophil granules, 167

Adipose tissue, 269

Adrenal, 285

Affixing sections, 24; frozen, 80; paraffin, 43

Agminated nodules, 273

Air-bubbles, 229

Albumen fixative, 11, 24

Albuminized water, 24

Alcohol, 7, 16, 17, 34, 231; absolute, 7; absolute, test for water in, 59; acetic-chloroform mixture, 230; acid, 8, 25, 52; alkaline, 53; and corrosive sublimate, 238; and ether, 8, 238; ethyl, 11; fixation, 34; grades of, 7; methyl, 12; renewing, 59; tertiary butyl, 67; wood, 12

Alcoholometer, 12

Algae, mounting medium for, 116

Alimentary canal, 272; of insects, to mount, 91

Alizarin Red S, 119

Allen, 55, 137, 138, 140, 143, 149, 156, 170

Allen's B-15 method, 170; gelatin method, 140; iron-hematoxylin modification, 55

Altman's acid fuchsin, 164, 250; mixture, 234; stain, 251

Alum-carmine, 244; dahlia, 268

Alum-cochineal, 9, 243

Ambystoma, 140; epidermal cells of, 161; mitosis in cells of, 158, 161; to study living cells in, 161

Ameba, 286; culturing, 286–88

Ammonia copper sulphate solution, 217

Amphibia: artificial fecundation in, 153; embryology of, 137–41

Amphioxus, 308

Amphiuma, 28

Amphophil granules, 267

Amyloid, 167, 245, 273

Anatomical material, preservation of, 32, 120; mounting in glycerin jelly, 120

Andrews, 142

Anemones, 299

Angular aperture, 206

Anilin, 20; anilin blue, orange G, and acid fuchsin, 244; dyes avoided in celloidin method, 72; oil, 23, 175; stain, acid, basic, 20; stains, 20, 244; stains, decolorizing, 244; water, 244

Antennae, to mount, 112, 114

Anthrax, 133, 136

Aorta, 268

Apathy's celloidin and paraffin method, 73

Apertometer, 209

Aphid, to kill and mount, 111

Aplanatism, 209

Apochromatic objective, 206, 213

Apparatus required, 1

Aqueous humor, 263

Arcella, 286

Archoplasm, stains for, 168

Areas of Cohnheim, 277

Areolar tissue, 269

Artemia, 306

315

Artery, 268
Arthropods, 306
Ascaris, polar bodies, fertilization and cleavage in, 154
Asphaltum, 107
Assheton, 150
Auerbach's fuchsin methyl-green stain, 250
Aurantia, 20, 250
Axial illumination, 216
Axis cylinder, 278
Axis of lens: principal, 198; secondary, 198
Axone, 278
Azo-carmine mixture, 87

Bacillus, 133; aerogenes capsulatus, 136; of anthrax, 133, 136; of bubonic plague, 136; of chancroid, 136; coli communis, 136; diphtheriae, 136; of dysentery, 136; of glanders, 136; of influenza, 136; of malignant edema, 136; mucosus capsulatus, 136; proteus, 136; pyano-cyaneus, 136; of tetanus, 136; of tuberculosis, 136; of typhoid, 136
Bacteria: cover-glass preparations of, 130; features to be observed in studying, 133; Gram's method of staining, 132; hanging-drop preparations of, 132, 133; material for demonstrating, 133; methylen-blue stain for, 132; mounting from fluid media, 130; mounting from solid media, 131; staining and mounting films, 131; spores, 133, 136; staining flagella, 136; stains for, 136; in tissues, 131, 133
Bacterial examination, 130
Baldwin, 293
Balsam, 10, 23; bottle, 4; mounting in, 36, 53; mounts, fading of, 62; neutral, 129; removing exuded, 61; xylol-, 11
Banta, 307
Bardeen, 180, 181
Barnacles, 306
Barker, 117, 293, 287, 300
Basophil granules, 168, 267
Batson, 104
Baumgartner, 48, 162
Bayberry wax for imbedding, 45

Beams, 47, 163, 165, 166, 176
Beetles, to mount as opaque objects, 111
Beggiatoa, 133
Benedict, 292
Bensley, 33, 162, 163, 170, 246; formol-bichromate sublimate, 239; freezing-drying method, 178; mitochondrial methods, 162, 164
Benzaldehyde for clearing and fixing whole objects, 119
Benzol, 23
Benzopurpurin, 20
Bergamot oil, 23, 175
Berlese's fluid, 113
Berlin blue, 97
Bethe's fluid, 255
Bibliography, 312
Bichloride of mercury, 237
Bichromate of potassium, 231; with dioxan, 66; dissociating fluid, 264; in various fixing mixtures, 232, 233, 234
Binocular loop, Hardy, 212
Binocular magnifier, 212
Binocular microscope, 209, 212
Bismarck brown, 22, 167, 246
Black pins in sections, 61
Bladder, 284
Blastoderm: of chick, 142; of fish, 146
Bleaching, 25, 47
Bleu de Lyon, 20, 254
Blocks for mounting celloidin, 71
Blood, 121–29; and blood-forming organs, 267; clinical examination of, 123; corpuscles, 124; corpuscles, living, 127; cover-glass preparation, 122, 267; crystals, 121, 122; currents, to observe, 126; dry preparations of, 122; effects of reagents on, 121; enumeration of corpuscles, 124; erythrocytes, 267; examination of fresh, 121; platelets, 121, 129; rapid method for, 123; Schultze's iodized serum, 264; serum, 263; serum, Loeffler's, 135; to study in sections, 127; tests for, 122; Wright's stain for 128
Blotting-paper method of reconstruction, 181
Blue color, to restore in injected tissues, 104

Bodine, 288
Bond, 306
Bone: corpuscles, 269; decalcifying, 95, 270; development of, 270; fibers of Sharpey, 270; grinding, 94; Haversian canals and lamellae, 270; isolation of corpuscles, 270; sectioning, 94; stains for developing, 245
Books on micro-technique, 312
Borax-carmine, 54; Grenacher's, 246
Borax-ferricyanide, 164
Bordeaux red, 168, 246
Born, 179
Bouin's picro-formol fixing fluid, 9, 30, 235, 242; fixing with, 30; in embryology, 157; Smith's modification for eyes, 242; use in cytology, 158
Boveri, 31, 231
Brain cells, 278; sand, 278
Breder, 127
Brittle objects: celloidin and paraffin method for, 73; paraffin-rubber method for, 45
Bronchi, 283
Brooks, 298
Brouha, 166
Brown, 103, 104
Brownian movement, 212
Bryozoa, 303
Buck, 173
Burckhardt, 149
Burr, 149
Burrows, 156
Butyl alcohol, 67

Cajal's method for neuro-fibrils, 85, 86
Calcification, test for, 168
Calibration of microscope, 212
Calkins, 293
Calleja's staining fluid, 247
Camera lucida, 212; Abbe, 212, 213; Wollaston, 212
Cannulae, glass, 103
Capillaries and small vessels, 268
Capillaries, Golgi method for, 84
Carbol-fuchsin for bacteria, 134
Carbol-xylol for clearing, 63
Carbolic acid, 22, 33
Carbon dioxide: for freezing, 77; to estimate minute quantities of, 176

Carborundum points, 1
Carchesium, 286, 290
Cards, record, 5
Carmalum Mayer's, 247
Carmine, 20; Beale's, 247; borax-, 246; muci-, 258; picric acid, and indigo-carmine, 247; picro-, 259; Schneider's acid, 248
Carnoy: fixing fluids, 230, 235; Lebrun fixing fluids, 158, 231
Carothers, 75
Carrel, 156
Cartilage, 270; capsule of, 270; connective tissue, and elastic fibers in, 270; elastic, 270; glycogen in, 270; hyaline, 270; white fibro-, 270
Cassia oil, 23
Caustic potash, caustic soda, 25, 264
Cedarwood oil, 22, 38, 44
Cell: dissection of living, 175; walls, to stain, 108
Cell-making, 107, 117; deep, 117
Celloidin, 11, 24; blocks for mounting, 71; bottle, 4; cement, 152; hardening, 71; imbedding a number of objects in, 72; removing from sections, 75; sections, affixing, 24; sections, to transfer from the knife, 73; serial sections, 74
Celloidin and paraffin methods compared, 72
Celloidin method, 68; clearing objects in the block, 73; in cytology, 174; dry-cutting method, 75; Gilson's rapid, 75; length of time, 71
Celloidin-paraffin method for brittle objects, 73
Cells of Paneth, 272
Cellular structures, tests for certain, 167–70
Celluloid: for corrosion preparations, 104; plating method, 74
Centering an object in a cell, 114
Centigrade to Fahrenheit to reduce, 311
Central illumination, 216
Central nervous system, 278
Centrosome, 168
Cerebellar cortex, 278
Ceruminous glands, 274
Cestodes, 115, 301
Chalkey's culture medium, 287

Chamberlain, 312
Chamber's micro-dissector, 176
Champy's mixture, 234
Champy-Kull triple stain, 234, 250
Chancroid, bacillus of, 136
Chen, 292
Chick: embryology of, 141–43; marking anterior end of young, 148; window method for, 156
Child, 139, 146
Chilomonas, 288, 294
Chi-Hsiun Chu, 34
Chinese black, 154
Chinolin blue, 248
Chipman, 307
Chitin, 114
Chloretone, 292
Chloride and acetate of copper, 235
Chlorine, Mayer's bleaching method, 47
Chloroform, 23; in celloidin method, 69; in paraffin method, 57
Cholera, spirillum of, 136
Choroid, 276; plexus, 279
Chrom-acetic-osmic acid, 235; acetic-formalin mixture, 236
Chromatin, tests for, 168
Chromic acid, 18, 235; decalcifying fluid, 265; washing out, 18
Chromosomes, stains for, 168, 262; to demonstrate quickly, 173
Chrom-silver method, Golgi's, 82
Cicatricula, 141
Cilia, 294
Ciliates, silver-line system, 294
Cinnamic aldehyde, 23
Circulatory system, 268
Clarke, 116, 117
Claus, 66, 177, 235, 246
Cleaning mixture, 61; microscope lenses, 226, 228; slides and covers, 60
Clearers, 22; for celloidin, 72
Clearing, 22, 106
Cleavage: in living material, 154; quick preparation of stages in teleosts, 147, 154
Clinging of paraffin sections to the knife, 50

Clitoris, 281
Clove oil, 23; in minute dissections, 92
Coccus, 133
Cochineal, 20; alum, 9, 243; and Lyons blue, 54
Cochlea, 274; nerve fibers and nerve endings of, 274
Coelenterates, 297
Cole, 256, 305, 306, 307
Collar cells of sponges, 296
Collodion, 24, 73; instead of celloidin, 73
Colostrum, 284
Colpoda, 288
Compound microscope, manipulation of the, 226
Compressors, 301, 302
Condenser, 213
Congo eel, 28
Congo glycerin, 92
Congo red, 10, 248
Conklin, 142, 146, 252
Conn, 21, 312
Connective tissues, 269, 271
Contractile animals killing, 16, 230
Convex or converging lens, 198
Cooler for paraffin microtome, 50
Coplin staining jars, 4
Copper chrome-hematoxylin method for mitochondria, 164
Coregonus, 160
Cornea, 276; silver-nitrate method for, 85
Corneal corpuscles and nerves, 276
Corneal spaces and canaliculi, 276
Corpus luteum, 281
Correction collar, 214
Corrosion of injected vessels, 25, 104
Corrosive sublimate, 8, 9, 18, 29, 237; with dioxan, 66; handling, 9; in various fixing fluids, 232, 237; washing out, 18, 29
Cort, 115
Cover-glass, 1; correction for thickness of, 214; standard, 214; to support, 92
Cowdry, 162, 165
Cox modification of the Golgi method, 84

Index

Crayfish, mitosis in testis of, 159; eye, 306
Creosote, 23
Crescents of Gianuzzi, 272
Crooked paraffin ribbons, 48
Crown glass, 204
Crumbling of objects or tissues, 45; in paraffin, 50
Crustacea: small, 110; to quiet, 306
Crystal violet, 22, 261
Curare, 126
Curtis, 308
Cyanin, 248; and erythrosin for spermatozoa, 248
Cylindrical end bulbs, 279
Cysticerci: to evert scolex, 301; to find, 301
Cytological methods, 158
Cytoplasmic stains, 20

Damar, gum, 23
Danchakoff's mixture, 174, 233
Danforth, 148
Daniels, 148
Daphnia, to rear, 306
Dark-field microscopy, 214
Davenport silver method, 86; celloidin substitute, 74
Dawson, 287
Daylite glass, 214
Dealcoholization, 23
Dealers, 4
Decalcification, 25, 95, 265; fluids for, 9, 265
Decolorization, 25
De Faure's fluid, 113
Dehydration, 18, 175
Delicate tissues, 45; paraffin method for, 57
Demilunes of Heidenhain, 272
Demonstration microscope, 214
Demonstration ocular, 214
Descemet, membrane of, 277
Desilicidation, 25
Developing bone, to stain, 245
Diapedesis, 126
Diaphragm, 215
Didinium, 290
Diehl, 115

Diethyl dioxide, 64
Difflugia, 286
Digestive organs, 272; blood vessels of, 272
Dilution, rules for, 7, 12
Dioxan method, 64; and freezing method, 67; in cytological work, 66; properties of, 65; repeated use of, 66
Diphtheria, 134, 136
Diplococci, 133
Diplococcus intra cellularis meningitidis, 136
Dipping tube, 116
Discoidal cleavage, 147
Dispersion, 202
Dissecting instruments, 1
Dissection of living cells, 175; of small embryos, 152
Dissections, minute, 91
Dissociation, 15, 26; digestion methods, 90, 302; fluids for, 264; general rules for, 92; by means of formaldehyde, 89, 264
Dogiel's methylen-blue method, 255
Doner, 92, 113
Double staining, 21, 54
Drawing, 183; in cytology, 189; in embryology, 188; in histology, 189; materials, 183; methods, 184; for publication, 190; in special courses, 188; table for reconstruction work, 181; use of camera lucida in, 191; use of colored crayons in, 189
Drawings: arrangement of, 187; arrangement of, for reduction, 194; combination, 193; fixing pencil, 186; for half-tone reproduction, 192; ink, 184; to keep clean, 187; labeling, 187; lettering for publication, 187, 194; pencil, 185; for publication, 190; reduction of, 191; for reproduction in color, 192, 193; reproduction of, by line process, 191; reproduction of pencil, 193; reproduction of wash, 193; shading, 185, 186; wash, 186
Drop method, 18, 19, 174
Dropping bottle, 62
Dry areas under the cover-glass, 61
Dry mounts, 112
Ducts, Golgi method for, 84
Duodenum, 272
Duval, 148
Dysentery, bacillus of, 136

Ear: external, 274; middle, 274
Earthworm, 303; alcohol method for, 303; chromic-acid method for, 304; to examine coelomic corpuscles of, 305; to immobilize, 305; to isolate setae of, 305; keeping alive in winter, 305; nephridia of, 305; ovary of, 305; sectioning, 304; testis of, 305
Eau de Javelle, 25
Echinoderms, artificial fecundation in, 153
Eggs: of insects in dry mounts, 112; of mouse, 148; notes on Amphibian, 139–41
Ehrlich: -Biondi triple stain, 248; triple stain, 249
Elastic elements of blood vessels, 268
Elastic fibers, 258, 271
Electrification of paraffin sections, 50
Embalming, 32
Embryograph, 216
Embryological methods, 137
Embryology: amphibia, 137–41; chick, 141–43, 147; frog, 137; mouse, 148, 149; pig, 151; rabbit, 149, 150; rat, 149–51; teleosts, 145, 147
Embryonic shield, 146
Embryonic tissues, cultivation of, removed, 155
Embryos: dissection of small, 152; early stages of mammalian, 149; human, 152; injection of, 101, 104; marking anterior end of young chick, 148; measuring length of, 145; older stages of mammalian, 151; shrinking and hardening of, 157; skeleton, to stain, 119; whole mounts and sections of teleost, 146
Enchytraeus, 305
Encircling fibers, 271
End bulbs, 279
Endothelial cells, 275
Endothelium of blood vessels, 268
Eosin, 10, 20, 54, 249; substitute, 67; Y, 249
Eosinophil granules, 267
Epidermis, under-surface view of, 283
Epididymus, 281
Epistylis, 290
Epithelial tissues, 275
Epithelium: of alimentary tract, 272; ciliated, 275; ciliated columnar, 275; columnar and glandular, 275; cubical, 275; isolation medium for, 234; pigmented, 275; squamous or pavement, 276; stratified, 276; transitional, 276
Erlicki's fluid, 233
Erythrocytes, 128, 267
Erythrosin, 249
Esophagus, 272
Ether: and alcohol, 8, 238; for freezing, 80
Euglena, 289, 290, 294
Euparal as a mounting and preserving medium, 23, 58, 176
Eustachian tube, 274
Evans, 103, 167
Evaporation from cells, to prevent, 116
Eycleshymer's clearer, 72; methods of orienting objects in celloidin, 144
Eye, 276
Eyelid, 276
Eyepiece. See Ocular
Eye-Point, 216

Fading of balsam mounts, 62
Fahrenheit to Centigrade, to reduce, 311
Fallopian tube, 281
Farrant's solution, 120
Fat: cells, 271; intestinal absorption of, 273; to remove from tissues, 169; tests for, 168, 169
Faust, 302
Fecundation, artificial, 153
Fenestrated membrane, 271
Fertilization in mammals, 148
Fetuses, 152
Feulgen's reaction, 168
Fibrillar connective tissue, 271
Fibrin, stained preparation of, 121
Filters, ray, 172
Fixation, 16, 30; by injection, 31
Fixing: with alcohol, 34; with Bouin's fluid, 30; with formalin, 30; general rules for, 28; tissues cut by the freezing method, 80; with Zenker's fluid, 29
Fixing and hardening agents, 230
Flagella of bacteria, 136; of protozoa, 293

Index

Flatness of field, 216
Flatworms, killing and mounting, 110, 115, 300
Flemming's fixing fluid, 235; use in cytology, 158
Flint, 104
Flint glass, 204
Flukes, 115
Focal point, 198
Foci, conjugate, 198
Foley, 87, 261
Foliate papillae, 274
Food vacuoles, to demonstrate, 289
Foot and Strobell, method of photomicrography, 172
Forceps, 1
Formalin, 8, 238; fixing with, 30; for frozen sections, 30; protecting the skin from, 33; to neutralize, 238; in various fixing mixtures, 238–40; with Zenker's fluid, 232
Formic glycerin, 150
Formol. *See* Formalin
Free acid, to detect, in tissues, 169
Free-hand sectioning, 35
Freezing attachment, 79, 80
Freezing method, 77; with dioxan, 67; for fixed tissues, 80; for fresh tissues, 79, 80; for objects alcohol injures, 80; rapid method, 81
Freezing-drying method, 178
French, 260
Fresh tissues, 26; sectioning by the freezing method, 79, 80; staining, 167
Friable objects, sectioning, 45, 73
Frog eggs, to secure out of season, 308
Frog embryos: fixing, staining, and mounting, 137–39; section method for, 137; whole mounts of, 138
Frozen sections, affixing, 80
Fry, 157
Fuchsin: acid, 20, 22, 250; basic, 250; and methyl-green stain, 250; and picric acid stain, 250; other combinations with, 134, 164, 243, 246, 250, 260

Gabbet's method for tubercle bacilli, 135
Gage, 26, 92, 169, 181, 194, 241; congo-glycerin, 92; formaldehyde dissociator, 264; stain for glycogen, 169
Gall bladder, 272
Gall duct, 272
Ganglion canaliculi, 279
Ganglion cells, 279
Gastric glands, 272
Gatenby, 165
Gay, 174
Geiman, 238
Gelatin capsules for small objects, 32
Gelatin injection mass, 104; to keep, 103
Gelatin method, Allen's, 140
Gelatinous coats of eggs, to remove, 139
Gentian violet, 20, 22, 134, 251
Germ ring, 136, 137
Germinal layers, 140
Gersh, 178
Geschichter, 81
Giemsa, stain, 129
Gilson's fixing fluid, 238; use in cytology, 158
Gilson's rapid celloidin process, 75
Gizzard of cricket or katydid, to mount, 91
Glanders, 134, 136
Glaser, 50
Glass marking, 1, 62
Glochidia, 308
Glucose bouillon, 135
Glycerin, 22, 37; mounting in, 108
Glycerin jelly, 22, 109; mounting in, 109
Glycogen, test for, 169
Gnat, to kill and mount, 111
Goblet cells, 272
Gold chloride, 252; for bleaching, 176; method for nerve endings, 88; toning solution, 87
Goldsmith, 115, 236, 302
Golgi apparatus, 16, 165, 166, 240
Golgi method, 82, 83; determining which elements will be impregnated in, 83; permanent preparations in, 84; Sevringhaus technique, 177
Gonococcus, 136
Graafian follicle, 281

Gradual change of liquids, 174; method for, 174
Graduated cylinders, 4
Gram's method, 132, 135, 251
Grandry's corpuscles, 279
Grantia, 296
Graphs, 192
Grasshopper, eggs: to cut, 176; mitosis in, 159
Graupner, Wiessberger, dioxan method, 64
Grave, 50
Green plants, mounting fluid for, 116
Gregarina, 291
Gregory, 103
Grenacher's borax-carmine, 246
Griffin, 299
Griggs, 62
Guild, 86
Guncotton, 24

Hair, 283; development of, 284; elements of, 284; follicle, 284; renewal, of, 284
Half-tones: from drawings, 192, 193; from photographs, 193
Hammond and Marshall, 151
Hamre, 95
Hance, 46, 55, 63, 171, 287
Hanging-drop preparations, 132
Hann, 156
Hanstein's rosanilin-violet, 257
Hard objects, 94–95
Hardening, 15, 17, 19
Harder's glands, 276
Hardesty, 46, 84, 312
Harris, 9, 252
Harrison, 156
Hayes, 294
Head of fly, to mount, 112
Heart, 269; valves of, 269
Heart-beat of insect, to see, 307
Heidenhain's iron-hematoxylin, 10, 54, 248; modification of Mallory triple stain, 87
Heliotype method of reproducing drawings, 194
Helly's fluid, 233
Helminthology, 300

Hemalum, Mayer's, 67
Hematein, 253; muci-, 258
Hematoidin crystals, 122
Hematoxylin, 9, 20, 52; bluing by means of tap-water or alkaline alcohol, 59; Conklin's picro-, 252; Delafield's, 9, 35, 52; Ehrlich's acid, 252; and eosin, double staining, 54, 71; Harris, 9, 252; Heidenhain's iron-, 10, 54, 253; Mallory's phosphotungstic, 253; precise staining in Delafield's, 60; ripening, 9, 20, 55
Hemin, crystals, 122
Hemocytometer, 124
Hemoglobin: crystals of, 121; tests for, 169
Hemolymph gland, 269
Hepatic interlobular connective tissue, 273
Hepatic lobules, 273
Herbst corpuscles, 279
Hermann's fluid, 243
Herpetomonas, 291
Hertwig macerating fluid, 90
Heuser, 103, 152, 182
Hinman, 104
His, 145, 181
Hjort, 81
Hogue, 287
Hollow organs, 31
Holmgren's canals, 240
Homogeneous immersion objective, 216
Hones, 46
Hoyer, 103
Hsw-Chuan Tuan, 56
Hubbert, 62
Huber, 84, 151, 180
Hucker, 135
Huettner, 168
Human embryos, 152
Huygenian ocular, 201
Hydra, 297; to discharge nematocysts of, 298; killing and mounting, 109; sectioning, 298; to secure spermaries and ovaries in, 297; staining and mounting, 109; staining nematocysts of, 298
Hydrofluoric acid, 26
Hydrozoa, 298
Hyman, 286

Index

Hypophysis, 279
Hyrox, 176

Illumination, 216; dark ground, 224
Images: defects in, 202; formation of, by lenses, 199
Imbedding, 23; in celloidin, 68; in gum and syrup, 77; in paraffin, 38
Impregnation methods, 21
Incubation of hen's egg, 141
Indifferent fluids, 26, 263
Indulin, 267
Indulinophil granules, 267
Infiltration methods, 23
Inflammation, 126
Influenza, bacillus of, 136
Infusoria, 288, 290–96; feeding, 292; macro- and micro-nucleus of, 292; quieting, 292; to see cilia of, 293; staining, 292
Injecting animals for preservation, 32, 33
Injection: by continuous air-pressure, 100; double, 100; through femoral artery, 103; by syringe, 98, 103; triple, 101
Injection mass, 97; blue, 97; cold fluid gelatin, 104; for gross anatomy, 105; to keep, 103; red, 97; yellow, 98
Injection methods, 26, 105; Locy's air method, 105
Ink: diamond, 62; for glass, 62
Insect larvae, quieting aquatic, 307
Insects: delicate, 114; having hard shells, 114; to mount, 113, 307; small, to dissect, 92; transparent and soft, 114
Intercellular bridges, 275
Intercolated disks in heart muscle, 269
Interpretation of prepared material, 15
Intestinal absorption of fat, 273
Intestine: large, 273; small, 273
Intra-vitam staining, 167
Iodine for removing corrosive sublimate, 29
Iris, 276
Iron-hematoxylin: Heidenhain's, 10; method, 54; with other stains, 56; use in cytology, 158
Isaacs, 126, 188
Isolation method, 15, 26, 89, 275

Jamming of paraffin sections, 48
Janse and Peterfi, 176
Janus green, 165, 167
Jelly of Wharton, 271
Jellyfish, 298
Jennings, 286, 303, 305
Johansen, 19, 67
Johnson, 45, 240; paraffin-rubber method, 45
Joris, 104
Juday, 296
Julin, 150

Kahle's fluid, 239
Kaufmann's Flemming-formol, 236
Keefe, 116
Keiller's fluid, 32
Kidney: blood vessels of, 284; cortex and medulla of, 285; epithelium of uriniferous tubules, 285; glomerulus and its capsule, 285; injection of, 101; isolation of uriniferous tubules, 285; medullary rays of, 285; nerves of, 285
Killing, 15; general rules for, 28
Kincaid, 33
King, 30, 176
Kirkham, 148, 149
Klein, fibrillar system of ciliates, 294, 296
Kleinenberg's picro-sulphuric, 242
Knower's injection method, 101–3
Koch-Ehrlich gentian violet, 134
Kofoid, 287
Kornhauser, 73, 158
Krichesky's modification of the Mallory triple stain, 245

Label, 53
Labeling, 63
Lachrymal gland, 276
Large objects, to cut in paraffin, 47
La Rue, 301
Larvae, soft, to mount, 113, 114
Larval stages of trematodes, 301
Larynx, 283
Lavdowsky's mixture, 239
Lecithin, tests for, 169
Lee, 38, 83, 248, 312

Leech, 306
Lefevre, 308
Legs of insects, to mount, 112
Leishman's Romanowsky's stain, 128
Lens, 276; capsule and epithelium of, 276; fibers, 277
Lenses, 198; doublets, 201; triplets, 201
Leprosy, bacillus of, 135
Leptothrix, 133
Leucocytes, 127, 268; ameboid movements in, 127; eosinophylic, 129; feeding, 127; granules of, 129, 267, 268; kinds of, 129; of lamellibranchs, 127; large mononuclear, 129; polynuclear neutrophylic, 129
Lewis, 156, 165, 182
Lichtgrün, 253
Ligament, 271
Ligamentum nuchae, 271
Light for microscopical work, 217
Light green, 253
Lillie, 243
Limnological apparatus, 296
Line-process, 191
Linstaedt, 74
Lip, 273
Lipman, 119
Lithography, 194
Little, 298
Liver, injection of, 101, 273
Liver fluke, 115; injecting, 300
Living cells, to study in Ambystoma, 161
Living tissue, staining, 167
Locke's solution, 155
Locy's injection methods, 105
Loeffler's alkaline methylen blue, 134; blood serum, 135
Long, 148
Ludford, 165
Lugol's solution, 254
Lund, 296
Lung, 283; blood vessels of, 283; elastic tissue of alveoli, 283; epithelium of, 283; fetal, 283; injection of, 101
Lymph capillaries, 109, 269
Lymph glands, 268
Lymphatics, injection of, 101

Lymphocytes, 128
Lyons blue, 10, 54, 254

MacCallum's macerating fluid, 9, 89, 265
McClung, 1, 12, 18, 23, 32, 43, 117, 143, 156, 158, 174, 312
MacLennan, 169
McNabb, 176
Macerated tissue, fixation of, 92
Macerating fluids, 264, 265
Maceration, 15, 26, 89, 90
Magenta S, 250
Magnifying power, 206, 218
Malarial parasite, 129, 268
Mallory and Wright, 312
Mallory's triple connective-tissue stain, 244
Mall's differential method for reticulum, 90
Malone, 86
Mammal, stages of maturation, fertilization, and segmentation in, 148
Mammary gland, 284
Manipulation of the compound microscope, 226
Mann, 253
Mann-Kopsch technique, 165
Manufacturers, 4
Marchi's method for degenerating nerve fibers, 279
Margolena, 168
Mark, 45, 148, 182
Marrow, red, 268
Martin, 95, 306
Mast, 290
Mast-cells, 268
Maturation in mammals, 148
Mayer, 11, 20, 47, 247, 258, 259; hemalum, 67
Mayo Clinic method for frozen sections, 80
Maximow's fluid, 233
Mealworms, rearing, 307
Mechanical stage, 218
Medulla oblongata, 279
Medullary sheath, 279
Medullated fibers of cord and brain, 280
Medullated nerve fiber, 280

Index

Medusae, 299
Medusoid forms, 298
Megakaryocyte nuclei, 236
Meissner's corpuscles, 280
Membrane of Descemet, 277
Membranes, silver-nitrate method for, 85
Menthol for slow killing, 298
Mercuric crystals, to remove, 29
Mesothelial cells, 275
Metagelatin, 109
Metal L's for imbedding, 47
Metallic substances for color differentiation, 20, 21, 82
Methods, general statement of, 15
Methyl green, 257
Methyl violet, 257
Methylated spirits, 12
Methylen azure, 129
Methylen blue, 20, 254; for bacteria, 132, 134; for impregnation of epithelia, 257; immersion methods with, 256; for intra-vitam staining, 254; Loeffler's alkaline, 134; for nerve cells, 256; for nerves and nerve terminations, 255; for non-striated muscle, 257; for ordinary sections, 257; polychromatic, 254
Metz, 174
Microchemical tests, 167–70
Micrococci, 133
Micrococcus tetrageneus, 136
Micro-dissections, 152
Micro-injection methods, 101–3
Micro-manipulator, 219
Micrometer, 219; ocular, 221; screw, 222; stage, 219; step, 222
Micron, 222
Microscope: compound with parts named, 203; demonstration, 214; dissecting, 215; manipulation of compound, 226; optical principles of the, 197; path of light rays through the compound, 205; principle of compound, 200, 201; principle of the simple, 200, 201; -tube, sectional view of, 204
Microscopes, representative compound, 207, 208
Microscopical material for general zoölogy, 286–309

Microscopical terms and appliances, 206–26
Microtome, 4, celloidin, 70; freezing, 78; Minot automatic rotary, 41; oiling the, 46; Spencer rotary, 42; well, 37
Microtome knife, sharpening the, 46
Milk, 284
Milky or hazy staining, 61
Miller, 101
Mineral ash, 169
Minot, 145, 152, 247
Minute dissections, 91: use of clove oil in, 92
Mirror, 222
Mites, 108, 113
Mitochondria, 16, 162–65, 170; Benda's method for, 163; Bensley's methods for, 163, 164; in living cells, 165; Regaud's method, 162; in tissue cultures, 165; Sevringhaus technique, 177
Mitosis: in Ascaris, 154; in cells of Ambystoma, 161; in eggs of white fish, 160; in grasshoppers, 159; material for demonstrating, 158, 159; in testis of crayfish, 159; in testis of Necturus, 160
Mollusks, 307
Monocystis, 291
Morrison, 104
Mosquito, to kill and mount, 111; larvae, 113
Mossman, 64, 239
Moulton, 81
Mounting, 23; general scheme of, 27; objects of general interest, 107
Mounts, temporary, 37
Mouse, use in embryology, 148–49
Mouth, epithelium of, 272
Mouth-parts of insects, to mount, 92, 114
Muci-carmine, Mayer's, 258
Muci-hematein, Mayer's, 258
Mucin, 258, 273; stains for, 170, 258
Mucoid connective tissue, 271
Mueller, drawing guide, 194
Müller's fluid, 234
Multiple, staining, 21
Muscae volitantes, 222

Muscle: areas of Conheim, 277; branched striated fibers, 277; cardiac, 277; ends of striated fibers, 277; fibrillae in striated, 278; of insects, 110; non-striated, 278; Purkinje fibers, 278; sarcolemma of, 278; striated, 278; to tendon, 278
Muscle-fibers, isolation of, 89
Muscular tissue, 277, 278
Mussel, 28, 307; clearing entire, 120; cross-section of gill, 307
Myelin, 279
Myelocytes, 129
Myers, 296

Nail, 284
Naphthol, alpha, 170
Naphthylamin yellow, 267
Narcotics in killing, 16
Necturus, mitosis in testis of, 160
Neisser's method for diagnosing diphtheria, 134
Nelson, 119
Nematodes, 115, 301
Nerve: plexuses of alimentary canal, 273; silver-nitrate method for, 84
Nerve cells, Golgi method for, 82
Nerve-endings, 280; gold-chloride method for, 88
Nerve-fiber bundles, 280
Nerve fibers: degenerating, 279; intra-epithelial, 279; medullated, 280; non-medullated, 280; stain for medullated, 263
Nerve-tissue, staining *in toto*, 119
Nervous system, 278; of insect, to mount, 91
Neubauer ruling, 124, 125
Neurofibrils, Cajal's method for, 85
Neuroglia, 280
Neurokeratin, 280
Neutral red, 258
Neutrophil granules, 268
Newton-gram method for chromatin, 168
Newton's rings, 125
Nigrelli, 127
Nigrosin, 20, 298
Nissl's granules, 170, 256, 280
Nissl's method of staining basophil substance in nerve cells, 256

Nitric acid, 17, 25; decalcifying fluid, 265
Nitrocellulose instead of celloidin, 74; for corrosion preparations, 104
Nodes of Ranvier, 280
Noland's fixative and stain, 293
Normal fluids, 26, 263
Normal saline, 8, 264
Nose, 281
Nuclear stains, 19
Numerical aperture, 222
Nymph of dragon fly, 307

Obelia, stain for, 298
Objective: achromatic, 204; apochromatic, 206, 213; immersion, 218; using the immersion, 228
Objectives: lens systems of, 202; rating of, 206
Ocular: compensating, 206, 213; demonstration, 213; Huygenian, 201; negative, 201; positive, 201; projection, 223; searching, 213; working, 213
Oculars, rating of, 206
Olfactory cells, 281; nerve processes of, 281
Oltman, 63
Oögenesis, 148, 154, 281
Opalina, 290
Opaque mounts, 111
Optical center of a lens, 198
Orange G, 10, 20, 22, 56, 258
Orcein, 258
Organs varying in density, 80
Orientation of objects in the imbedding-mass, 144
Orienting serial sections, 143
Origanum, oil of, 23
Orth's lithium carmine, 132
Osmic acid, 18, 169, 240; test for fat, 169; vapor, 241; washing out, 18
Osmium-bichromate mixture in Golgi method, 83
Otoliths, 275
Ova, 282
Ovary, 282
Over-correction, 225
Oviduct, 282

Oxidase reaction, 170
Oxyphil granules, 267

Pacinian corpuscles, 280
Painter, 171
Pancreas, 273, granules of, 273
Paper boxes, preparing, 39
Paracarmine, Mayer's, 259
Paraffin, 11, 24, 38; baths, 12, 13, 38; to clean, 44; ribbon carrier, 43
Paraffin method, 24, 38, 52; compared with celloidin method, 72; for delicate objects, 57; difficulties likely to be encountered in, 48
Paraffin-rubber method of Johnson, 45
Paraffin sections: of large objects, 47; staining and mounting, 52
Paramecium, 288, 290, 294
Parfocal, 223
Parlodion, 24, 73
Parmenter, 161
Parotid gland, 273
Pathogenic bacteria, **136**
Patton, 144
Payne, 162
Pearl, 240
Peaslee, 115
Pectinatella, **303**
Pedesis, 212, 223
Pedicellaria, 303
Penetration, 223
Penis, 282
Peter, 181
Petrunkevitch's fixing fluid, **34**
Peyer's patches, 273
Pflüger's egg-tubes, 282
Phloroglucin method of decalcification, 265
Photographing cellular structures, 172
Photographs, reproduction of, 193
Photomicrography, 223
Pickerel, mitosis in egg of, 154
Picric acid, 18, 241; for destaining iron-hematoxylin, 56; decalcifying fluid, 266; as a stain, 259; washing out, 18; with dioxan, 66
Picric-alcohol, 241
Picro-acetic, 241; formalin, 242
Picro-anilin blue, 259

Picro-carmine, 259
Picro-sublimate, 242
Picro-sulphuric, 242; for chick embryos, 127
Pig, embryology of, 151
Pipette, egg, 139
Pituitary, injecting anterior lobe of, 308
Placenta, 282
Planaria: killing and mounting, 110, 299; to kill with pharynx protruded, 300
Plankton, 296
Plasmosomes, stains for, 170, 257
Platelets, blood-, 121, 129
Plating celloidin sections, 74
Platino-aceto-osmic mixture, 243
Plumatella, 303
Pneumococcus, 136
Pohlman, 45
Pointer ocular, 223
Polar bodies, 149, 153, 154
Polariscope, 223
Polypoid forms, 298
Potash clearing method, 119
Preservation, 19, 31, 32; in paraffin, **45**
Prickle-cells of skin, 284
Prostate, 282
Protoplasmic currents, to demonstrate, 174
Protozoa, 286–96; permanent preparations of, 292
Pulsating vacuoles, to slow, 289
Purkinje cells, 280
Purkinje fibers, 269
Pyridine-silver method, 86
Pyrogallol, 260
Pyroligneous acid, 260
Pyroxylin, 24, 133

Quince jelly for Euglena cultures, 290
Quinoline blue, 248

Rabbit, ovulation and early embryology of, 149–51
Rabinowitz, 168
Rabl's picro-sublimate, 242
Radula of snail, 115
Ranson, 86, 260

Ranvier's one-third alcohol, 265; cross of, 85
Rat, embryology of, 149–51
Rath's picro-sublimate, 242
Ray of light, 197
Razor, section 1, 35
Razor blades, 42
Reagan, 103
Reagents, preparation of, 7
Reconstruction: blotting paper, 181; geometrical, 181; practical exercise in, 180; reconstruction methods, 179–82; use of photography in, 182; wax, 179, 182
Record cards, 5
Rectified spirits, 12
Reduction: of drawings, 191, 194; test for intracellular, 169
References, 312
Refraction, 197
Refractory materials, to section, 176
Regaud's method, 162
Relative merits of paraffin and of celloidin, 72
Remak's fibers, 280
Reproductive organs, 281
Resolving power, 224
Resorcin-fuchsin, 260
Respiratory organs, 283
Reticular connective tissue, 271
Reticulum, Mall's method for, 90
Retina, 277
Rhigolene, for freezing, 80
Rice, 181
Riddle, 169
Ringer's solution, 264
Ripart and Petit, liquid of, 235
Robertson, 159
Rolling of paraffin sections, 48
Rosanilin-violet, 257
Ross board, 191
Rotifers, 303
Rubbing sections off the slide, to avoid, 59
Rubin S, 250
Rules, general, 5

Sabin, 103, 117; modification of the Spalteholz method, 117
Safranin, 20, 260; -gentian-orange preparations, 161, 261; and gentian violet, 161, 261; use in cytology, 158, 161
Salivary glands: granules of, 273; of insect, to mount, 92
Sandal-wood oil, 23
Sarcinae, 133
Sarcolemma, 278
Scale insects, 113
Scales of insects, 112
Scalpel, 1
Scammon, 63, 181
Schaefer, 66
Scharlach R, 262; stain for fat, 169
Schaudium's fluid, 238
Schneider, 248
Schulemann, 167
Schultze's iodized serum, 264
Sciara, 174
Scissors, dissecting, 1
Sclera, 277
Scraping of microtome knife, 49
Scratches across paraffin sections, 49
Scratching noise of microtome knife, 48
Sealing bottles, etc., 32
Sealing cover-glasses, 108
Sebaceous glands, 284
Secretion antecedents, 170
Section methods, 15, 27
Section razor, 33, 35
Sectioning: by the freezing method, 77; injected organs, 100
Sections: affixing, 24; cutting celloidin, 69; cutting frozen, 78; cutting paraffin, 40; flooding with the dye, 62; free-hand, 35; preserving frozen, 35; washing-off of, 61
Segmentation in mammals, 148
Semicircular canals, 275
Seminal vesicle, 282
Seminiferous tubules, 282
Serial sections, orienting, 143
Sevringhaus, 177; technique for Golgi apparatus and mitochondria, 177
Sex-chromosomes, 173
Shading drawings, 186
Shadows, in drawing, 186
Sheet method for celloidin sections, 74

Sheib, ameba cultures, 287
Shellac for cell-making, 117
Shipley, 117
Shop skeleton letters for hand labeling, 187
Siedentopf, 225
Sihler's hematoxylin method, 119
Silver nitrate, 82, 85, 262; methods, 84
Skin and its appendages, 283; blood vessels of, 284
Slide, box, 1; marking, 1, 62, 63; standard, 1
Slifer, 176
Slow-killing of aquatic forms, 307
Small intestine, epithelium of, 273
Small objects, transferring, 31
Small pieces, sectioning free-hand, 36
Smear preparations: bacteria, 130; blood, 122; to show mitosis in, 159
Smegma bacilli, 135
Smith, 75, 147, 286, 291; bichromate-acetic mixture, 232
Snails, to kill expanded, 290, 308
Sobotta, 149
Sodium-chloride dissociator, 265
Solutions, rules regarding, 5
Spaeth, 154
Spalteholz method of clearing total specimens, 117
Spawning fish, 153
Spectrum: secondary, 206, 209; tertiary, 206, 209
Spermatogenesis, 282
Spermatozoa, 282, 293
Spinal cord, 280
Spinal ganglia, 281
Spindles, 159, 170
Spiracles, 114
Spirillum, 133; of Asiatic cholera, 136
Spleen, 268; blood vessels of, 269
Splitting of paraffin sections, 49
Sponges, 296
Spores of bacteria, staining, 136
Sporozoa, 291
Sputum, to examine for tubercle bacilli, 135
Staining, 19, 20; in bulk, 56; causes of failure in, 60–61; celloidin sections, in hematoxylin and eosin, 71; and mounting paraffin sections, 52; progressive, 25; rack, 62; regressive, 25
Stains, 243–63; classes of, 20; renewing, 62; standardized, 21, 22
Staphylococci, 133
Staphylococcus pyogenes: alba, 136; aureus, 136
Starfish, 303
Stenders, 4
Stentor, 286
Stephenson, 290
Sting, to mount, 91
Stipple-board, 191
Stitt, 302
Stomach, 274; cardiac end of, 274; epithelium of, 272; fundus of, 274; pyloric end of, 274
Stoppers, to remove, 12
Streeter, 153
Streptococci, 133
Streptococcus capsulatus, 136; pyogenes, 136
Strop for microtome knife, 46
Stylonychia, 294
Sublingual glands, 274
Submaxillary glands, 274
Sudan III, 169, 262; stain for fat, 169
Sugar in fixing, 17, 74
Sulphalizarinate of soda, 163
Supplies required, 1
Supporting tissues, 269–72
Supporting vials, 32
Suprarenal glands, 285
Swank, 74
Sweat glands, 284
Sympathetic ganglia, 281
Synovial villi, 272
Syphilis, treponema of, 135

Table: imbedding-, 11; warming, 11, 14
Table of equivalent weights and measures, 310
Table of tissues and organs with methods of preparation, 267–85
Tactile corpuscles, 281
Tactile menisci, 281
Tandler, 104
Tapeworms, 115

Tap-water for washing sections stained in hematoxylin, 59
Tashiro's apparatus for measuring carbon-dioxide output, 176
Taste-buds, 274
Teasing, 15, 26, 89
Teichmann's crystals, 122
Teleosts, artificial fecundation in, 153; care of eggs, 146; embryology of, 146; to preserve eggs of, 147; to study living eggs, 146
Tellyesnicky's fluid, 232
Temperature, effect on paraffin sectioning, 45
Tendon, 272; cells, 272; to muscle, 272
Terminal bars, 276
Tertiary butyl alcohol, 67
Testis, 282; for avian and mammalian spermatogenesis, 175
Tests for certain cellular structures, 167–70
Tetanus, bacillus of, 136
Tetracocci, 133
Thermometers, 311
Thin sections: compared with thick, 36; to cut, 45; treatment of, 59
Thionin, 262
Thoma mixing pipettes, 125
Thymus gland, 268, 283
Thyroid gland, 283
Ticks, 113
Tigroid granules (substance), 170, 256
Tilt of microtome knife, 49
Tissues: which crumble, 45, 50; cultivation of removed, 155; length of time required for staining, 60
Toisson's solution, 125
Toluidin blue, 20
Toluol, 23
Tongue, 274; papillae and folliculi of, 274
Tonsil, 274
Tooth, decalcified, 94, 274; development of, 274; enamel prisms of, 274; grinding, 94; ground, 94; odontoblasts, 274; sectioning decalcified, 94
Trachea, 283
Tracheae of insects: to demonstrate, 114; Golgi's method for, 84
Transparent larvae, 108

Trematodes, 300; larval stages of, 301
Trichinella, 301; to demonstrate living, 302
Trichlorethylene, 63
Trichocysts, to demonstrate, 289
Trypan blue, 167
Trypanosoma, 291
Tube-length, 204, 205, 206, 224
Tubercle bacilli, 135
Turner, 290
Turning cells, 107
Turntable, 108
Turpentine, 23, 59
Typhoid, bacillus of, 136

Ultra-microscopy, 224
Umbilicus, 282
Under-correction, 225
Unna, 257, 258
Urea, use in cytology, 17, 174
Ureter, 285
Urethra, 282, 285
Urinary organs, 284
Uterus, 282

Vagina, 282
Van Beneden, 150
Van Gehüchten, 243
Van Giesen's stain, 250
Variation in thickness of paraffin sections, 50
Vas deferens, 282
Vein, 269
Vertebrata, 308
Vision, conventional distance of, 206
Visual purple, in rods, 277
Volvox, aureus, 291; globator, 291
Von Ebner's decalcifying fluid, 266
Vorticella, 290
Vulcanite, 71

Wagner, 32, 105
Walker, 81, 97
Walls, 76
Walton, 31, 117
Warming table, 14
Washing, 17; devices, 33
Watch-glass, Syracuse, 4

Water method for affixing sections, 24
Water-mites, 108
Wax plates: anchorage by twisted wires, 182; methods for rapidly cutting, 182; preparing for reconstructions, 179, 180
Weber, 181
Weigert-Pal stain for medullated nerve fibers, 263
Weigert's borax-ferricyanide, 164; elastic tissue stain, 260; method for bacteria, 132
Weights and measures, table of equivalent, 310
Weissberger, 64
Welch, 48
Wenrich-Geiman, 238
Wetmore, 72
Whelks, to kill expanded, 308
White blood corpuscles, 268
White objects, to orient, 45
Whitman, 101, 139
Whole objects, mounting, 27, 107, 109, 110

Wildman, 163
Window method for chick embryo, 156
Wings of insects, 112, 114
Winiwarter, 236
Wintergreen, oil of, 23
Wisconsin paraffin-melting oven, 13
Wood's metal, 104
Worcester's fluid, 240
Working distance, 226
Wright, 58, 128–29; stain for blood, 128
Wrinkles in paraffin sections, 48
Wu, 31, 165, 287, 289, 297, 299

X-element, 173
Xylol, 23, 38; for removing paraffin, 59

Zebrowski mixture, 113
Zenker's fluid, 8, 232; fixing with, 29; formalin mixtures, 232
Ziehl-Neelson carbol-fuchsin, 134
Zoölogy, materials for a course in general, 286–309
Zsigmondy, 225

ENNIS AND NANCY HAM LIBRARY
ROCHESTER COLLEGE
800 WEST AVON ROAD
ROCHESTER HILLS, MI 48307